Encyclopaedia of Mathematical Sciences

Volume 18

Editor-in-Chief: R. V. Gamkrelidze

A. I. Kostrikin I. R. Shafarevich (Eds.)

Algebra II

Noncommutative Rings
Identities

With 10 Figures

Springer-Verlag
Berlin Heidelberg New York
London Paris Tokyo
Hong Kong Barcelona
Budapest

Scientific Editors of the Series:
A. A. Agrachev, E. F. Mishchenko, N. M. Ostianu, L. S. Pontryagin
Editors: V. P. Sakharova, Z. A. Izmailova

Title of the Russian edition:
Itogi nauki i tekhniki, Sovremennye problemy matematiki,
Fundamental'nye napravleniya, Vol. 18, Algebra 2
Publisher VINITI, Moscow 1988

Mathematics Subject Classification (1980):
16-02, 08B99, 17A01

ISBN 3-540-18177-6 Springer-Verlag Berlin Heidelberg New York
ISBN 0-387-18177-6 Springer-Verlag New York Berlin Heidelberg

Library of Congress Cataloging-in-Publication Data
Algebra 2. English. Algebra II / A. I. Kostrikin, I. R. Shafarevich, eds. p. cm. –
(Encyclopaedia of mathematical sciences; v. 18)
Translation of: Algebra 2, issued as v. 18 of the serial: Itogi nauki i tekhniki.
Seriia͡ sovremennye problemy matematiki. Fundamental'nye napravleniia͡.
Includes bibliographical references and index.
ISBN 0-387-18177-6 (U.S.)
1. Noncommutative rings. 2. Algebraic varieties. I. Kostrikin, A. I. (Alekseĭ Ivanovich)
II. Shafarevich, I. R. (Igor' Rostislavovich), 1923–. III. Title. IV. Title: Algebra two. V. Series.
QA251.4.A313 1991 512–dc20 90-45761

Printed in the United States of America
41/3140-543210 - Printed on acid-free paper

List of Editors, Contributors and Translators

Editor-in-Chief

R. V. Gamkrelidze, Academy of Sciences of the USSR, Steklov Mathematical Institute, ul. Vavilova 42, 117966 Moscow, Institute for Scientific Information (VINITI), ul. Usievicha 20 a, 125219 Moscow, USSR

Consulting Editors

A. I. Kostrikin, I. R. Shafarevich, Steklov Mathematical Institute, ul. Vavilova 42, 117966 Moscow, USSR

Contributors

Yu. A. Bakhturin, Moscow State University, Leninskie Gory, 119899 Moscow, USSR

L. A. Bokhut', Institute of Mathematics of the Siberian Branch of the Academy of Sciences of the USSR, 630090 Novosibirsk, USSR

V. K. Kharchenko, Institute of Mathematics of the Siberian Branch of the Academy of Sciences of the USSR, 630090 Novosibirsk, USSR

I. V. L'vov, Institute of Mathematics of the Siberian Branch of the Academy of Sciences of the USSR, 630090 Novosibirsk, USSR

A. Yu. Ol'shanskij, Moscow State University, Leninskie Gory, 119899 Moscow, USSR

Translator

E. Behr, Department of Mathematics, Illinois State University, Normal, IL 61761, USA

Contents

I. Noncommutative Rings

L.A. Bokhut', I.V. L'vov, V.K. Kharchenko

Translated from the Russian
by E. Behr

Contents

Introduction

The algebra of square matrices of size $n \geq 2$ over the field of complex numbers is, evidently, the best-known example of a non-commutative algebra[1]. Subalgebras and subrings of this algebra (for example, the ring of $n \times n$ matrices with integral entries) arise naturally in many areas of mathematics. Historically however, the study of matrix algebras was preceded by the discovery of quaternions which, introduced in 1843 by Hamilton, found applications in the classical mechanics of the past century. Later it turned out that quaternion analysis had important applications in field theory. The algebra of quaternions has become one of the classical mathematical objects; it is used, for instance, in algebra, geometry and topology.

We will briefly focus on other examples of non-commutative rings and algebras which arise naturally in mathematics and in mathematical physics.

The exterior algebra (or Grassmann algebra) is widely used in differential geometry – for example, in geometric theory of integration. Clifford algebras, which include exterior algebras as a special case, have applications in representation theory and in algebraic topology. The Weyl algebra (i.e. algebra of differential operators with polynomial coefficients) often appears in the representation theory of Lie algebras. In recent years modules over the Weyl algebra and sheaves of such modules became the foundation of the so-called microlocal analysis. The theory of operator algebras (i.e. subalgebras of the algebra of bounded operators on Hilbert spaces) – in particular, the W^*-algebras or von Neumann algebras, and C^*-algebras – emerged thanks to the stimulating influence of quantum mechanics, and became an important part of functional analysis, with numerous connections with representation theory and other areas of mathematics. Universal enveloping algebras of Lie algebras are important in particular for the theory of representations of Lie groups and Lie algebras. Group algebras of finite groups are fundamental objects in the representation theory of finite groups. They are also used, for example, in cryptography. Group algebras of topological groups are important in the theory of infinite-dimensional representations of groups. The modern theory of finite-dimensional algebras lies on the boundary of such disciplines as algebraic K-theory, the theory of algebraic groups, and algebraic geometry. For $n \geq 3$ there is a correspondence between n-dimensional projective spaces and the (associative) division rings coordinatizing them, which allows one to reduce many questions about such spaces to algebraic questions about division rings. The so-called continuous geometries, introduced by von Neumann, are coordinatized by regular rings.

On the contents of this chapter. In § 1 examples of rings are given, as well as the basic definitions, statements of problems, and some motivations.

[1] It is recommended that the reader become familiar with Chapter 11, which contains the basic concepts of algebra, before reading this chapter. (The Editors)

The theory of finite-dimensional algebras is a classic part of ring theory. § 2 is devoted to it. An important notion in the structure theory of rings is the concept of a module. § 3 is devoted to modules and some classes of rings which are defined in terms of modules over them. The primary task of ring theory is the classification of rings up to isomorphism. Even though this goal in its full generality has not been achieved, it nevertheless leads to numerous results on the structure of rings. This is discussed in § 4. § 5 describes fragments of theories being developed within the framework of ring theory and disciplines closely related to it. Annotated bibliography is included at the end of this chapter.

While writing the chapter, the authors received help in one form or another from many mathematicians. I.R. Shafarevich and A.I. Kostrikin carefully reviewed the first draft of the chapter, and made many suggestions as to concrete questions as well as the general plan and character of the exposition. The following mathematicians responded to our request and wrote for us materials which we then used: V.I. Arnautov (as co-author with M.I. Ursul'), A.A. Bovdi, Yu.A. Drozd, A.S. Merkurjev and V.I. Yanchevskij. A.V. Mikhalov and L.A. Skornyakov read the chapter and offered their remarks. Most helpful were meetings with A.Z. Anan'in, O.K. Babkov, K.I. Beidar, I.H. Bekker, A.A. Boyarkin, G.M. Brodskij, A.I. Valitskas, V.N. Gerasimov, R.I. Grigorchuk, A.G. Grigorian, E.I. Zel'manov, Yu.I. Kafiev, A.R. Kemer, L.A. Koifman, S.F. Krendelev, Yu.N. Mal'tsev, V.T. Markov, S.Yu. Prishchepionek, A.M. Stepyn, V.A. Ufnarovskij and A.S. Stern (many of them also prepared for us materials which, regretfully, we could not use directly in view of the limited scope of the chapter).

We are deeply grateful to all of them for their assistance and collaboration.

§ 1. Basic Definitions and Examples

1.1. A *ring* R is a set R with operations of addition $+$ and multiplication $\cdot$, such that $< R, + >$ is an abelian group, and the distributive laws $a(b+c) = ab + ac$ and $(b+c)a = ba + ca$ hold. We will consider only associative rings, i.e. rings with associative multiplication operation: $(ab)c = a(bc)$. In addition, unless stated otherwise, it will always be assumed that the ring has a unity element, i.e. a distinguished element $1 = 1_R$ such that $1a = a1 = a$ for all $a \in R$. Ring axioms do not imply the following property, well-known from arithmetic: if $ab = 0$ then $a = 0$ or $b = 0$. Elements for which this statement fails are called *zero divisors.* If there are no zero divisors in the ring, it is then called an *integral domain* or a *ring without zero divisors.*

A ring R is *commutative* if the identity $ab = ba$ is valid for all $a, b \in R$.

As the title of this section implies, our attention will be primarily focused on non-commutative (more precisely: not necessarily commutative) rings – the theory of commutative rings has its own peculiarities, tight connections with algebraic geometry, deeply developed methods, and is regarded as a separate theory. Nevertheless, many concepts of the general theory of non-commutative rings originate in commutative ring theory, and we will occasionally illustrate this by considering commutative rings.

1.2. Ideals, Factor Ring. An *ideal* of the ring R is a non-empty subset $I \subseteq R$, closed under subtraction, and stable under multiplication – both on the left and on the right – by all elements of the ring (notation: $I \triangleleft R$). In every ring the sets $\{0\}$ and R are ideals. These two ideals are called *improper*, while the others (if any) are *proper*. A ring which contains only improper ideals is called *simple*.

Let I be an ideal of the ring R. We will define on the set R a relation $a \sim b \Leftrightarrow a - b \in I$. It is easy to verify that this is an equivalence relation, compatible with the operations of addition and multiplication: $a \sim b, c \sim d \Rightarrow ac \sim bd,\ a + c \sim b + d$. Therefore R decomposes into equivalence classes, $R = \bigcup_r (r + I)$, where $r + I$ is the equivalence class containing the element r: it consists of elements of the form $r + i$ with $i \in I$. On the set of equivalence classes we define the operations $(r + I)(s + I) = rs + I$ and $(r + I) + (s + I) = (r + s) + I$. As a result, the set of equivalence classes is made into a ring, which is denoted by R/I and called the *factor ring of R modulo the ideal I*. The factor ring construction in a sense transforms the proper ideal I into an improper (i.e. zero) one.

Ideals of a ring can be multiplied and added: if $I, J \triangleleft R$ then $I + J = \{a + b \,|\, a \in I, b \in J\}$ and $IJ = \{\sum_i a_i b_i \,|\, a_i \in I, b_i \in J\}$ are again ideals. Evidently, one can in this way obtain improper ideals from proper ones. As usual, one also defines powers of an ideal: $I^n = I \cdot I \cdots I$ (n times). In addition, intersection of any family of ideals is an ideal, and it is easy to see that $IJ \subseteq I \cap J$.

1.3. The Ring of Integers $\mathbb{Z}$. This ring is known to the reader from school courses. $\mathbb{Z}$ is an example of a commutative ring. The set $(n) = n\mathbb{Z}$ of all integral multiples of some fixed integer n, is an ideal. Using Euclid's division algorithm it can be shown that all ideals of $\mathbb{Z}$ are of the form $n\mathbb{Z}$ ($n \geq 0$). The factor ring $\mathbb{Z}/n\mathbb{Z}$ consists of residue classes $k + n\mathbb{Z}$. All numbers in each class are characterized by the property that they yield the same remainder k on division by n. The factor ring $\mathbb{Z}/n\mathbb{Z}$ can be identified, for $n > 0$, with the ring $\mathbb{Z}_n$ of residues modulo n. The ring $\mathbb{Z}_n$ consists of n elements $0, 1, \ldots, n-1$, on which operations of addition $a \oplus b$ and multiplication $a \odot b$ are defined as remainders on division of the integers $a + b$ and ab, respectively, by n. From the algebraic point of view the rings $\mathbb{Z}_n$ and $\mathbb{Z}/n\mathbb{Z}$ are indistinguishable or, as we say, isomorphic.

1.4. Homomorphism, Isomorphism, Automorphism, Subring. Let R and S be two rings. A mapping $\varphi : R \to S$ is called a *homomorphism* if it preserves the operations: $\varphi(x - y) = \varphi(x) - \varphi(y)$, $\varphi(xy) = \varphi(x)\varphi(y)$ and $\varphi(1) = 1$ (if the rings are being considered as rings without unity, the last condition is omitted from this definition). The *kernel* of a homomorphism φ is the set $\ker\varphi = \{x \in R \mid \varphi(x) = 0\}$. It is easy to see that $\ker\varphi$ is an ideal of R. The *image* of a homomorphism φ is the image of the mapping φ: $\operatorname{Im}\varphi = \{y \in S \mid \exists x \in R \text{ with } \varphi(x) = y\}$. A subset of the ring R, which is closed under the ring operations and contains the unity element, is called a *subring.* It is easily seen that an image of a homomorphism is a subring. A homomorphism whose kernel is zero will be called *injective* (or an *embedding*, or a *monomorphism*). A homomorphism $\varphi : R \to S$ is called *surjective* (or a homomorphism *onto*, or an *epimorphism*) if $\operatorname{Im}\varphi = S$. A homomorphism which is one-to-one (both injective and surjective) is called an *isomorphism.*

In the example of the preceding section, the mapping $k \mapsto k + n\mathbb{Z}$ will be a homomorphism from $\mathbb{Z}$ onto $\mathbb{Z}/n\mathbb{Z}$, while the mapping $k + n\mathbb{Z} \mapsto k - [\frac{k}{n}]n$ (where $[\frac{k}{n}]$ is the integer part of $\frac{k}{n}$) will be an isomorphism of $\mathbb{Z}/n\mathbb{Z}$ onto $\mathbb{Z}_n$. This example has a general character. Namely, if I is an ideal, then the mapping $r \mapsto r + I$ is a homomorphism of R onto R/I. Conversely, if $\varphi : R \to S$ is a homomorphism, then one can consider the homomorphism $\psi : R \to R/\ker\varphi$ defined as above. In fact, $\varphi(r + i) = \varphi(r)$ if and only if $i \in \ker\varphi$, i.e. φ maps each equivalence class onto a single element; we can thus define a mapping $\bar{\varphi} : R/\ker\varphi \to S$. This mapping turns out to be an isomorphism of $R/\ker\varphi$ onto $\operatorname{Im}\varphi$.

An *automorphism* $\varphi : R \to R$ is an isomorphism of the ring onto itself.

1.5. Fields, Rings of Algebraic Integers. A commutative ring F with $1 \neq 0$ is called a *field* if, for any $a \neq 0$ in F, the equation $ax = 1$ has a solution in F. The ones most often encountered in mathematics are: the field of rational numbers $\mathbb{Q}$, the field of real numbers $\mathbb{R}$, and the field of complex numbers $\mathbb{C}$. Classical Galois theory has sprung from the investigation of finite extensions of the field $\mathbb{Q}$ (i.e. fields K whose elements are complex numbers and whose dimensions as vector spaces over $\mathbb{Q}$ are finite – we assume that the reader is familiar with the basics of linear algebra) and their automorphisms. Its development led to the modern Galois theory of both commutative and non-commutative rings.

Together with the integers, a fundamental role is played in number theory by algebraic integers, i.e. (complex) solutions of equations $x^n + a_{n-1}x^{n-1} + \ldots + a_0 = 0$ with integer coefficients. For example, numbers of the form $a + b\sqrt{c}$ $(a, b, c \in \mathbb{Z})$ are algebraic integers (this follows from Viète's theorem covered in elementary mathematics courses). It can be shown that the difference and the product of algebraic integers are such numbers themselves, i.e. that algebraic integers form a ring. Algebraic integers lying in a given finite extension K of the field $\mathbb{Q}$ form a ring, called a *ring of algebraic integers.* The theory of factorization in rings of algebraic integers differs from factorization theory of integers in that an algebraic integer cannot be, generally speaking, uniquely presented as a product of algebraic integers which are indecomposable in the given ring. For example, $9 = 3 \cdot 3 = (2 + \sqrt{-5})(2 - \sqrt{-5})$ are two factorizations into numbers which are indecomposable in the ring of algebraic integers of the form $a + b\sqrt{-5}$ (with $a, b \in \mathbb{Z}$). In order to establish uniqueness in a precise sense, 'ideal multipliers' were introduced, which led to the notion of an ideal of a ring (factorization theory of algebraic integers has been investigated in the works of Kummer, Dedekind and D.I. Zolotarev). Classical number theory studies factorization theory, or the theory of ideals, in rings of algebraic integers. Thorough understanding of the properties of these rings led to the notion of a commutative Dedekind ring and to the transfer of factorization theory from rings of algebraic integers to commutative Dedekind rings. This was accomplished by E. Noether and proved to be one of the origins of modern algebra.

We will remark (without making any use of it later) that the field of real numbers $\mathbb{R}$ is obtained from the field $\mathbb{Q}$ by the operation of topological completion, if the topology on $\mathbb{Q}$ is defined by the standard norm $\|x\| = |x|$. The field $\mathbb{Q}$ can also be equipped with other topologies, in such a manner that ring operations will be continuous. These are topologies determined by the p-adic norm $\|p^m \cdot \frac{a}{b}\| = 2^{-m}$, where p is a fixed prime number and the integers a, b are not divisible by p. Completions of $\mathbb{Q}$ with respect to p-adic norms produce fields $\mathbb{Q}_p$ of p-adic numbers, which are of importance in algebraic number theory and in the so-called p-adic analysis.

One of the important properties of the field of complex numbers $\mathbb{C}$ lies in the fact that this field is algebraically closed, i.e every equation $x^n + a_{n-1}x^{n-1} + \ldots + a_0 = 0$ with complex coefficients a_i has a solution in $\mathbb{C}$.

When p is prime, the residue ring $\mathbb{Z}_p$ discussed above is also a field, which has the property that the sum $\underbrace{1 + \ldots + 1}_{p \text{ times}} = p \cdot 1 = 0$. A field in which this equality holds has, by definition, *characteristic* p. Every field of characteristic $p > 0$ contains a minimal subfield, which is isomorphic to $\mathbb{Z}_p$. We say that a field F is of *characteristic zero* if $p \cdot 1 \neq 0$ for every positive integer p. It is clear that a field of characteristic zero contains a subfield isomorphic to $\mathbb{Q}$.

1.6. Algebras Over a Field and Over a Commutative Ring. A ring R is called an *algebra over a field* F (or an F-*algebra*) if R is a vector space over F and the vector space structure is linked with multiplication in R by the law $\alpha(ab) = (\alpha a)b = a(\alpha b)$, where $\alpha \in F$ and $a, b \in R$. The rings $\mathbb{R}$ and $\mathbb{C}$ can be considered as algebras over $\mathbb{Q}$. The field $\mathbb{C}$, in turn, is an algebra over $\mathbb{R}$. Every field of characteristic $p > 0$ is an algebra over $\mathbb{Z}_p$.

If R is an F-algebra then the mapping $\alpha \mapsto \alpha 1_R$ (where $\alpha \in F$) of the field F into R preserves all operations and its kernel is null: if $\alpha \neq 0$ and $\alpha 1_R = 0$ then $1_R = (\alpha^{-1}\alpha)1_R = \alpha^{-1}(\alpha 1_R) = \alpha^{-1} \cdot 0 = 0$. F can therefore be identified with the subring $F \cdot 1_R$, and $\alpha a = \alpha(a1) = a(\alpha 1) = a\alpha$ (where $\alpha \in F, a \in R$), i.e. F is contained in the center of the algebra R. The *center* of a ring (or algebra) R is the set $C(R) = \{c \in R \mid \forall a \in R,\ ca = ac\}$. The converse is also true: if there is an embedding φ of the field F into the center $C(R)$, then R naturally becomes an algebra over F: $\alpha r := \varphi(\alpha) r$. An algebra R over a field F is *central* if $F = C(R)$. This point of view allows one to define algebras over an arbitrary commutative ring K. A ring R is called an *algebra over the commutative ring* K if there is a homomorphism $\varphi: K \to C(R)$; then the elements $\varphi(\alpha) r$ and $r\varphi(\alpha)$ are customarily written as, respectively, αr and $r\alpha$.

The examples presented below are essentially procedures which allow one to construct a non-commutative ring from a ring or a field; if the original ring is commutative then, as a rule, an algebra over it is obtained.

1.7. The Skew Field of Quaternions. A ring R is called a *skew field* (or a *division ring*) if for every non-zero $a \in R$ there exists an element $a^{-1} \in R$ such that $aa^{-1} = a^{-1}a = 1$ (i.e. R is a "non-commutative field"). If a division ring is at the same time an algebra over some field, then it is called a *division algebra*. We will consider the set $\mathbb{H}$ of sums of the form $a + b\mathbf{i} + c\mathbf{j} + d\mathbf{k}$, where $a, b, c, d \in \mathbb{R}$ and $\mathbf{i}, \mathbf{j}, \mathbf{k}$ are symbols ("imaginary units"). We will define addition on $\mathbb{H}$ coordinate-wise, and multiplication by the distributive law and the 'clockwise rotation' rule: $\mathbf{ij} = \mathbf{k}$, $\mathbf{jk} = \mathbf{i}$, $\mathbf{ki} = \mathbf{j}$, $\mathbf{kj} = -\mathbf{i}$, $\mathbf{ik} = -\mathbf{j}$, $\mathbf{ji} = -\mathbf{k}$ and $\mathbf{i}^2 = \mathbf{j}^2 = \mathbf{k}^2 = -1$. As a result we obtain a 4-dimensional algebra over $\mathbb{R}$, called the *quaternion algebra*. Even though the ring $\mathbb{H}$ contains the field $\mathbb{C}$ (sums of the form $a+b\mathbf{i}$), it is not an algebra over $\mathbb{C}$ because $\mathbb{C}$ is not contained in the center of $\mathbb{H}$. The ring $\mathbb{H}$ is a division ring: if $\alpha = a + b\mathbf{i} + c\mathbf{j} + d\mathbf{k}$ then $\alpha^{-1} = |\alpha|^{-2}\bar{\alpha}$, where $\bar{\alpha} = a - b\mathbf{i} - c\mathbf{j} - d\mathbf{k}$ and $|\alpha| = \sqrt{a^2 + b^2 + c^2 + d^2}$. The norm defined by the last equality makes $\mathbb{H}$ into a complete normed $\mathbb{R}$-algebra (see Sect. 5.5).

1.8. The Matrix Algebra $M_n(F)$ over a Field F. This algebra consists of ordinary $n \times n$ matrices $A = (a_{ij})$ with entries from the field F, $(a_{ij}) + (b_{ij}) = (a_{ij} + b_{ij})$, $\alpha(a_{ij}) = (\alpha \cdot a_{ij})$ and $(a_{ij}) \cdot (b_{ij}) = \left(\sum_{k=1}^{n} a_{ik} b_{kj}\right)$. The identity matrix $1_n = (\delta_{ij})$ (where δ_{ij} is the Kronecker delta symbol) has the usual property $1_n \cdot A = A \cdot 1_n = A$. As a basis of $M_n(F)$ over F one can take the matrix units e_{ij} (e_{ij} is the matrix whose $(i,j)^{\text{th}}$ entry is 1 and all the others equal 0). Consequently, $M_n(F)$ is a finite-dimensional algebra, and its dimension (over F) is n^2. The multiplication table of the algebra $M_n(F)$ relative to the basis $\{e_{ij}\}$ has the form $e_{ij}e_{ks} = \delta_{jk}e_{is}$. The elements e_{ii} are *idempotents*, i.e. $e_{ii} \neq 0$ and $e_{ii}^2 = e_{ii}$. An idempotent e is called *primitive* (or *minimal*) if it cannot be written as a sum of orthogonal idempotents, i.e. $e \neq f + h$ for any idempotents f, h such that $fh = hf = 0$. The idempotents e_{ii} are primitive. Equality $1_n = e_{11} + \ldots + e_{nn}$ gives a decomposition of the unity element of the algebra $M_n(F)$ into a sum of orthogonal primitive idempotents. It is easy to see that $M_n(F)$ is a central simple (see Sect. 1.2 and 1.6) algebra over F. In addition, $M_n(F)$ contains many *one-sided ideals*. A subset $I \neq \emptyset$ of a ring R is called a *right (left) ideal* if it is closed under subtraction and is invariant under right (left) multiplication by elements of the ring: $a, b \in I$, $r \in R \Rightarrow a - b \in I$ and $ar \in I$ ($ra \in I$, respectively). For example, $e_{ii}M_n(F)$ is a right ideal (the set of 'i-th rows' in $M_n(F)$), while $M_n(F)e_{ii}$ is a left ideal (the set of 'i-th columns' in $M_n(F)$).

One of the most important properties of the ring $M_n(F)$, and every other finite-dimensional algebra, is the *finiteness condition on descending chains of right (left) ideals*, or *descending chain condition on right (left) ideals*: every strictly descending chain $I_1 \supset I_2 \supset \ldots$ of right (left) ideals contains only finitely many elements. This condition is equivalent to the *minimality condition on right (left) ideals*: every family of right (left) ideals of the ring has an element which is minimal with respect to inclusion. A ring which satisfies the minimality condition on right (left) ideals is called *right (left) Artinian.* A ring is *Artinian* if it is both left and right Artinian. Occasionally, right (left) Artinian rings are called *Artinian on the right (left).* The Artinian condition for a ring is in many cases a succesful substitute for the notion of finite dimensionality of an algebra (in the sense that many theorems about finite-dimensional algebras generalize to Artinian rings – see Sect. 2.14 below).

If R is any ring then the matrix ring $M_n(R)$ is defined analogously to the algebra of matrices over a field. $M_n(R)$ inherits many properties of the ring R which are formulated in terms of (left, right) ideals and homomorphisms. For example, if R is a simple or a left (right) Artinian ring, then $M_n(R)$ will also satisfy the corresponding property. Thorough understanding of the reason behind this phenomenon leads to the concept of Morita equivalence of rings (see 1.20, 1.22 and 4.8 below), and the link between properties of R and $M_n(R)$ is explained by the fact that R and $M_n(R)$ are Morita equivalent. Characteristics of R which are formulated in terms of elements of the ring do not always carry over to $M_n(R)$. And so $M_n(R)$ for $n > 1$ always contains zero divisors (e.g. $e_{12} \cdot e_{12} = 0$), even when R has no zero divisors (see Sect. 1.1). If R is a field or a ring with no zero divisors, then in $M_n(R)$ a product of non-zero ideals (see Sect. 1.2) is non-zero. This suggests the proper substitute, in the non-commutative case, for the notion of a ring without zero divisors. A ring R is called *prime (semiprime)* if a product of non-zero ideals of R is non-zero (respectively, a square of a non-zero ideal is non-zero). The properties of primeness and semi-primeness carry over from a ring to the ring of matrices over it (and, in general, are preserved by Morita equivalence). This partly justifies the fact that in the theory of non-commutative rings these notions are more important and useful than the property of having no zero divisors.

Matrix rings naturally arise in the theory of simple finite-dimensional algebras and simple Artinian rings. F.E. Molin (1893) showed that a simple finite-dimensional algebra over the field of complex numbers is isomorphic to an algebra of matrices over $\mathbb{C}$. Generalizing this and later results, E. Artin in 1927 established – in essence – that a simple left (right) Artinian ring is isomorphic to a matrix ring over a division ring.

1.9. The Group Algebra FG of a Group G Over a Field F. Let, for simplicity, $G = \{e = g_0, g_1, \ldots, g_{n-1}\}$ be a finite group. We will consider a vector space FG with the basis $\{g_i\}$ and we will define on it multiplication induced by the product $g_i g_k = g_k$ in the group G. FG becomes a finite-dimensional F-algebra with basis $\{g_i\}$ and multiplication table $g_i g_j = g_k$. The algebra FG is a classical example of a finite-dimensional algebra. If G is a non-trivial group, FG is not a simple algebra because it contains, for example, the *augmentation ideal* $A(FG) = \{\sum \alpha_i g_i \mid \sum \alpha_i = 0\}$. In this case FG is not a central algebra either, because for every element $g \neq e$ the element $C(g) = \sum_{x \in G} x^{-1} g x$ lies in the center of the algebra FG (since when $y \in G$, $x^{-1}gxy = yx_1^{-1}gx_1$ for a suitable $x_1 \in G$). It turns out that the center of FG is spanned (as a vector space) by the elements $C(g)$ $(g \in G)$. We analogously define a group algebra FG of an infinite group over a field F, or RG – over a ring R (as the set of all finite sums $\sum_{g \in G} \alpha_g g$, where $\alpha_g \in F$ or $\alpha_g \in R$ respectively; finiteness of the sum is understood to mean that only finitely many of its coefficients are non-zero).

1.10. The Algebra of Polynomials $F[x]$ in one indeterminate x over a field F consists of the usual polynomials $f(x) = \sum_{i=0}^{n} a_i x^i$ (where $a_i \in F$), with the usual operations of addition, multiplication and multiplication by elements of the field F. This algebra is a commutative integral domain (see Sect. 1.1). As in the case of $\mathbb{Z}$, every ideal of $R = F[x]$ is principal, i.e. has the form aR (this again follows from Euclid's division algorithm). In this sense, R is a commutative *principal ideal domain.* If R is any ring (or algebra) then the *ring (algebra) of polynomials* $R[x]$ is defined in a completely analogous manner. If R is an integral domain then it follows from properties of the degree function on $R[x]$ that $R[x]$ is itself an integral domain. The algebra of polynomials $F[x_1, \ldots, x_n]$ in indeterminates $x_1, \ldots, x_n$ over a field F is defined by induction as the algebra $R[x_n]$, where $R = F[x_1, \ldots, x_{n-1}]$. Polynomial algebras $F[x_1, \ldots, x_n]$ have fundamental significance in algebraic geometry. Namely, an algebraic variety in a projective space is a set of zeroes of a certain collection of polynomials. Hilbert's basis theorem (see below) shows that in this situation it is always possible to consider only a finite set of polynomials. The requirements of geometry led to the creation of the theory of ideals in polynomial algebras (M. Noether, D. Hilbert, E. Lasker).

The construction of the polynomial ring $R[x]$ does not, as a rule, preserve properties of the ring R (for instance, simplicity of the ring). Nevertheless, there is one important property (aside from the property of being an integral domain) which is preserved while passing to the ring $R[x]$. A ring R is called *right Noetherian* if every right ideal I of the ring R is *finitely generated*, i.e. there exists a finite set $\{\alpha_1, \ldots, \alpha_n\}$ of elements from R such that $I = \alpha_1 R + \ldots + \alpha_n R = \{\sum \alpha_i b_i \mid b_i \in R\}$. *Left Noetherian* rings are defined similarly. In the commutative case they are referred to simply as Noetherian rings. It turns out that if R is a right (left) Noetherian ring then so is $R[x]$. It then follows that $F[x_1, \ldots, x_n]$ is a Noetherian ring (*Hilbert's basis theorem*).

The standard construction of fractions is applicable to the ring $F[x_1, \ldots, x_n]$. As a result, we obtain the *field* $F(x_1, \ldots, x_n) = \{f/g \mid f, g \in F[x_1, \ldots, x_n],$

$g \neq 0\}$ *of rational functions* in $x_1, \ldots, x_n$ over F. Finite extensions of the field $F(x_1, \ldots, x_n)$ are called fields of algebraic functions over F. The theory of those fields is a part of classical algebra and algebraic geometry.

1.11. The Algebra of Formal Series $F[[x]]$ in one indeterminate x over a field F consists of formal power series $f(x) = \sum_{i=0}^{\infty} a_i x^i$ with formally defined operations of addition and multiplication. This ring is an algebraic analogue of the ring of analytic functions. Properties of the least degree function: $o(f) = \min\{i \mid a_i \neq 0\}$ for $f \neq 0$ and $o(0) = \infty$, show that $F[[x]]$ is a commutative integral domain. All ideals of the algebra $R = F[[x]]$, as in the case of the algebra $F[x]$, are obtained as principal ideals $(f) = fR$, i.e. R is a principal ideal domain. The main singularity of $F[[x]]$, as contrasted with $F[x]$, lies in the fact that if f is a series with a non-zero constant term, i.e. $o(f) = 0$, then f has an inverse element. Indeed, we will write f as $f = \alpha(1-g)$, where $\alpha \in F$ and $o(g) > 0$. Then $f^{-1} = \alpha^{-1}(1 + g + g^2 + \ldots)$. It follows that all non-zero ideals of $F[[x]]$ form the chain $F[[x]] = (1) \supset (x) \supset (x^2) \supset \ldots$. In particular, there exists a unique maximal (largest) ideal, namely $\mathbf{m} = (x)$. A commutative ring which contains a unique maximal ideal is called *local*. A commutative ring is local if all of its non-invertible elements form an ideal. In the non-commutative case this last statement is adopted as definition of a local ring. The study of local rings is the central task of commutative ring theory (since rings of (germs of) regular functions at a point of an algebraic variety are local).

If R is any ring, then the ring $R[[x]]$ is defined in a completely analogous way. If R is an integral domain then $R[[x]]$ is a domain as well. If R is a right (left) Noetherian ring then $R[[x]]$ is also a right (left) Noetherian ring. The algebra $F[[x_1, \ldots, x_n]]$ of formal series in many indeterminates is defined inductively as $R[[x_n]]$, where $R = F[[x_1, \ldots, x_{n-1}]]$.

1.12. Algebras of Skew Polynomials and Series. Let R be a ring, α – an *endomorphism* (i.e. a homomorphism of R into itself) and δ – an α-*derivation*, i.e. an additive mapping of R into R which satisfies $\delta(ab) = \delta(a)b + \alpha(a)\delta(b)$ for all $a, b \in R$. If, on the set of polynomials $\sum a_i x^i$ over R, a (non-commutative) multiplication is defined by the rule $xa = \alpha(a)x + \delta(a)$ (for $a \in R$), we obtain a ring denoted by $R[x; \alpha, \delta]$ (a *skew polynomial ring*). For example, the *Weyl algebra* $A_1(F)$ over a field F is defined as the algebra $F[y][x; 1, ']$, where $'$ is the usual derivative in $F[y]$ (i.e. the elements x and y are linked in $A_1(F)$ by the relation $xy = yx + 1$). Skew polynomial rings provide examples of non-commutative *principal ideal rings* (i.e. rings R in which every left ideal has the form Ra and every right one – aR). Namely, let D be a division ring, α – an automorphism and let δ be an α-derivation. Then $D[x; \alpha, \delta]$ is a principal ideal ring. This follows from the fact that Euclid's division algorithm is valid in $D[x; \alpha, \delta]$.

If R is a ring and α an endomorphism, then the ring $R[[x; \alpha]]$ of *skew power series* is defined as the set of formal series with an operation of multiplication induced by the rule $xa = \alpha(a)x$ (for $a \in R$). In this way one constructs the so-called Hilbert division rings. Let D be a division ring and α an automorphism. We will consider formal Laurent series $\sum_{i \geq n} a_i x^i$, where n is any integer, possibly negative. Introducing on these series an operation induced by the rule $xa = \alpha(a)x$ for $a \in R$ (whence $x^{-1}a = \alpha^{-1}(a)x^{-1}$), one obtains a ring denoted by $D[[x, x^{-1}; \alpha]]$. In fact, it turns out to be a division ring (called a *Hilbert division ring*). In particular, if $\Phi = F(y)$ is the field of rational functions and α – the automorphism of the F-algebra Φ defined by $\alpha(y) = y + 1$, then the center of the division ring $\Phi[[x, x^{-1}; \alpha]]$ coincides with F, and we obtain an example of a division ring which is infinite-dimensional over its center.

1.13. Free Algebras, Systems of Generators and Defining Relations, Identities. Free algebras are a non-commutative analog of polynomial rings. Let F be a field, $X = \{x_i \mid i \in I\}$ – a collection of (non-commuting) indeterminates (letters). A word of length $k \geq 0$ in X will mean a sequence $x_{i_1} \cdots x_{i_k}$ (when $k = 0$, we get the trivial word 1). Two words are equal, if their lengths are identical and the corresponding letters coincide (first – with the first one, etc). We will consider an (infinite-dimensional) vector space $F\langle X \rangle$, whose basis consists of all words in X. Multiplication in $F\langle X \rangle$ is induced by juxtaposition of words: $(\sum \alpha_u u)(\sum \beta_v v) = \sum \alpha_u \beta_v uv$. The algebra $F\langle X \rangle$ is called the *free algebra over F with (free) generating set X*. If R is any algebra over F, then it is easy to see that every mapping of X into R can be uniquely extended to a homomorphism of $F\langle X \rangle$ onto R. From this, we conclude that every algebra over F is a *homomorphic image* of a free algebra, i.e. there exists a surjective homomorphism (see Sect. 1.4) $\varphi : F\langle X \rangle \to R$. In order to see this, we will introduce the notion of a system of generators of an algebra. If V is a subset of R, then $\langle V \rangle$ will denote the smallest subalgebra of R, which contains V – i.e. the smallest subring (see Sect. 1.4), containing V and being a vector subspace of R over F (this subalgebra exists, as it equals the interection of all subalgebras of R which contain V). The elements of $\langle V \rangle$ have the form $\sum \alpha_I v_I$, where $v_I = v_{i_1} \cdots v_{i_n}$ with $v_i \in V$ (here multiplication is understood to be the product in R, rather than the formal one). We say that V is a *system of generators* of R, if $R = \langle V \rangle$. Let X be some collection of indeterminates, whose cardinality equals that of a system of generators V of the algebra R (for example, $V = R$). Then any one-to-one mapping $X \to V$ induces the desired surjective homomorphism $\varphi : F\langle X \rangle \to R$. Once the mapping $X \to V$ has been fixed (e.g. every letter $x \in X$ is represented by the same symbol as $\varphi(x) \in V$), then elements of the kernel $\operatorname{Ker} \varphi$ are called *relations of the algebra R with respect to the system of generators V*. If $\operatorname{Ker} \varphi$ is generated as an ideal by elements f_i $(i \in I)$, i.e. $\operatorname{Ker} \varphi = \sum_{i \in I} F\langle X \rangle f_i F\langle X \rangle$ as a sum of vector subspaces, then we say that $\{f_i \mid i \in I\}$ is a *system of defining relations of R with respect to the generating set V*, and the algebra $F\langle X \rangle / \operatorname{Ker} \varphi$, isomorphic (see Sect. 1.4) to the algebra R, is denoted as follows: $\langle X;\ f_i = 0\,(i \in I) \rangle$ (in this notation, reference to the base field is missing). For example, the Weyl algebra $A_1(F)$ can be defined as the algebra $\langle\, x, y;\ xy - yx - 1 = 0\,\rangle$ with two generators x, y and a single relation $xy - yx - 1 = 0$.

Similar constructions can be performed for algebras over a commutative ring k (see Sect. 1.6). In this case, the free k-algebra $k\langle\, X\,\rangle$ consists of formal finite sums $\sum \alpha_v v$, where $\alpha_v \in k$ and the v's are words in X. In precisely the same way, every k-algebra has a presentation $\langle\, X;\ f_i = 0\,(i \in I)\,\rangle$. In particular, every ring can be regarded as an algebra over $\mathbb{Z}$ (see Sect. 1.6), and therefore every ring has a presentation in terms of generators and defining relations.

Presentation of algebras by generators and relations makes it easy to define the free product. Let $R = \langle\, X;\ f_i = 0\,(i \in I)\,\rangle$ and $S = \langle\, Y;\ g_j = 0\,(j \in J)\,\rangle$. Then $R*_k S = \langle\, X \bigcup Y;\ f_i = 0,\ g_j = 0\,(i \in I, j \in J)\,\rangle$ is called the *free product of R and S over k*. Naturally, it is assumed here that $X \cap Y = \emptyset$. If R and S are algebras over a field F, then there is a basis of the free product $R *_F S$ consisting of 1 and words of the form $r_1 s_1 r_2 s_2 \ldots r_n s_n$, where r_i and s_i are elements, distinct from 1, of certain fixed bases of R and S respectively, and the factors r_1 and s_n can even be absent (this easily follows from the composition lemma – see Sect. 1.18 below). A non-zero element $f(x_1, \ldots, x_n) \in F\langle\, X\,\rangle$ (as well as the equation $f(x_1, \ldots, x_n) = 0$) is called an *identity*, or *identity relation*, of an algebra R, if $f(a_1, \ldots, a_n) = 0$ for any elements $a_1, \ldots, a_n \in R$. For example, the algebra of matrices $M_2(F)$ satisfies Hall's identity $[[x, y]^2, z]$, where $[x, y] = xy - yx$ is the *commutator* of the elements x and y. Every n-dimensional algebra satisfies the standard identity $S(n+1)$ of degree $n+1$:

$$\sum (-1)^{\operatorname{sgn} \pi} x_{\pi(1)} \ldots x_{\pi(n+1)} = 0\,,$$

where π runs over all permutations from the group S_{n+1}. Commutative rings satisfy the identity $[x, y] = 0$. The most important class of rings with identities are the matrix rings $M_n(K)$ over commutative rings K – they satisfy the identities $S(2n) = 0$ (Amitsur, Levitzki).

One of the important tasks of ring theory is to determine all identities of a given algebra. All identities in countably many indeterminates $X = \{x_1, x_2, \ldots\}$ of an algebra R, together with 0, form the *fully characteristic ideal* (or T-ideal) $T(R)$ of $F\langle\, X\,\rangle$ (i.e. an ideal invariant under all endomorphisms of $F\langle\, X\,\rangle$). The problem of describing all identities of the algebra R is equivalent to determining generators of $T(R)$ as a T-ideal. Thus far, every known T-ideal has finitely many generators. Of special interest is the question of finite generation of T-ideals for algebras over a field of characteristic 0 (it was raised in this case by Specht, and is referred to as *Specht's problem* [1]). For more details on identities, see Sect. 5.4; identities of algebras are specifically discussed in Part II.

---

[1] Recently, an affirmative solution of Specht's problem was obtained by A.R. Kemer (appeared in Algebra and Logic, Novosibirsk, vol.26, no.5 (1987) pp. 597 - 641).

Let $R = F[X]$ or $R = F\langle X \rangle$, and let R_n be the subspace of all homogeneous elements of degree n (and 0). Then $R = \bigoplus_{n \geq 0}$ (direct sum of subspaces), and $R_n R_m \subseteq R_{m+n}$. Any algebra R with subspaces R_m, $m \geq 0$, which satisfies the above relations, is called *graded.* Any graded algebra R can be completed by considering all, generally speaking infinite, sums $\sum_{n \geq 0} r_n$ (where $r_n \in R_n$), and inducing on them natural operations of addition and multiplication (where multiplying two elements one does not, as is easily seen, obtain infinitely many like terms). Completion of the algebra of polynomials $F[x_1, \ldots, x_n]$ is nothing else but the algebra of commutative formal series $F[[x_1, \ldots, x_n]]$. Completion of the free algebra $F\langle X \rangle$ is called the algebra of *non-commutative formal power series*, and is denoted $F\langle \langle X \rangle \rangle$.

1.14. Weyl Algebra $A_n(F)$ (or the algebra of differential operators in n variables) is generated over F by elements $x_1, \ldots, x_n$ and $y_1, \ldots, y_n$, which are subject to defining relations $[x_i, y_j] = \delta_{ij}$, $[y_i, y_j] = [x_i, x_j] = 0$. Elements of the form $x_{i_1} x_{i_2} \ldots x_{i_k} y_{i_1} y_{i_2} \ldots y_{i_m}$ (where $i_1 \leq \ldots \leq i_k$ and $j_1 \leq \ldots \leq j_m$), constitute a basis of $A_n(F)$. This follows from the 'composition lemma' (see Sect. 1.18). In other words, the elements of $A_n(F)$ are linear differential operators $\sum a_{\alpha,\beta} x^\alpha \partial^\beta$ (where $\alpha = (i_1, \ldots, i_k)$, $i_1 \leq \ldots \leq i_k$, $x^\alpha = x_{i_1} \cdot \ldots \cdot x_{i_k}$, $\beta = (j_1, \ldots, j_m)$, $j_1 \leq \ldots \leq j_m$, $\partial^\beta = \frac{\partial}{\partial x_{j_1}} \cdots \frac{\partial}{\partial x_{j_m}}$), with polynomial coefficients from the algebra $F[x_1, \ldots, x_n]$.

1.15. The Exterior Algebra (or *Grassman algebra*) G_n of rank n is generated over a field F by the variables $x_1, \ldots, x_n$, subject to defining relations $x_i^2 = 0$, $x_i x_j = -x_j x_i$ (such variables are called anti-commuting). As a basis of the algebra G_n one can take all words $x_{i_1} \cdot \ldots \cdot x_{i_k}$, where $i_1 < \ldots < i_k$. This again follows from the composition lemma. Consequently, dimension of the algebra G_n equals $1 + \binom{n}{1} + \binom{n}{2} + \ldots + \binom{n}{n-1} + 1 = 2^n$. Multiplication in G_n is often denoted by the 'wedge' $\wedge$: $x_i \wedge x_j = -x_j \wedge x_i$. If $f_i = \sum_{j=1}^n a_{ij} x_i$ $(1 \leq i \leq n)$ are n-linear forms, then $f_1 \wedge f_2 \wedge \ldots \wedge f_n = \det(a_{ij})\, x_1 \wedge \ldots \wedge x_n$ in the algebra G_n – where $\det(a_{ij})$ is the determinant of the matrix (a_{ij}). This equation may be adopted as a definition, and then used as a root of the whole theory of the determinant. One frequently considers Grassman algebras of countable rank: $G = \langle x_1, x_2, \ldots ; x_i^2 = 0, x_i x_j + x_j x_i = 0 \rangle$. We will note that this algebra satisfies the identity $[[x, y], z] = 0$, since homogeneous elements of even degree lie in its center. The algebras G_n and G are examples of graded algebras (see Sect. 1.13).

1.16. Clifford Algebra $C(n, f)$ **of a Quadratic Form** f. Let $f(x_1, \ldots, x_n) = \sum a_{ij} x_i x_j$ (with $a_{ij} = a_{ji}$) be a quadratic form in n variables over a field F of characteristic $\neq 2$. The algebra $C(n, f)$ is given by generators $e_1, \ldots, e_n$ and defining relations $e_i e_j + e_j e_i - 2a_{ij} = 0$. When $f = 0$, the algebra $C(n, f)$ becomes G_n. The basis of G_n considered above is also a basis of the algebra $C(n, f)$. This, once again, is a consequence of the composition lemma.

1.17 Universal Enveloping Algebra $\mathcal{U}(L)$ of a Lie Algebra L. Let L be a *Lie algebra* over F – finite-dimensional, for the sake of simplicity. This means that L is a finite-dimensional vector space over F, and the (non-associative) multiplication in L satisfies two identities: $x^2 = 0$ (anti-commutativity) and $(xy)z+(yz)x+(zx)y = 0$ (Jacobi identity). If, for example, A is an associative algebra, then defining on it the commutation operation $[a,b] = ab - ba$ we obtain a Lie algebra, denoted by $A^{(-)}$. For a given Lie algebra L we will construct an associative algebra $\mathcal{U}(L)$ such that L will be a Lie subalgebra of $\mathcal{U}(L)^{(-)}$. Let $a_1, \dots, a_n$ be a basis of the algebra L as a vector space over F. We will consider the multiplication table of the algebra L in this basis: $a_i a_j = \sum_{k=1}^n \alpha_{ij}^k a_k$ (with $\alpha_{ij}^k \in F$). In view of the equality $a_i a_j = -a_j a_i$, it can be assumed here that $i > j$. We will use $\mathcal{U}(L)$ to denote the (associative) algebra defined by the generators $a_1, \dots, a_n$ and relations $a_i a_j - a_j a_i - \sum \alpha_{ij}^k a_k = 0$, where $n \geq i > j \geq 1$ (see Sect. 1.13). It turns out, that words of the form $a_{i_1} \dots a_{i_k}$ (with $i_1 \leq \dots \leq i_k$) constitute a basis of the algebra $\mathcal{U}(L)$. This statement is known as the *Poincaré-Birkhoff-Witt theorem*, and it can also be derived from the composition lemma. It follows that the elements $a_1, \dots, a_n$ are linearly independent in $\mathcal{U}(L)$, and hence in $\mathcal{U}(L)^{(-)}$ they generate a subalgebra isomorphic to L. If L is an *abelian Lie algebra*, i.e. a Lie algebra with null multiplication, then $\mathcal{U}(L)$ is isomorphic to the algebra of polynomials in n variables.

1.18. Composition Lemma. Let $R = F\langle x_1, \dots, x_n \rangle$ be a free associative algebra (see Sect. 1.13). We will assume that the variables are ordered: $x_i > x_j$ whenever $i > j$. We will order the set of words in $\{x_i\}$ first according to length ($u > v$ if the lengths $l(u) > l(v)$), while words of equal length will be ordered lexicographically (first comparing, as in a dictionary, first letters; then, if they are identical – the second ones etc.) For example, $x_2x_3x_4 > x_2x_2x_5 > x_5$. If $f \in R \setminus \{0\}$, then f is a linear combination of pairwise distinct words, and $\bar{f}$ will denote the leading word of f (i.e. the largest word appearing in f with a non-zero coefficient). Two words u, v will be called linked by a word w, if $w = ux = yv$ for some x and y and the sub-words u and v 'intersect' in w (that is, $l(u) + l(v) > l(w)$). We will now introduce the notion of composition of elements in a free algebra. Let f and g be non-zero (possibly equal) elements of R, with $f = \alpha\bar{f} + \dots$ and $g = \beta\bar{g} + \dots$; assume moreover, that $\bar{f}$ and $\bar{g}$ are linked by some word $w = \bar{f}x = y\bar{g}$. Then the *composition* of f and g relative to w is the following element of the algebra R: $(f,g)_w = \beta f x - \alpha y g$. It is clear that the leading word of the composition is smaller than w (since if $u \geq v$ and $t \geq s$, then $ut \geq vs$).

Let now S be a subset of R. We say that S is closed under composition when the following conditions are satisfied:

1) if $f, g \in S$ and $(f,g)_w$ is their composition, then in the algebra $F\langle X \rangle$ the element $(f,g)_w$ can be written in the form $\sum \alpha_i a_i s_i b_i$, where $s_i \in S$, a_i, b_i are words, $\alpha_i \in F$, and the leading word of each summand $a_i s_i b_i$ is strictly smaller than w, i.e. $\overline{a_i s_i b_i} = a_i \bar{s_i} b_i < w$;

2) if s_1 and s_2 are distinct elements of S, then $\bar{s_1}$ does not contain $\bar{s_2}$ as a sub-word.

Composition Lemma. *Let S be a subset of the algebra $F\langle x_1, \ldots, x_n \rangle$, which is closed under composition. Then the set of all those words in $x_1, \ldots, x_n$ which are not sub-words of larger words $\bar{s}$ of elements $s \in S$, constitutes a basis of the algebra $A = \langle x_1, \ldots, x_n;\ s = 0\,(s \in S) \rangle$.*

Now in order to find bases of the algebras A_n, G_n, $C(n, f)$ and $\mathcal{U}(L)$ it suffices to verify, that the sets of their defining relations are closed under composition. From a technical point of view, while computing a composition $(f, g)_w$ one may discard those terms of the form αasb $(s \in S)$, for which $a\bar{s}b < w$, replacing the equality sign with congruence $\equiv$. In this notation, closure under composition can be expressed as $(f, g)_w \equiv 0$.

1.19. Localizations, Ore Condition, Classical Ring of Fractions. Let S be a subset of a ring (or an algebra) R, not containing zero. *Localization of R relative to S* is the ring (or algebra) $S^{-1}R = \langle R, s^{-1}\,(s \in S);\ ss^{-1} = s^{-1}s = 1\,(s \in S)\rangle$, i.e. the ring (algebra) obtained by adjoining to R (that is, to its system of generators and relations – see Sect. 1.13) new generators s^{-1} for all $s \in S$, and new defining relations $ss^{-1} = s^{-1}s = 1$. It is easy to deduce from general facts that the algebra $S^{-1}R$ does not depend on the choice of generators and relations of R. Elements of $S^{-1}R$ are all linear combinations of monomials $r_0 s_1^{-1} r_1 s_2^{-1} r_2 \ldots s_k^{-1} r_k$ $(s_i \in S,\ r_i \in R)$, with relations of equality between such elements not yielding, in practice, to analysis. This analysis becomes considerably more clear when a matrix construction of $S^{-1}R$ is used (see Sect. 5.3).

Of particular interest is the case when all elements of $S^{-1}R$ have the form $s^{-1}r$ $(s \in S,\ r \in R)$ and the mapping $r \mapsto r = r \cdot 1^{-1}$ from R to $S^{-1}R$ is an injection. In this situation we say that $S^{-1}R$ is the *classical left ring of fractions of R relative to S*, denoting it by $Q_{\text{cl}}(R, S)$.

We will say that the *left Ore condition relative to S* holds in R, when for any $s \in S$ and $r \in R$ there exist $s_1 \in S$ and $r_1 \in R$ such that $s_1 r = r_1 s$.

Theorem. *The classical left ring of fractions $Q_{\text{cl}}(R, S)$ exists if and only if elements of S are not zero divisors (see Sect. 1.1) and the left Ore condition relative to S is satisfied in R.*

If S is the set of all regular elements (i.e. non zero divisors) of R and if $Q_{\text{cl}}(R, S)$ exists, then we say that R is a *left Ore ring*, or a *ring with left Ore condition*. In this case the ring of fractions $Q_{\text{cl}}(R, S)$ is denoted by $Q_{\text{cl}}(R)$.

1.20. Modules. Let R be a ring. An abelian group $\langle M, + \rangle$ is called a *left R-module* (denoted by ${}_RM$), if a left action of elements of R on elements of M is defined: $rm \in M$ for $r \in R$, $m \in M$, with all axioms of a vector space being satisfied (including $1 \cdot m = m$). A *right R-module* M_R is defined in a symmetric way (R acts on M on the right: $(r, m) \mapsto mr$ for $r \in R$ and

$m \in M$). The concept of a module over a ring is a natural generalization of the notion of a vector space, and has a fundamental importance not only in algebra, but in all of mathematics. Modules over a ring R are essentially nothing but representations (homomorphisms) of R in the endomorphism rings of abelian groups. More precisely, if M is a left module over the ring R, then to every element $r \in R$ one can assign the homotethy $r_M : M \to M$,

$$r_M(m) = rm \quad (m \in M),$$

which is an endomorphism of the abelian group M. This defines a homomorphism of rings $R \to \operatorname{End} M$, $r \mapsto r_M$ (where $r \in R$). Conversely, if A is any additive abelian group and φ – a homomorphism of the ring R into the endomorphism ring of A, then A has a structure of a left R-module according to the formula $ra := \varphi(r)(a)$ (where $r \in R$ and $a \in A$).

One naturally associates with modules the concepts of a homomorphism, isomorphism, submodule and factor module. A mapping of left R-modules $\varphi : M \to N$ is called a *homomorphism*, if it preserves the module operations: $\varphi(m_1+m_2) = \varphi(m_1)+\varphi(m_2)$ and $\varphi(rm) = r\varphi(m)$ (with $m, m_1, m_2 \in M$ and $r \in R$). A subset V of M is a *submodule* if it is an additive subgroup of M, and $rv \in V$ for all $v \in V$, $r \in R$. Every homomorphism $\varphi : M \to N$ is associated with two submodules: the kernel $\operatorname{Ker} \varphi = \{m \in M \mid \varphi(m) = 0\} \subseteq M$, and the image $\operatorname{Im} \varphi = \{n \in N \mid \exists m \in M \text{ with } \varphi(m) = n\} \subseteq N$. A homomorphism φ is called an *embedding* (or *injection*) if $\operatorname{Ker} \varphi = \{0\}$. A bijective (injective and onto) homomorphism is called an *isomorphism.*

For any submodule $V \subseteq M$ one can naturally define a *factor module* M/V. As an abelian group it is isomorphic to the factor group M/V, while multiplication by elements of R is defined by the formula $r(m+V) = rm+V$. In this setting, the following isomorphism theorem holds: $M/\operatorname{Ker} \varphi \simeq \operatorname{Im} \varphi$, where $\varphi : M \to N$ is a homomorphism. Every module has two improper submodules $\{0\}$ and M. A module $M \neq \{0\}$ which has no proper submodules, is called *simple* (or *irreducible*). Such modules play an important part in many areas of ring theory and representation theory.

One of the simplest examples of a module over a ring R is the ring itself, together with its multiplication. This module is called *regular* (left – ${}_RR$, or right – R_R, depending on the context). We can now view left (or right) ideals as submodules of the regular module ${}_RR$ (or R_R, respectively).

We will now describe several natural types of modules. A G-module of a group G is a vector space M, on which elements $g \in G$ act in such a way that $(g_1g)m = g_1(gm)$, $em = m$ and $g(\alpha m + \beta m_1) = \alpha gm + \beta gm_1$ (where $g, g_1, e \in G$, $m, m_1 \in M$ and α, β are scalars). Every given G-module can be naturally made into a module over the group algebra FG according to the formula $(\sum \alpha_i g_i)m = \sum \alpha_i g_i m$. On the other hand, a G-module structure on M is equivalent to a homomorphic representation of G by non-singular linear transformations of the vector space M. Because of this, the description of modules over group algebras is the central goal of the theory of *group representations*. As an example, we will remark that the theorem on the Jordan normal form of matrices over $\mathbb{C}$ can be interpreted as a statement about the structure of all finite-dimensional modules over the polynomial algebra $\mathbb{C}[x]$ (see Sect. 1.25 below).

A *Lie module* M over a Lie algebra L is an abelian group together with an action $am \in M$ (for $a \in L$ and $m \in M$), for which $(ab)m = a(bm) - b(am)$ for all $a, b \in L$ and $m \in M$. Every Lie module M induces an ordinary left module M over the universal enveloping algebra $\mathcal{U}(L)$ of the algebra L, and conversely. Hence the theory of Lie modules (or, in other words, the representation theory of Lie algebras) is equivalent to the theory of modules over universal enveloping algebras.

A module ${}_R M_S$ (i.e. a left R-module and right S-module) is called a *bimodule*, if $(rm)s = r(ms)$ for all $r \in R$, $m \in M$ and $s \in S$. The bimodule ${}_R R_R$ will be called the regular bimodule; its subbimodules are ideals of R. The notion of a bimodule is related to an important idea of a Morita context (R, V, W, S) and the ring $\begin{pmatrix} R & V \\ W & S \end{pmatrix}$ which corresponds to it. We will now describe this concept. Let R, S be rings and ${}_R V_S$, ${}_S W_R$ – bimodules, and let $(v, w) \in R$, $(w, v) \in S$ ($v \in V$, $w \in W$) denote binary operations which obey the 'associative laws' $(v, w)v_1 = v(w, v_1)$, $w_1(v, w) = (w_1, v)w$ and are bilinear: $(x_1 + x_2, y) = (x_1, y) + (x_2, y)$, $(x, y_1 + y_2) = (x, y_1) + (x, y_2)$, $(ax, y) = a(x, y)$, $(xb, y) = (x, by)$ and $(x, ya) = (x, y)a$ (where, depending on the operation, either $x, x_i \in V$, $y, y_i \in W$, $a \in R$ and $b \in S$, or $x, x_i \in W$, $y, y_i \in V$, $a \in S$ and $b \in R$). Then the four-tuple (R, V, W, S) is called a *Morita context* (for the reader familiar with category theory, we will remark that a Morita context is precisely a preadditive category with two objects E_1 and E_2, where $R = \mathrm{Hom}(E_1, E_1)$, $S = \mathrm{Hom}(E_2, E_2)$, $V = \mathrm{Hom}(E_2, E_1)$ and $W = \mathrm{Hom}(E_1, E_2)$). The ring associated with the Morita context is $\begin{pmatrix} R & V \\ W & S \end{pmatrix}$, in which operations are derived from the usual matrix operations and operations in the Morita context (this ring is the endomorphism ring of the object $E_1 \oplus E_2$ in the generated additive category). A Morita context is called *surjective* if $(V, W) = R$ and $(W, V) = S$. A *standard Morita context* has the form $(R, R^n, {}^nR, M_n(R))$, where $R^n = V$ is the module consisting of row vectors of length n, ${}^nR$ is the module of column vectors of length n and (v, w), (w, v) are the usual matrix multiplications on $v \in V$ and $w \in W$. A standard Morita context turns out to be surjective, which implies – as we will see below – that rings R and $M_n(R)$ are Morita equivalent (see Sect. 4.8).

An *exact sequence* $\ldots \to A_{n-1} \to A_n \to A_{n+1} \to \ldots$ of module homomorphisms is a sequence in which the image of each homomorphism is equal to the kernel of the succeeding one. For example, the sequence $0 \to A \xrightarrow{\varphi} B$ is exact if and only if φ is a monomorphism. Exactness of $B \xrightarrow{\varphi} C \to 0$ is equivalent to the statement that φ is an epimorphism.

1.21. Free, Projective and Injective Modules. Let R be a ring, X – a set. The *free left module* $M_R(X)$ over R with basis X consists of all finite formal sums $\sum r_i x_i$, in which the element $1x_i$ is identified with x_i. Operations of addition and multiplication by elements of R are defined coordinatewise and by the distributive law. A free module with basis X is characterized by the property that it is generated by the set X, and every mapping $\varphi : X \to L$ into a left R-module L uniquely extends to a homomorphism from $M_R(X)$ into L. If $X = \{x\}$ is a singleton, then by mapping the element $rx \in M_R(x)$ to r we obtain an isomorphism $M_R(x) \simeq {}_R R$. Structure of any free module is made more clear by the construction of a direct sum of modules.

We will first define a direct product of modules. Let M_i $(i \in I)$ be a family of left R-modules. In the Cartesian product of sets $\prod_{i \in I} M_i$ we will define module operations component-wise: $(m_i) + (n_i) = (m_i + n_i)$, $a(m_i) = (am_i)$, where $m_i \in M_i$ and $a \in R$. The module obtained in this way is called the *direct product*, and is denoted by $\prod_{i \in I} M_i$. *Direct sum* $\bigoplus M_i$ is the submodule of $\prod M_i$ consisting of those sequences $(m_i)_{i \in I}$ in which all but finitely many components equal zero. Each module M_i can be identified with the submodule of the direct sum, containing those sequences whose all but the i-th components are zero. With this identification we see that the module $\bigoplus M_i$ is generated by the submodules M_i. This allows to give an internal characterization of decomposability of a module into a direct sum of its submodules M_i $(i \in I)$. Namely, $M = \bigoplus M_i$ if and only if M is generated by the submodules M_i (i.e. equals their sum), and the intersection $M_j \cap \sum_{i \neq j} M_i$ is zero for all $j \in I$. In particular, M decomposes into a direct sum of two submodules M_1 and M_2 if and only if $M_1 + M_2 = M$ and $M_1 \cap M_2 = (0)$.

We can now assert that a free module $M_R(X)$ coincides with the direct sum of submodules $M_R(x)$ $(x \in X)$, each of which is isomorphic to the left regular module R, i.e. $M_R(X) \simeq \bigoplus {}_R R$.

Free modules have the following fundamental property (which, unlike their definition, does not depend on the choice of a basis X): every homomorphism φ of a free module F into a factor module V/U can be 'lifted' to V, i.e. there is a homomorphism $\psi : F \to V$, whose composition with the natural homomorphism $V \to V/U$ equals φ (if $\varphi(x) = v_x + U$ then we can let $\psi(x) = v_x$, where $x \in X$). This property defines the class of *projective* modules. A definition of a projective module P can be given in the language of diagrams: every diagram with an exact row

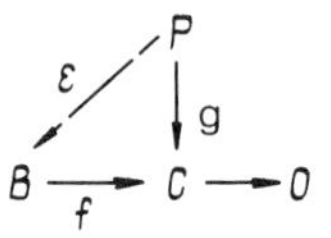

can be completed to a commutative diagram, i.e. $\exists \varepsilon$ such that $g = f\varepsilon$.

Theorem 1. *A module P is projective if and only if it is a direct summand of a free module:* $P \oplus N \simeq M_R(X)$.

One may get the impression that every projective module, as a direct summand of a direct sum of regular modules, is also a direct sum of regular modules (i.e. is free). This, however, is far from the truth – a regular module itself may decompose into a direct sum of modules which are not free (but necessarily projective, according to Theorem 1). Let, for example, $R = M_n(F)$ be the ring of $n \times n$ matrices over a field F. Then ${}_RR = V_1 \oplus \ldots \oplus V_n$, where V_i is the left ideal of matrices whose i-th column may be non-zero, and none of the submodules V_i is free.

Another fundamental concept of the theory of modules is obtained by reversing all arrows in the diagram used to define a projective module. Namely, a left module Q is *injective*, if every diagram

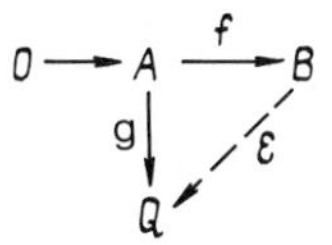

with exact top row can be completed to a commutative diagram: $\exists \varepsilon$ such that $g = \varepsilon f$. Among the examples of injective $\mathbb{Z}$-modules we have the *divisible* (or *complete*) *abelian groups*, i.e. abelian groups in which every equation $nx = a$ ($a \in A$, $n \in \mathbb{Z}$) can be solved for x. It can be shown that every injective $\mathbb{Z}$-module is a divisible abelian group. It is well known that every abelian group can be embedded in a divisible one – a similar situation is valid in the case of modules.

Theorem 2. *Any left R-module can be embedded in an injective module.*

Among all injective modules containing M there exists the smallest one, which is determined up to isomorphism and called the injective hull (or envelope) of the module M (for details, see Sect. 3.5).

Injective modules are characterized by the fact that they separate as direct summands in all of their module extensions: if $Q \subseteq N$, then $Q \oplus M = N$ for some $M \subseteq N$. The regular module may not be injective (e.g. if $R = \mathbb{Z}$ – the ring of integers, or $R = F[x]$ – the algebra of polynomials). In the case when $R = M_n(F)$ is the ring of matrices over a field, the regular module is always injective; moreover, in this situation all modules are both projective and injective. In general, if the left regular module ${}_RR$ is injective, the ring R is called *(left) self-injective.* For finite-dimensional algebras, self-injectivity is equivalent to the algebra being quasi-Frobenius – a notion arising in the representation theory of finite groups (see Sect. 3.12). For a somewhat more detailed treatment of projective and injective modules, see Sections 3.4 and 3.5.

1.22. Categories and Functors. We will recall that a *category* K consists of a class of objects $a, b, c, \ldots$ and a class of morphisms (or transformations) $\alpha : a \to b$ etc. between them. Morphisms $\alpha : a \to b$ and $\beta : b \to c$ having a common codomain and domain respectively, can be 'composed': $\beta\alpha : a \to c$ is a morphism. In addition, the composition operation (which, as we can see, is not always defined) is associative, and for every object a there exists an 'identity' morphism $1_a : a \to a$ such that $\alpha 1_a = \alpha$ and $1_b \alpha = \alpha$ for any $\alpha : a \to b$. A morphism $\alpha : a \to b$ is called an isomorphism in a category K if there exists a morphism $\beta : b \to a$ for which $\alpha\beta = 1_b$ and $\beta\alpha = 1_a$. As examples of categories we list the categories R–Mod (left R-modules and their homomorphisms), Mod-R (right R-modules), Ring (rings and their homomorphisms), Φ-Alg (Φ-algebras), Set (sets and mappings between them) and (R, S)-Bimod ((R, S)-bimodules). The category $\mathbb{Z}$-Mod is simply the category of abelian groups (and is denoted by Ab). An object a in a category K is called *initial (terminal)*, if for every object x from K there exists a unique morphism $a \to x$ (resp. $x \to a$) in the category K. Every initial (and terminal) object is unique up to isomorphism. A *(covariant) functor* $F : K \to K_1$ from category K to category K_1 is a pair of mappings: $a \mapsto F(a)$ (from objects of K to objects of K_1) and $\alpha \mapsto F(\alpha)$ (from morphisms of K to morphisms of K_1), for which $\alpha : a \to b$ implies $F(\alpha) : F(a) \to F(b)$, and $F(1_a) = 1_{F(a)}$, $F(\alpha\beta) = F(\alpha)F(\beta)$. A *contravariant functor* $F : K \to K_1$ 'reverses the arrows', i.e. if $\alpha : a \to b$ then $F(\alpha) : F(b) \to F(a)$, and $F(1_a) = 1_{F(a)}$, $F(\alpha\beta) = F(\beta)F(\alpha)$.

We will introduce a natural definition of isomorphism of categories. A functor $F : K \to K_1$ is called an *isomorphism of categories*, if there exists a functor $G : K_1 \to K$ such that $FG = 1$ and $GF = 1$, where 1 is the identity functor (on K_1 and K respectively). The categories Φ-Mod and Mod-Φ, where Φ is a commutative ring, provide an example of isomorphic categories (every left Φ-module M can be made into a right module by setting $ma := am$ for $a \in \Phi, m \in M$, and conversely). If rings R and S are isomorphic, then the categories R-Mod and S-Mod are also isomorphic. More useful is the notion of *equivalence of categories.* Two functors $F, G : K \to K_1$ are called equivalent ($F \sim G$) if for every object $a \in K$ the objects $F(a)$ and $G(a)$ are naturally isomorphic, i.e. if for each a there exists an isomorphism $\alpha_a : F(a) \to G(a)$ such that whenever $f : a \to b$ is a morphism, we have $G(f)\alpha_a = \alpha_b F(b)$. Categories K and K_1 are called *equivalent*, if there exist functors $F : K \to K_1$ and $G : K_1 \to K$ such that the functors GF and FG are equivalent to the identity functors 1_K and 1_{K_1} ($1_K(a) = a$, $1_K(\alpha) = \alpha$). Categories which are equivalent as defined above are indistinguishable in the sense of category theory, and can be regarded as 'identical'. Categories R-Mod and $M_n(R)$-Mod are among the important examples of equivalent categories. Rings R and S for which the module categories R-Mod and S-Mod are equivalent, are called *Morita-equivalent* or *equivalent in the sense of Morita* (and denoted $R \sim S$). This way, the rings R and $M_n(R)$ are Morita-equivalent. We will later see (cf.

Sect. 4.8) that this notion is left-right symmetric, and that $R \sim S$ if and only if there exists a surjective Morita context (R, V, W, S). We will remark that centers of Morita-equivalent rings are isomorphic. In particular, commutative rings are Morita-equivalent if and only if they are isomorphic.

1.23. Functors Ext **and** Tor. Let M, N be left R-modules. We will denote by $\mathrm{Hom}(M, N)$ the abelian group of all homomorphisms from M to N: if $\alpha : M \to N$ and $\beta : M \to N$, then $(\alpha + \beta)(x) = \alpha(x) + \beta(x)$. With the first argument fixed we obtain a functor, covariant in the second argument, $\mathrm{Hom}_R(-, N) : R - \mathrm{Mod} \to \mathrm{Ab}$. Analogously, by fixing the second argument we obtain a contravariant functor $\mathrm{Hom}_R(-, N) : R - \mathrm{Mod} \to \mathrm{Ab}$. Moreover, we are dealing here with a *bifunctor* Hom, covariant in the second argument and contravariant in the first one. This means that for any left R-modules M, N, M', N' and module homomorphisms $f : M' \to M$, $g : N \to N'$ there is a homomorphism $\mathrm{Hom}(f, g) : \mathrm{Hom}_R(M, N) \to \mathrm{Hom}_R(M', N')$, which satisfies the properties: $\mathrm{Hom}(ff', g'g) = \mathrm{Hom}(f', g')\mathrm{Hom}(f, g)$ (where $f' : M'' \to M'$, $g' : N' \to N''$), and $\mathrm{Hom}(1, 1) = 1$. Such homomorphism is defined by the formula $\mathrm{Hom}(f, g)(\alpha) = g\alpha f$ (for $\alpha \in \mathrm{Hom}_R(M, N)$).

Homology theory, with use of the so-called derived functors, allows to construct from the bifunctor Hom_R a sequence of bifunctors $\mathrm{Ext}^n_R(-, -)$ (for $n \geq 1$) of the same 'variance' as Hom_R (see Vol. 11, Sect. 21). We cannot describe that construction here, but in proofs it is usually sufficient to use properties of these bifunctors and the fact of their existence. We will quote the more important properties: for any module X and an exact sequence of homomorphisms $0 \to A \to B \to C \to 0$ there is an exact sequence

$$
\begin{aligned}
0 \to \mathrm{Hom}(X, A) \to \mathrm{Hom}(X, B) \to \mathrm{Hom}(X, C) \to \mathrm{Ext}^1(X, A) \to \\
\to \mathrm{Ext}^1(X, B) \to \mathrm{Ext}^1(X, C) \to \mathrm{Ext}^2(X, A) \to \dots
\end{aligned}
$$

and a similar one (with the letters A and C interchanged) when X appears in the second argument. In this way, $\mathrm{Ext}^1_R(-, -)$ provides a measure of 'right non-exactness' of the bifunctor $\mathrm{Hom}(-, -)$.

Theorem. *For projective modules P (injective modules Q), and those modules only, the functor $\mathrm{Hom}_R(P, -)$ (respectively, $\mathrm{Hom}_R(-, Q)$) is right exact, i.e. $\mathrm{Ext}^1_R(P, Y) = 0$ (resp. $\mathrm{Ext}^1_R(X, Q) = 0$).*

The abelian group $\mathrm{Hom}(M, M)$, where M is a right R-module, has a natural operation of multiplication (composition), which gives it a ring structure. $\mathrm{End}_R(M, M)$ is called the *endomorphism ring* of the module M. In addition, the module M then becomes a left module over the ring $\mathrm{End}_R M$: if $\varepsilon \in \mathrm{End}_R M$ and $m \in M$, then $\varepsilon m = \varepsilon(m)$, and the associative law $\varepsilon(ma) = (\varepsilon m)a$ shows that M is also an $(\mathrm{End}_R M, R)$-bimodule.

Let now M be a right and N a left R-module. We will define their tensor product. Let $\mathbb{Z}(M \times N)$ be a free abelian group generated by the set of all ordered pairs $m \times n$ ($m \in M$, $n \in N$). In it, we will consider the subgroup U generated by elements of the form $(m + m_1) \times n - m \times n - m_1 \times n$, $m \times (n + n_1) - m \times n - m \times n_1$ and $ma \times n - m \times an$, where $m, m_1 \in M$, $n, n_1 \in N$ and $a \in R$. The abelian group $\mathbb{Z}(M \times N)/U$ is called the *tensor product* $M \otimes_R N$. Denoting the coset $m \times n + U$ by $m \otimes n$ we see that the tensor product $M \otimes N$ is generated by the tensors $m \otimes n$, and that the identities $(m + m_1) \otimes n = m \otimes n + m_1 \otimes n$, $m \otimes (n + n_1) = m \otimes n + m \otimes n_1$ and $ma \otimes n = m \otimes an$ hold in it. The tensor product also turns out to be a bifunctor, covariant in both arguments: for homomorphisms $f : M_R \to M'_R$ and $g : {}_RN \to {}_RN'$ the product $f \otimes g$ is defined by the formula $(f \otimes g)(m \otimes n) = f(m) \otimes g(n)$, and extended by linearity to sums of tensors.

In homology theory one also constructs for the bifunctor $- \otimes -$ a sequence of covariant bifunctors Tor_n, which measure the 'inexactness on the left' of the tensor product: for any right module X and any exact sequence of left module homomorphisms $0 \to A \to B \to C \to 0$, there is an exact sequence $\ldots \to \mathrm{Tor}_2(X, C) \to \mathrm{Tor}_1(X, A) \to \mathrm{Tor}_1(X, B) \to \mathrm{Tor}_1(X, C) \to X \otimes A \to X \otimes B \to X \otimes C \to 0$. A similar statement is valid with respect to the second argument. We will see in Sect. 3 that the functors Ext^n and Tor_n are used in the theory of rings and modules, primarily with the aim of introducing to this theory the geometric idea of dimension of a module and a ring.

1.24. Baer and Jacobson Radicals. Let A be a finite-dimensional commutative algebra over the field of complex numbers $\mathbb{C}$. An element a is called *nilpotent*, if $a^n = 0$ for some n. A classical theorem of Weierstrass asserts that if A (as above) contains no nilpotent elements, then A is isomorphic to a direct sum of a certain number of copies of the field $\mathbb{C}$. In order to give this theorem a form needed by us, we will introduce the concept of the radical of an algebra A (this notion did not yet exist in Weierstrass' time). Namely, the radical $\mathrm{Rad}\, A$ will denote the set of all nilpotent elements of the algebra A (again, A is the algebra described above). Then $\mathrm{Rad}\, A$ is an ideal, and $A/\mathrm{Rad}\, A$ does not contain nilpotent elements – i.e. its structure is described by Weierstrass' theorem. In addition, we have a decomposition of A into the sum $\mathrm{Rad}\, A + B$, where B is a subalgebra of A isomorphic to $A/\mathrm{Rad}\, A$, $B \cap \mathrm{Rad}\, A = (0)$ (as we say, $\mathrm{Rad}\, A$ splits in A as a direct summand) and the subalgebra B is unique. If A is any commutative ring, $\mathrm{Rad}\, A$ can also be defined as the set of nilpotent elements (i.e. the set of 'radicals of zero': $\sqrt{0} = \mathrm{Rad}\, A$), and turns out to be an ideal of A.

Let now A be an arbitrary (non-commutative) finite-dimensional algebra over the field $\mathbb{C}$. In this case, generally speaking, the set of nilpotent elements does not form an ideal. The radical of such an algebra was first defined by F.E. Molin. In contemporary terms, the definition can be formulated as follows. An ideal (ring, algebra) I is called *nilpotent* if $I^n = 0$ for some n. The sum of all nilpotent ideals of A is again a nilpotent ideal, containing every nilpotent ideal of A, i.e. it constitutes the largest nilpotent ideal of A. This ideal is called the *radical* $\mathrm{Rad}\, A$ of the algebra A (here it is not necessary for the base field to be the field of complex numbers). The radical of $A/\mathrm{Rad}\,A$ is zero or, in other words, the algebra $A/\mathrm{Rad}\,A$ is *semisimple with respect to* Rad (i.e. *semiprime*). A theorem of Molin asserts that $\mathrm{Rad}\,A$ splits in A as a direct summand, and that any semisimple algebra over the field $\mathbb{C}$ is a direct sum of matrix algebras $M_n(\mathbb{C})$. Further results of this nature for finite-dimensional algebras over the field $\mathbb{R}$ were obtained by E. Cartan and, in their final form (for an arbitrary base field), by Wedderburn (see Sect. 2). Uniqueness – up to isomorphism – of the semisimple algebra B in the decomposition $A = \mathrm{Rad}\,A \oplus B$ was demonstrated much later by A.I. Mal'tsev.

We will remark – without going into details – that there is a certain analogy between the results described above, and the theory of symmetric bilinear forms on finite-dimensional vector spaces. If $g : V \times V \to F$ is such a form (with F – a field and V – a finite-dimensional space over F) then the radical of V, relative to g, is the subspace $V^{\perp} = \{v \in V \mid g(v, V) = 0\}$. In this situation, $V/V^{\perp}$ is equipped with a non-degenerate bilinear symmetric form $\bar{g}$, induced by the form g. In that sense, the formula $\mathrm{Rad}\,(V/\mathrm{Rad}\,V) = 0$ holds in this case as well. There are interesting connections between the notions of the radical of a symmetric bilinear form and the radical of a finite-dimensional algebra. For example, let $A = C(n, f)$ be the Clifford algebra of a quadratic form $f(x_1, \ldots, x_n) = \sum a_{ij}x_ix_j$ (where $a_{ij} = a_{ji} \in F$). We will consider the corresponding bilinear form $g(\sum x_ie_i, \sum y_je_j) = \sum a_{ij}x_iy_j$. It turns out that the radical of the algebra $C(n, f)$ is generated, as an ideal, by the radical of the form g. If A is a finite-dimensional algebra over a field F of characteristic 0, then its radical coincides with the radical of the bilinear form given on the vector space A by $(a, b) = \mathrm{tr}\,(ab)$ $(a, b \in A)$, where for an element $a \in A$ its trace $\mathrm{tr}\,(a)$ is defined as the trace of the left multiplication $x \mapsto ax$ $(x \in A)$ by a. This last statement is connected with the fact that a matrix from $M_n(F)$ is nilpotent if and only if the traces of its powers are all zero.

The concept of a radical can be generalized to the class of infinite-dimensional algebras or arbitrary rings, but the way of doing so will be far from unique. We will concentrate on two of the most important ones – the (lower nil) Baer radical, and the Jacobson radical.

A ring R is called *semiprime* (see Sect. 1.8) if it contains no non-zero nilpotent ideals. The *Baer radical* is the smallest ideal $\mathrm{Rad}\,R$ of R such that $R/\mathrm{Rad}\,R$ is semiprime. $\mathrm{Rad}\,R$ can be constructed by means of a transfinite inductive process: Let N_1 be the sum of all nilpotent ideals of the ring R, N_2 – the sum of all ideals nilpotent modulo N_1, etc. Setting $N_\alpha = \bigcup_{\beta<\alpha} N_\beta$ for limit ordinals α, we find that this transfinite procedure stabilizes, and its result is the Baer radical. Baer radical clearly may not be nilpotent, but it is *locally nilpotent* as a ring, i.e. any finite set of its elements generates a nilpotent ring (without 1). In terms of elements, the Baer radical has the following convenient description. An element x of the ring R is called *strongly nilpotent*, if for any sequence of elements $a_1, a_2, \ldots, a_n, \ldots$ of R the sequence $x_1 = x, x_2, x_3, \ldots$, where $x_{n+1} = x_na_nx_n$, contains zero. It turns out that $\mathrm{Rad}\,R$ consists precisely of all strongly nilpotent elements.

In the finite-dimensional situation, one can look at $\operatorname{Rad} A$ from a module-theoretic point of view. Let ${}_AV$ be a simple left A-module. Then $(\operatorname{Rad} A)V$ is a submodule of V, which has to equal zero (since otherwise $(\operatorname{Rad} A)V = V$, and so $(\operatorname{Rad} A)^n V = V$, which contradicts $(\operatorname{Rad} A)^n = (0)$ for some n). This implies that $\operatorname{Rad} A$ is contained in the intersection of *annihilators* $\operatorname{Ann} V = \{x \in A \,|\, xV = (0)\}$ of all simple left A-modules V. It isn't difficult to show that $\operatorname{Rad} A$ is in fact equal to this intersection (and equals the intersection of annihilators of all simple right A-modules as well). Let now R be any ring. The *Jacobson radical* $J(R)$ is the intersection of annihilators of all simple left R-modules. $J(R)$, as it turns out, has the following properties:

1) $J(R)$ equals the intersection of all maximal left ideals of the ring R, that is, proper left ideals which are not contained in any other proper ideals of R;

2) $J(R)$ is the set of those $a \in R$ for which $1 - xay$ is invertible for any $x, y \in R$ (such elements are called quasi-regular, and the radical $J(R)$ itself is sometimes called the quasi-regular radical of R);

3) $J(R)$ coincides with its right analogue (i.e. with the intersection of annihilators of all right simple R-modules, as well as with the intersection of all maximal right ideals of the ring).

Significance of the Jacobson radical is clearly seen in the example of a commutative *Banach algebra* A over the field $\mathbb{C}$ (i.e. A is a normed complete space with norm $\|x\| \geq 0$, where $\|xy\| \leq \|x\| \cdot \|y\|$ and, consequently, the algebra operations in A are continuous relative to that norm). If $\mathbf{m}$ is a maximal ideal of A, then it is easy to see that $A/\mathbf{m}$ is a Banach field extension of $\mathbb{C}$, i.e. by the Gelfand-Mazur theorem $A/\mathbf{m} \simeq \mathbb{C}$. Let $\chi_{\mathbf{m}} : A \to A/\mathbf{m} \simeq \mathbb{C}$ be the canonical epimorphism. If $x \in A$, then one can assign to it a function $\hat{x}$ on the set $\mathcal{M}$ of maximal ideals: $\hat{x}(\mathbf{m}) = \chi_{\mathbf{m}}(x)$. The space $\mathcal{M}$ is endowed with the so-called Zariski topology, in which the mapping $x \mapsto \hat{x}$ is a continuous homomorphism of A into the ring of continuous complex-valued functions $\mathbb{C}_c(\mathcal{M})$ on $\mathcal{M}$. The kernel of this mapping happens to be $J(A)$. We see that in this case $J(A)$ measures the deviation of A from a subring of the ring of continuous functions on a compact topological space.

Finally, we will consider the example of local rings. Recall that a ring R is *local* if all of its non-invertible elements form a two-sided ideal P. Since a proper one-sided ideal of R contains no invertible elements, it must be contained in P, i.e. P is the unique maximal one-sided ideal and so, by property (1) of the Jacobson radical, $J(R) = P$. Aside from this, in the factor ring $R/J(R)$ all non-zero elements are invertible, i.e. $R/J(R)$ is a division ring.

It can be shown that the radical of a ring of matrices over a ring R is equal to the ring of matrices over the radical of R: $J(M_n(R)) = M_n(J(R))$.

We also note that $\operatorname{Rad} R \subseteq J(R)$, and the inclusion may be proper – e.g if $R = F[[x]]$ is the algebra of power series, R contains no nilpotent elements and hence $\operatorname{Rad} R = 0$. On the other hand, it is local with a unique maximal ideal $xF[[x]]$, which means that $J(R) = xR$. We will add that in general the structure of $J(R)$ can be made arbitrarily complex; for example, there exist simple rings (without 1) equal to their Jacobson radical.

1.25. Some Classes of Modules. We will describe a few important classes of modules. Among them we count the already introduced simple modules. In many situations simple modules are 'building blocks' from which other modules are constructed. For example, let R be an algebra over a field F. We will consider an arbitrary R-module V, which is a finite-dimensional vector space over F (i.e. a finite-dimensional R-module). Then in V one can construct (not uniquely, in general) a sequence of submodules $V_0 = 0 \subset V_1 \subset V_2 \subset \ldots \subset V_n = V$, such that V_{i+1}/V_i are simple R-modules for all $i = 0, \ldots, n-1$. Indeed, for V_1 one can take any non-zero submodule of V whose dimension over F is the smallest possible; for V_2 one can choose any submodule, strictly containing V_1, with the smallest possible dimension, etc.

It turns out that the simple modules V_{i+1}/V_i are invariants of the module V, i.e. the set of these modules does not depend on the choice of the original chain of submodules (this is a statement of the Jordan-Hölder theorem, see Sect. 3). In this fashion with finite-dimensional modules (in general, with so-called modules of finite length) one can associate simple modules which, to a large extent, determine their structure.

Another important class consists of the so-called completely reducible modules. A module M is *completely reducible* if it is isomorphic to a direct sum of simple modules. Clearly, completely reducible modules are built of simple modules in a particularly 'nice' way. It is sometimes the case (which is an especially important one) that all modules (right and left) over a ring R are completely reducible. Such rings are called *classically semisimple.* This term has its roots in the theory of finite-dimensional algebras. It was already noticed by the founders of this theory that any module over a finite-dimensional semisimple algebra is completely reducible. Rings with this property, i.e. those rings over which every module is completely reducible, became known as *semisimple* and, in time, classically semisimple. A theorem of Maschke asserts that a group algebra FG of a finite group G over a field F of characteristic 0, or whose characteristic does not divide the order of G, is semisimple. FG is therefore in this case a clasically semisimple ring. Above (Sect. 1.8) we introduced the notion of a left (right) Artinian ring. It turns out that every Jacobson semisimple left (right) Artinian ring is classically semisimple. Moreover, every classically semisimple ring is Artinian (both left and right), i.e. the classes of semisimple Artinian and classically semisimple rings coincide. In some cases not all, but only finite-dimensional modules over an algebra R are completely reducible. Cartan's theorem on complete reducibility in the theory of Lie algebras states that every finite-dimensional module over a semisimple (relative to the radical in the class of Lie algebras) Lie algebra over a field of characteristic 0 is completely reducible (as a Lie algebra module). This in particular implies that every finite-dimensional module over the universal enveloping algebra of a semisimple Lie algebra over a field of characteristic 0 is completely reducible.

We conclude with yet another important class of modules – the indecomposable modules. A module V is called *indecomposable* if it is not isomorphic to a direct sum of two non-zero modules. Indecomposable modules (just as the simple ones) are encountered in classical mathematics. For instance, the well-known Jordan's theorem about the normal form of matrices essentially states that any finite-dimensional module over the polynomial algebra $\mathbb{C}[x]$ is a direct sum of indecomposable modules, while every indecomposable module has a basis $e_1, \ldots, e_k$ in which left multiplication by x is given by the Jordan block $xe_1 = \alpha e_1 + e_2$, $xe_2 = \alpha e_2 + e_3$, ..., $xe_k = \alpha e_k$, and the resulting Jordan matrix, for any given module, is unique up to the order of blocks. In the general case, every finite-dimensional module over an algebra is isomorphic to a direct sum of indecomposable modules, and such decomposition is unique modulo the order of summands. This follows from the Krull-Schmidt theorem which, in its full generality, also applies to modules of finite length (see Sect. 3). For any finite-dimensional algebra A there are only finitely many non-isomorphic simple A-modules and, in principle, their structure is understood well enough. In contrast, indecomposable modules over a finite-dimensional algebra A can be of very complex nature (e.g. they can have arbitrarily large dimension) and it is those modules – not the simple ones – which turn out to carry in them fundamental information about all other A-modules (naturally, when A is not a semisimple algebra, for otherwise complete reducibility implies that simple and indecomposable modules are identical). Study of indecomposable modules over finite-dimensional algebras became the core of a program for investigation of modules over such algebras, proposed by Brauer. That program is being successfully realized at this time (see Sect. 2).

1.26. Bibliographical Notes. Initial familiarity with the concepts of ring theory can be gained from university courses such as Kostrikin [1977], Kurosh [1962], Skornyakov [1983], van der Waerden [1967, 1971] and Jacobson [1985]. The broadest textbook on ring theory, Jacobson [1964], covers: the Jacobson radical and semisimple rings, irreducible modules and primitive rings with minimal condition on one-sided ideals, tensor products and the Brauer group, Galois theory of rings of linear transformations and fields, Baer, Levitzki and nil radicals, the space of primitive ideals of a ring, applications to commutativity theorems, PI-algebras and to the Kurosh problem. A much later monograph by Herstein [1968] includes, apart from classical material, Goldie's theorems and the method of Golod-Shafarevich. The textbook by Lambek [1966] contains an introduction to homological algebra, functors Tor and Ext, an exposition on the construction of the full and classical rings of quotients. As an indroduction to ring theory we also recommend Bokut' [1977, 1981] which describe, in particular, the composition method and Cohn's theorem on embeddability of universal enveloping algebras into division rings. Dixmier

[1977] is specifically devoted to universal enveloping algebras of (mainly finite-dimensional) Lie algebras.

For the knowledge of general constructions of rings of quotients one can recommend, aside from Lambek [1966], the textbook by Stenström [1975] as well as the survey by Elizarov [1973].

Cohn [1985] can be regarded as a handbook on the theory of free rings and embeddings of rings in skew fields. It contains both ring- and module-theoretic material, which is related to lattice theory and commutative algebra.

Categorical approach to the foundations of ring theory is adopted in Faith [1973, 1976], which can serve as a reference guide to a major fragment of modern theory of non-commutative rings and modules over them. Functors Tor and Ext are treated in detail in the classic monograph on homological algebra by Cartan and Eilenberg [1956]. Textbook by MacLane [1971] is the most popular guide to category theory.

§ 2. Finite-Dimensional Algebras

2.1. Introduction. Foundations of the theory of (finite-dimensional) associative algebras were laid in the work of F.E. Molin (1893), E. Cartan (1898) and Wedderburn (1908). In the 30's of this century thanks to the efforts of A.A. Albert, E. Artin, R. Brauer, E. Noether and H. Hasse this theory "reached its zenith" (according to Albert's statement made in 1939) by determining all division algebras over the field of algebraic numbers. The theory of algebras received its next boost in the past 15 yeras, when many problems posed in the 30's were solved.

2.2. Algebras of Small Dimension. The natural goal of any theory is the classification of the objects it studies. Obviously, any field which is a finite extension of the base field is a finite-dimensional algebra. Leaving aside the problem of classifying finite extensions of fields (which is a part of field theory), it is natural to consider the case of an algebraically closed base field. We will try to classify at least algebras of small dimension over an algebraically closed field F. We will denote the dimension of A over F by $\dim_F A$. If $\dim_F A = 1$ then $A = F \cdot 1_A$, where 1_A is the unit element, i.e. $A \simeq F$, so that the first interesting case is that of $\dim_F A = 2$. We will choose in a two-dimensional algebra a basis, taking 1 as the first basis element: $A = F \cdot 1_A + Fa$, $a \notin F \cdot 1_A$. Then multiplication in A is uniquely determined by the value of the product $a^2 = \alpha a + \beta 1_A$, where – as is readily seen – the associative law will be automatically satisfied. We will consider the polynomial $g(x) = x^2 - \alpha x - \beta$. The element a is a root of this polynomial (i.e. $g(a) = 0$). Two cases are possible:

1) Roots x_1 and x_2 of the polynomial $g(x)$ are distinct. Then $x_1 + x_2 = \alpha$ and $x_1x_2 = -\beta$. Let $b = a - x_1(x_2 - x_1)^{-1} \cdot 1_A$. Since $b \notin F \cdot 1_A$, $\{1, b\}$ is a basis of A, and an easy calculation shows that $b^2 = b$.

2) $g(x)$ has one double root, $g(x) = (x - x_1)^2$. Setting $b = a - x_1 \cdot 1_A$ we obtain a basis $\{1, b\}$ for which $b^2 = (a - x_1 \cdot 1_A)^2 = g(a) = 0$. We conclude that any two-dimensional algebra over an algebraically closed field has a basis $\{1, b\}$ such that either $b^2 = b$ or $b^2 = 0$. This means that, up to isomorphism, only two such algebras exist over an algebraically closed field.

Analogous, only slightly more complicated reasoning shows that there are only five (up to isomorphism) three-dimensional algebras over an algebraically closed field: two of them are spanned by the basis $\{1, e, a\}$ with multiplication table $e^2 = e$, $a^2 = 0$, $ea = \varepsilon a$ and $ae = 0$, where $\varepsilon = 0$ or $\varepsilon = 1$. Two more have bases $\{1, a, b\}$ with multiplication given by $a^2 = \varepsilon b$, $b^2 = ab = ba = 0$, where $\varepsilon = 0$ or $\varepsilon = 1$. The fifth one has a basis $\{1, e, f\}$ and multiplication $e^2 = e$, $f^2 = f$, $ef = fe = 0$.

This situation may inspire a conjecture that in any given dimension there are only finitely many non-isomorphic algebras. Such conjecture, however, is already negated by the case of four-dimensional algebras. We will consider a four-dimensional algebra A_t with basis $\{1, a_1, a_2, b\}$ with multiplication table $a_1^2 = a_1a_2 = b$, $a_2a_1 = 0$, $a_2^2 = tb$ and $b^2 = 0$, where $t \in F$. We will show that when $t \neq t'$, the algebras A_t and $A_{t'}$ are not isomorphic. Suppose that in the algebra A_t there is a basis $\{1, a_1', a_2', b'\}$ with multiplication equivalent to that of the algebra $A_{t'}$. Let $a_i' = \sum \alpha_{ij}a_j + \alpha_i b$, $b' = \beta_1 a_1 + \beta_2 a_2 + \beta b$. We have $b' = a_1'a_2' = \gamma b$ (since $A_t^2 = Fb$), i.e. $\beta_1 = \beta_2 = 0$ and $\beta = \gamma \neq 0$. Equalities $a_1'a_1' = a_1'a_2' = b'$, $a_2'a_1' = 0$ and $a_2'a_2' = t'b'$ yield, in the basis $\{1, a_1, a_2, b\}$, the equalities which may be written in matrix form as

$$C\begin{pmatrix} 1 & 1 \\ 0 & t \end{pmatrix}C' = \begin{pmatrix} 1 & 1 \\ 0 & t' \end{pmatrix}\gamma, \qquad (1)$$

where $C = (\alpha_{i,j})$ and $C' = (\alpha_{j,i})$ is its transpose. In particular, equating determinants of both sides, we get $(\det C)^2 = \gamma t'$. On the other hand, subtracting the transpose of (1) from (1), we obtain $C\begin{pmatrix} 0 & 1 \\ -1 & 0 \end{pmatrix}C' = \begin{pmatrix} 0 & 1 \\ -1 & 0 \end{pmatrix}\gamma$, and comparing determinants again we find that $(\det C)^2 = \gamma$ – meaning that $t = t'$, as required.

We now switch to the case of an arbitrary base field, and consider several constructions which play an important rôle in the theory of finite-dimensional algebras.

2.3. Crossed Products. Let G be a finite group of automorphisms of a field K, and let F be the subfield of elements fixed by G: $F = \{a \in K \mid \forall g \in G\ a^g = a\}$. In other words, K is a normal separable extension of the field F whose Galois group is G (one can also say that 'K is a Galois extension of the field F with Galois group G'). We will consider the right vector space

(K, G) over the field K with basis $\{u_g \mid g \in G\}$, where u_g are symbols. Then every element of (K, G) has the form $\sum_{g \in G} u_g \alpha_g$ (where $\alpha_g \in K$). We will define on this space a multiplication operation by means of the distributive law and conditions (2) and (3) below:

$$\alpha u_g = u_g \alpha^g \ (\alpha \in K), \tag{2}$$

$$u_g u_h = u_{gh} \sigma(g, h), \tag{3}$$

where $\sigma : G \times G \to K^*$ is some fixed function of two variables from the group G with values in the multiplicative group of the field K. In this way we obtain an algebra (K, G, σ) over F which, generally speaking, is non-associative, with multiplication given by

$$\Big(\sum_{g \in G} u_g \alpha_g\Big)\Big(\sum_{h \in G} u_h \beta_h\Big) = \sum_{g, h \in G} u_{gh} \sigma(g, h) \alpha_g^h \beta_h \ .$$

For the associative law to hold, it is necessary and sufficient to guarantee that equations $(u_g u_h) u_f = u_g (u_h u_f)$ are valid. Rewriting this condition using (2) and (3) we obtain the following conditions on the mapping σ, equivalent to associativity of the algebra (K, G):

$$\sigma^f(g, h) \sigma(gh, f) = \sigma(g, hf) \sigma(h, f), \tag{4}$$

where $\sigma^f(g, h) = (\sigma(g, h))^f$.

The function $\sigma : G \times G \to K^*$ satisfying relations (4) is called the *(Noether) factor system of* (G, K). The algebra (K, G, σ) determined by such mapping is called the *crossed product of the field K and the group G with a system of factors σ*. We have the following important

Theorem. *The crossed product (K, G, σ) is a central* (see Sect. 1.6) *simple* (see Sect. 1.2) *algebra over the field $F = K^G$. Dimension of this algebra equals the square of the order of the group G (or, equivalently, the square of the degree of K over F).*

If σ and τ are two systems of factors corresponding to a single Galois extension $K \supset F$ with Galois group G then, multiplying sides of their respective relations (4), we see that the mapping $\lambda : (g, h) \to \sigma(g, h)\tau(g, h)$ is also a factor system. Mappings $(g, h) \to 1$ and $(g, h) \to \sigma(g, h)^{-1}$ are factor systems of (G, K) as well. It is now clear that the set of factor systems of (G, K) forms an abelian group.

Next we turn our attention to the fact that the algebra (K, G, σ) can be described (up to isomorphism) as a crossed product (K, G, σ') by means of another system of factors. Indeed, let μ be any map from the group G into the multiplicative group K^* of the field K. We will let $u'_g = u_g \mu(g)$. Then every element of (K, G, σ) is a linear combination $\sum_{g \in G} u'_g \alpha_g$; moreover, relations

$$\alpha u'_g = u'_g \alpha^g \ ,$$

$$u'_g u'_h = u'_{gh}\mu(gh)^{-1}\sigma(g,h)\mu^h(g)\mu(h)$$

are satisfied. This means that $(K, G, \sigma) \simeq (K, G, \sigma')$, where

$$\sigma'(g,h) = \sigma(g,h)\mu(gh)^{-1}\mu^h(g)\mu(h)\,. \tag{5}$$

This fact allows us to identify factor systems which differ by the factor in formula (5).

Definition. The quotient group of the group of (G, K)-factors modulo the subgroup of factor systems of the form $(g, h) \to \mu(gh)^{-1}\mu^h(g)\mu(h)$ is denoted by $H^2(G, K)$ and called the *second homology group* of G over the field K.

The group $H^2(G, K)$ corresponds to the set of all crossed products of K with G; namely, as can be shown, two factor systems σ and τ are equal in $H^2(G, K)$ if and only if their corresponding crossed products are isomorphic: $(G, K, \sigma) \simeq (G, K, \tau)$.

As an example we will consider the case when $K = \mathbb{C}$ is the field of complex numbers and $F = \mathbb{R}$ is the real field. Let $G = \{e, \varphi\}$ be the group of automorphisms of $\mathbb{C}$ consisting of two elements, where e is the identity automorphism (the neutral element of G) and φ is the conjugation mapping: $c^\varphi = \bar{c}$ (clearly, $\varphi^2 = e$). We will define a factor system σ, setting $\sigma(g, h) = 1$ for all $g, h \in G$. Then $(\mathbb{C}, G, \sigma)$ has basis $\{u_e = 1, \mathbf{i}, u_\varphi, u_\varphi\mathbf{i}\}$ over $\mathbb{R}$. It is easy to see that these basis elements multiply in the same way as the matrices

$$\begin{pmatrix} 1 & 0 \\ 0 & 1 \end{pmatrix}, \begin{pmatrix} 0 & 1 \\ -1 & 0 \end{pmatrix}, \begin{pmatrix} 0 & 1 \\ 1 & 0 \end{pmatrix} \text{ and } \begin{pmatrix} -1 & 0 \\ 0 & 1 \end{pmatrix},$$

respectively. By assigning to an element $1\alpha + \mathbf{i}\beta + u_\varphi\gamma + u_\varphi\mathbf{i}\delta$ the matrix $\begin{pmatrix} \alpha-\delta & \beta+\gamma \\ \gamma-\beta & \alpha+\delta \end{pmatrix}$ we can then conclude that the crossed product of $\mathbb{C}$ and G with a unitary factor system is isomorphic to the ring of all 2×2 matrices over $\mathbb{R}$.

With the same group G, we will now consider another factor system: $\sigma_1(g, h) = 1$ if $g \neq \varphi$ or $h \neq \varphi$, while $\sigma_1(\varphi, \varphi) = -1$. Then $(\mathbb{C}, G, \sigma_1)$ has an $\mathbb{R}$-basis $\{1, \mathbf{i}, u_\varphi, u_\varphi\mathbf{i}\}$ and its multiplication table coincides with that of the division ring of quaternions (see Sect. 1.7), i.e. we have found that the algebra of quaternions is a crossed product. Moreover, it is easy to show that $|H^2(G, \mathbb{C})| = 2$, meaning that there are no other crossed products of the field $\mathbb{C}$ with the group of order two.

2.4. Cyclic Algebras. Let K be a cyclic extension of the field F. This means, by definition, that K is a normal and separable extension of F whose Galois group $G = \mathrm{Aut}\,(K/F)$ is cyclic: $G = \{1, s, \ldots, s^{n-1}\}$ (where n equals the degree $[K : F]$). Let γ be an element of the field F. We will consider an F-algebra $(K, G, \gamma) = u_1K \oplus u_sK \oplus \ldots \oplus u_{s^{n-1}}K$ defined just like the crossed product, only with multiplication in it given by the relations $u_{s^i}u_{s^j} = u_{s^{i+j}}$

(when $i+j<n$) and $u_{s^i}u_{s^j} = \gamma u_{s^{i+j-n}}$ (if $i+j \geq n$). The algebra (K,G,γ) is called a *cyclic* algebra. It is isomorphic to the crossed product of K and G relative to a suitable factor system, namely $\sigma(s^i,s^j) = \begin{cases} 1 & \text{for } i+j<n, \\ \gamma & \text{for } i+j \geq n. \end{cases}$ The cyclic algebra (K,G,γ) is therefore a central simple F-algebra.

Example. We will consider an arbitrary crossed product (K,G,τ) with a cyclic group G. We will show that one can choose a basis for it in such a way that it will turn into a cyclic algebra. Let $G = \{1 = s^0, s, \ldots, s^{n-1}\}$ be a cyclic group. For brevity we will write $\sigma_{i,j}$ instead of $\sigma(s^i,s^j)$. Then the condition for associativity takes the form $\sigma_{i,j+k}\sigma_{j,k} = \sigma_{i+j,k}\sigma_{i,j}^{s^k}$. When $i=j=0$, we have $\sigma_{0,k}\sigma_{0,k} = \sigma_{0,k}\sigma_{0,0}^{s^k}$, i.e. $\sigma_{0,k} = \sigma_{0,0}^{s^k}$. Similarly, when $j=k=0$ we obtain $\sigma_{0,0} = \sigma_{i,0}$. These equalities show, in particular, that the element $v_1 = \sigma_{0,0}^{-1}u_1$ is the unity of this crossed product. Next we will let $v_s^i = (u_s)^i$, where $i = 1,2,\ldots,n-1$. Then $v_{s^i}v_{s^j} = v_{s^{i+j}}$ when $i+j<n$ and $v_{s^i}v_{s^j} = \gamma v_{s^{i+j-n}}$ when $2n > i+j \geq n$ and where $\gamma \in K$ is defined by the equation $(u_s)^n = \gamma v_1$. Moreover, $\gamma \in F$ because $u_s\gamma^s = \gamma u_s = \gamma v_1 u_s = u_s^{n+1} = u_s\gamma v_1 = u_s\gamma$, and hence $\gamma^s = \gamma$. Noting that the law of commutation between the elements v_{s^i} and elements of the field K remains valid, $v_{s^i}\alpha^{s^i} = \alpha v_{s^i}$, we conclude that every crossed product with a cyclic group is a cyclic algebra.

2.5. Direct Sums and Tensor Products of Finite-Dimensional Algebras. In this section we will desribe two of the more important constructions of algebras. We begin with a direct sum. Let $A_1,\ldots,A_n$ be any algebras over a field F. A will denote the direct sum of vector spaces A_i $(1 \leq i \leq n)$. Elements of A can be viewed as formal sums $a_1 + \ldots + a_n$, where $a_i \in A_i$. The vector space A becomes an algebra if one introduces in it multiplication defined by the formula $(a_1+a_2+\ldots a_n)(a_1'+a_2'+\ldots a_n') = a_1a_1' + a_2a_2' + \ldots + a_na_n'$. This algebra is called the *direct sum of algebras* A_i: $A = \bigoplus_{i=1}^n A_i$. Algebras A_i can be regarded as subalgebras of A (without common 1), identifying an element $a_i \in A_i$ with the sum $0+\ldots+a_i+\ldots+0$. In this context the A_i's turn out to be ideals of A, and these ideals annihilate each other: $A_iA_j = 0$ for $i \neq j$ (we assume here that $0 = 0+\ldots+0$). Moreover, the intersection of each of them with the sum of all the remaining ones is zero. These properties, as it turns out, characterize decomposability of an algebra into a direct sum. We also note that dimension of A equals the sum of dimensions of the A_i's.

We now present the definition of a tensor product $A \otimes B$ of two F-algebras A and B. Let $\{a_i \mid 1 \leq i \leq n\}$ be a basis of the algebra A and $\{b_i \mid 1 \leq i \leq m\}$ – a basis of B. We will consider an nm-dimensional vector space $A \otimes B$ over F, with a basis consisting of elements denoted by $a_i \otimes b_j$ $(1 \leq i \leq n$ and $1 \leq j \leq m)$. We define multiplication on this space by the distributive law and the formula $(a_i \otimes b_j)(a_k \otimes b_l) = a_ia_k \otimes b_jb_l$. The right-hand-side expression can be transformed into a linear combination of basis elements according to the relations $(a+b)\otimes c = a\otimes c + b \otimes c$, $a \otimes (b+c) = a\otimes b + a \otimes c$ and

$\alpha a \otimes b = a \otimes \alpha b = \alpha(a \otimes b)$, where $\alpha \in F$. The algebra $A \otimes B$ obtained this way is the *tensor product of algebras* A and B. If we now identify an element $a \in A$ with the element $a \otimes 1_B$ and $b \in B$ with $1_A \otimes b$, then we will see that A and B are subalgebras of their tensor product with common unity element $1_A \otimes 1_B$. Moreover, elements of A commute with those of B: $(a \otimes 1)(1 \otimes b) = a \otimes b = (1 \otimes b)(a \otimes 1)$. We also see that the dimension of $A \otimes B$ is the product of dimensions of A and B. It turns out that these properties fully characterize the tensor product of finite-dimensional algebras. Namely, let C be an algebra with 1, containing subalgebras A and B which have the same 1 as C, and which generate C. If elements of A commute with elements of B and the dimension of C equals the product of dimensions of A and B, then C is isomorphic to the tensor product $A \otimes B$. We also remark that the above definition of a tensor product can be easily generalized to the case of infinite-dimensional algebras A and B.

Example 1. Let A be any F-algebra and let $B = M_n(F)$ be the ring of all $n \times n$ matrices over F. In this case $A \otimes M_n(F) \simeq M_n(A)$, the ring of $n \times n$ matrices with entries from A. Indeed, we will choose as a basis of the algebra $B = M_n(F)$ the set of matrix units $\{e_{i,j}\}$. We see that every element of the algebra $A \otimes M_n(F)$ can be uniquely expressed as $x = \sum a_{i,j} \otimes e_{i,j}$, and we can identify it with the matrix $(a_{i,j})$, since the product $(\sum a_{i,k} \otimes e_{i,k})(\sum b_{k,j} \otimes e_{k,j}) = \sum a_{i,k}b_{k,j} \otimes e_{i,j}$ agrees with the matrix product $(a_{i,j})(b_{i,j}) = (\sum_k a_{i,k}b_{k,j})$.

Example 2. Extension of the base field. If A is an algebra over a field F and a field K is an extension of F, then one can consider the tensor product $A \otimes_F K$ which, by definition, will be an F-algebra. This tensor product, however, can also be regarded as an algebra over K by defining $(\sum a_i \otimes k_i)k = \sum a_i \otimes k_i k$ (for $k_i, k \in K$). Since a basis of A over F is also a basis of $A \otimes K$ over K, we say that the K-algebra $A \otimes K$ is obtained from A by extension of the field of scalars.

One of the most important results on tensor products is the following

Theorem. *Tensor product of central simple algebras* (see Sections 1.2 and 1.6) *is a central simple algebra.*

A proof of this fact can be easily obtained by means of the following useful property of linearly independent elements of any central simple algebra (over a field F): if $d_1, \ldots, d_n$ are linearly independent (over F) elements of A then there exist elements $s_i, t_i \in A$ $(1 \le i \le m)$ such that $\sum_i s_i d_1 t_i = 1$ and $\sum_i s_i d_k t_i = 0$ for all $2 \le k \le n$. This statement follows from the density theorem (see Sect. 4.2, Theorem 2). More precisely, one can view the algebra A as an irreducible (A, A)-bimodule with the ring of endomorphisms F (i.e. as an irreducible left module over the algebra $A \otimes_F A^{\text{op}}$, where A^{op} is the opposite algebra of A – the algebra canonically anti-isomorphic to A).

Let now x be a non-zero element of the tensor product $A \otimes B$ of the central simple algebras A and B. We will show that the ideal (x) generated by it contains 1. We will write x as a sum $\sum_i d_i \otimes b_i$, where d_i are basis elements (possibly not all) of the algebra A, while b_i are non-zero elements of B. By the property of linearly independent elements which was quoted above, we can find in the ideal (x) an element $b = \sum_i (s_i \otimes 1)x(t_i \otimes 1) = 1 \otimes b_1$. Since B is a simple algebra, there exist elements $v_i, w_i \in B$ $(1 \le i \le k)$ such that $\sum_i v_i b_1 w_i = 1$. It follows that the ideal (x) contains the element $\sum_i (1 \otimes v_i) b (1 \otimes w_i) = 1 \otimes 1 = 1$, as claimed.

Next let $x = \sum_{i=1}^n d_i \otimes b_i$ (d_i, b_i chosen as above) be an element of the center of $A \otimes B$. We will assume that the bases of both algebras A and B contain the respective unity elements. If one of the elements d_i – say, d_1 – is not equal to 1, then the same property used again yields elements $s_i, t_i \in A$ $(1 \le i \le m)$ such that

$$\sum_i s_i d_1 t_i = 1\,, \quad \sum_i s_i 1 t_i = 0\,, \quad \sum_i s_i d_k t_i = 0 \ (2 \le k \le n)\,.$$

We see that because x is central, the sum $\sum_i (s_i \otimes 1)x(t_i \otimes 1) = x \sum_i (s_i \otimes 1)(t_i \otimes 1)$ equals 0. On the other hand, this sum is equal to $1 \otimes b_1$ – a contradiction. Hence the element x has the form $1 \otimes b$, i.e. corresponds to a central element of B, and so $x = (1 \otimes 1)\beta$, where $\beta \in F$ (because $b = \beta \cdot 1_B$).

In Sections 2.6 and 2.7 below we will present the fundamental structure theorems of the theory of finite-dimensional algebras.

2.6. Frobenius' Theorem. Historically, the first theorem in the theory of algebras was the following

Theorem (Frobenius (1886)). *Every finite-dimensional division algebra over the field* $\mathbb{R}$ *is isomorphic to* $\mathbb{R}$, $\mathbb{C}$ *or* $\mathbb{H}$.

The idea on which the proof is based uses the fact that every element x of a finite-dimensional division algebra A is a root of a quadratic trinomial with coefficients in $\mathbb{R}$: by finite-dimensionality, the elements $1, x, x^2, \ldots$ are linearly dependent over $\mathbb{R}$, i.e. $x^n + r_1 x^{n-1} + \ldots + r_n = 0$. Since every polynomial over the field of real numbers decomposes into a product of quadratic and linear factors, x must be a root of one of them, and hence $x \in \mathbb{R}$ or x is a root of a quadratic trinomial $x^2 + ax + b$. We will remark that a similar approach also turns out to be effective in the study of certain non-associative division algebras over $\mathbb{R}$.

2.7. Structure of Finite-Dimensional Algebras. Over the field of complex numbers C, structure of an arbitrary algebra was described in the dissertation of F.E. Molin (1893). Those results were then rediscovered and extended to the real case by E. Cartan (1898). Finally, Wedderburn (1908) proved analogs of the above theorems in the case of an arbitrary base field. Theorems about the structure of finite-dimensional algebras have become known to researchers as "Wedderburn's theorems". For complete formulation of these results it is necessary to introduce several definitions. We will recall that an algebra (without 1) is called *nilpotent* if $A^n = 0$ for some n (i.e. $a_1 \cdot \ldots \cdot a_n = 0$ for any $a_i \in A$). The smallest n of this property is called the *index of nilpotency* of A. For example, the algebra $T_n^0(F)$ of strictly upper triangular $n \times n$ matrices over F is nilpotent with index n. Indeed, its square $\left(T_n^0(F)\right)^2$ consists of upper triangular matrices whith zeros on both the main diagonal and the parallel one above it, its cube contains upper triangular matrices with zeros on three diagonals etc. We will recall that the radical $\operatorname{Rad} A$ of a finite-dimensional algebra A is the largest nilpotent ideal of A (which contains all nilpotent ideals of that algebra; see Section 1.24). The factor algebra $A/\operatorname{Rad} A$ is semisimple, i.e. its radical is zero.

Theorem 1. *Every semisimple finite-dimensional algebra A decomposes in a unique way into a direct sum of finitely many simple algebras: $A = B_1 \oplus \ldots \oplus B_k$.*

Theorem 2. *Every simple finite-dimensional algebra A is isomorphic to a matrix algebra $M_n(D)$ over some division algebra D, where the integer n and the algebra D (up to isomorphism) are uniquely determined by the algebra A.*

These theorems are valid also for algebras which are not *a priori* assumed to have unity element.

Let A be a central simple finite-dimensional algebra $A \simeq M_n(D)$, where D is a central division algebra. The integer n is called the *reduced degree* of A. It can be shown that the dimension of a central division algebra is always a square, $\dim_F D = i^2$. The integer $i = i(A)$ is called the *index of the algebra* A. We have $\dim_F A = (ni)^2$.

A semisimple finite-dimensional algebra A is called *separable* if for any field extension $K \supset F$ the algebra $A \otimes_F K$ is also semisimple. This is equivalent to the statement that the centers C_i of simple summands B_i in the decomposition $A = B_1 \oplus \ldots \oplus B_k$ are all separable field extensions of the field F. Recall that an algebraic field extension $C \supset F$ is *separable* if for every $z \in C$ the polynomial of minimal degree with coefficients in F whose roots include z does not have multiple roots in the algebraic closure of the field F. This latter statement is automatically valid when the field has characteristic 0 (which is to say that F contains the field of rational numbers as a subfield). Every semisimple algebra over a field of characteristic 0 is therefore separable.

Theorem 3. *If the factor algebra* $A/\operatorname{Rad} A$ *of a finite-dimensional algebra A over a field F is separable, then A decomposes (as a vector space) into a direct sum of its radical* $\operatorname{Rad} A$ *and some semisimple subalgebra* $B \simeq A/\operatorname{Rad} A$, *i.e.* $A = \operatorname{Rad} A \oplus B$.

The question of uniqueness of the above decomposition was settled by A.I. Mal'tsev (1943):

Theorem 4. *If $A = \operatorname{Rad} A \oplus B = \operatorname{Rad} A \oplus B'$ are two decompositions of a finite-dimensional algebra A into a sum of its radical and a separable subalgebra, then there exists an invertible element $x \in A$ for which $B' = x^{-1}Bx$ (that is, the semisimple component is unique up to an inner automorphism).*

We now describe the idea on which proof of Theorem 2 is based. We will consider in the simple algebra A a right ideal V of minimal (non-zero) dimension. V can be viewed as a vector space over the field F. In the ring of all linear transformations of this vector space into itself we will single out the set D of all those mappings which commute with right multiplications by elements of A, i.e. $d \in D \Leftrightarrow \forall a \in A\ d(xa) = d(x)a$. A fundamental observation states that D is a division algebra (*Schur's lemma*). Indeed, it is quite clear that D is closed under the vector space operations and multiplication, so it is only necessary to verify that any non-zero transformation from D has an inverse which belongs to D. If $0 \neq d \in D$ then the set $d(V)$ is a right ideal of the algebra A, contained in V: $d(V)A \subseteq d(VA) \subseteq d(V)$. Since V has minimal dimension, $d(V) = V$ which means that there exists an inverse linear transformation $d^{-1} : V \to V$ such that $d^{-1}(d(v)) = v$. It is clear that $d^{-1}(va) = d^{-1}(v)a$ for all $a \in A$, and hence $d^{-1} \in D$. We can now view V from a different perspective. The operation of left multiplication by elements of the division ring D is defined on V, so that V is a left vector space over D. Moreover, right multiplications by elements of the ring A are compatible with the D vector space structure of V, i.e. right multiplications are D-linear transformations of V. The ring $E = \operatorname{End}({}_D V)$ of all such transformations (relative to right multiplication $v(fg) = (vf)g$ for $v \in V$, $f, g \in E$) is isomorphic to the ring of matrices $M_n(D)$, where $n = \dim_D V$. This means that we have a homomorphism $A \to M_n(D)$. Since the kernel of a homomorphism is an ideal, and A is a simple algebra, this (non-zero) homomorphism must be injective. It is only a little more difficult to show that this mapping is an isomorphism.

Proof of Theorem 1 is based on the following simple observation. Consider a non-zero two-sided ideal I of the algebra A whose dimension is smallest possible. It is easy to see that I is a simple algebra (without 1, generally speaking). By Theorem 2, I is isomorphic to an algebra of matrices over a division ring. In particular, this algebra contains a unity element e which, obviously, will not be the 1 of A when $A \neq I$. We will consider the set $W = \{x - xe \mid x \in A\}$. It turns out that W is an ideal, and $A = I \oplus W$. The argument is completed by an obvious induction on the dimension of A.

Finally, among the fundamental results of the theory of algebras we also have the following

Theorem 5 (T. Skolem, E. Noether). *If f, g are two non-zero homomorphisms of a simple finite-dimensional algebra B into a central simple finite-dimensional algebra A, then A contains an invertible element a such that $g(x) = a^{-1}f(x)a$ for all $x \in B$.*

This theorem implies a number of important properties of central simple finite-dimensional algebras, e.g. that any two isomorphic subalgebras of a central simple finite-dimensional algebra are in fact conjugate. Another interesting consequence states that every automorphism of a full matrix algebra $M_n(F)$ is *inner*, i.e. has the form $x \mapsto a^{-1}xa$ for some invertible element a.

2.8. The Brauer Group. If A and B are central simple (finite-dimensional) algebras over a field F then their tensor product $A \otimes B$ is also a central simple algebra. Since tensor product is associative and commutative, the set of all central simple algebras is a commutative semigroup relative to that operation.

Definition 1. Central simple finite-dimensional algebras A and B are called *similar*, $A \sim B$, if their underlying division rings are isomorphic, i.e. $A \simeq M_n(D)$, $B \simeq M_k(D)$ for some integers n, k and a division ring D.

It can be shown that two central simple finite-dimensional algebras are similar if and only if they are Morita-equivalent (see Sect. 4). We will denote by $[A]$ the class of all central simple finite-dimensional algebras which are similar to A.

Theorem 2. *The set* $\mathrm{Br}\,(F)$ *of all equivalence classes of central simple finite-dimensional algebras over* F *is an abelian group under the operation* $[A] + [B] = [A \otimes B]$. *This group is called the Brauer group of the field* F.

The rôle of the neutral element in the Brauer group is played by the class $[F]$. The inverse element of a class $[A]$ is the class $[A^{\mathrm{op}}]$ of the opposite algebra of A. This follows from the fact that $A \otimes A^{\mathrm{op}} \simeq M_n(F)$, where n is the dimension of A. Indeed, the dimension of $A \otimes A^{\mathrm{op}}$ equals n^2. On the other hand, assigning to the element $\sum a_i \otimes b_i$ the mapping $x \mapsto \sum b_i x a_i$ of the vector space A, we obtain a homomorphism from $A \otimes A^{\mathrm{op}}$ into the algebra of all linear transformations (with right composition) of the space A over F, which is isomorphic to the matrix algebra $M_n(F)$. The kernel of this homomorphism is zero since the algebra $A \otimes A^{\mathrm{op}}$ is simple; the dimension of the algebra $M_n(F)$ is n^2 as well, so that the homomorphism we constructed will be an isomorphism. It can be shown that order of a class $[A]$ in the Brauer group divides the index $i(A)$ of the algebra A, with the order and index having the same prime factors (we will recall that order of $[A]$ is the smallest integer $m > 0$ such that $m[A] = 0$).

Let now K be an extension of the field F. If A is a central simple algebra over F then $A \otimes_F K$ is a central simple algebra over K. If $A \sim B$ then $A \otimes K \sim B \otimes K$, i.e. $[A \otimes K] = [B \otimes K]$ in the group $\mathrm{Br}\,(K)$. Consequently, the mapping $[A] \mapsto [A \otimes K]$ will be a homomorphism from $\mathrm{Br}\,(F)$ into $\mathrm{Br}\,(K)$, whose kernel is denoted by $\mathrm{Br}\,(K/F)$.

Definition 3. We say that a central simple algebra A is *split* by a field $K \supset F$ (and K is called a *splitting field of* A) when $A \otimes_F K \simeq M_n(K)$.

The group $\mathrm{Br}(K/F)$ is therefore determined by the central simple algebras over F which are split by the field K. Since a division algebra over an algebraically closed field coincides with the field itself, the Brauer group $\mathrm{Br}(\bar{F})$ of the algebraic closure of a field F is trivial and, consequently, $\mathrm{Br}(\bar{F}/F) = \mathrm{Br}(F)$ – because every central simple algebra has a splitting field. It can be shown that every maximal subfield of the component of a central simple algebra turns out to be its splitting field.

In the case when $K \supset F$ is a Galois extension, as we saw above, the crossed product construction provides us with examples of central simple algebras over F. All those algebras are split over the field K, so that there is a natural inclusion $H^2(G, K) \to \mathrm{Br}(K/F)$, where G is the Galois group of the extension $K \supset F$ (we will remark that all crossed products of the field K with the group G have identical dimension n^2, and hence similarity between them means isomorphism). Furthermore, it turns out that this inclusion is an isomorphism, meaning that every central simple algebra which splits in a finite Galois extension of F is similar to a crossed product. One can show that any central simple algebra has some finite Galois extension as its splitting field. We therefore have

Theorem 4. *Every central simple algebra is similar to a crossed product.*

We will now consider a few concrete examples. Frobenius' theorem shows that the algebra of quaternions is the unique non-one dimensional central division algebra over the field of real numbers, i.e. $\mathrm{Br}(\mathbb{Z}) \simeq \mathbb{Z}_2$. If K is a finite field then all finite-dimensional K-algebras are also finite. *Wedderburn's theorem* on finite division rings asserts that every such ring is commutative. This means that over a finite field there are no finite-dimensional central division algebras (other than the field itself), hence the Brauer group of such a field is trivial. In the case of arbitrary fields of finite characteristic $p > 0$, a theorem of Witt states that the Brauer group is p-divisible (i.e. for any element a of the group, the equation $px = a$ has solutions for x).

2.9. Algebras Over an Algebraic Number Field. We will recall that an *algebraic number field* is any finite extension of the field of rational numbers $\mathbb{Q}$. The following theorem has been obtained, as we already mentioned in Sect. 2.1, thanks to efforts of many mathematicians.

Theorem. *Let F be an algebraic number field, and let A be a central simple F-algebra. Then A is a cyclic algebra, and the order of $[A]$ in the Brauer group* $\mathrm{Br}(F)$ *is equal to its index $i(A)$.*

Number-theoretic methods are essential in the proof of this theorem.

2.10. Examples of Non-Crossed Products. Albert constructed the first example of a central division algebra (of dimension 16) which was a crossed product, but was not cyclic. Moreover, every division algebra of dimension 16 is a crossed product. The question whether every central division algebra must be a crossed product remained open. It was answered in the negative by Amitsur. A counterexample is provided by the algebra $\mathbb{Q}(X_1, \ldots, X_m)$, where $X_1, \ldots, X_m$ are so-called generic $n \times n$ matrices and n is subject to certain restrictions (it is divisible by 8 or by a square of an odd prime integer). We will pause to describe a construction of this algebra in more detail. Let $X_i = (x^i_{jk})$ (for $1 \leq i \leq m$ and $1 \leq i, k \leq n$) be 'generic' matrices in algebraically independent, commuting indeterminates $x^i_{j,k}$. Further, let $\mathbb{Q}[X_1, \ldots, X_m] = \mathbb{Q}_{n,m}$ be the algebra generated over $\mathbb{Q}$ by the matrices $X_1, \ldots, X_m$ (this is the *algebra of generic matrices over* $\mathbb{Q}$. It turns out that this algebra has no zero divisors, and satisfies the (left and right) Ore condition (see Sect. 1.19). We will denote its classical division ring of fractions by $\mathbb{Q}(X_1, \ldots, X_m)$. Let C be the center of $\mathbb{Q}(X_1, \ldots, X_m)$. Then $\mathbb{Q}(X_1, \ldots, X_m)$ is a central division algebra of dimension n^2 over C.

There also exist central simple division algebras over a field of prime characteristic, which are not crossed products. Moreover, there exist p-algebras with this property (a central simple algebra A over a field F is a p-algebra if F has characteristic p and the reduced degree of A equals p^k for some k).

2.11. The Brauer Group and the Functor K_2. We begin this section by considering a concrete example of a central simple algebra. Let F be a field containing an n-th primitive root ζ of 1 (i.e. the polynomial $t^n - 1$ decomposes in F into linear factors: $t^n - 1 = \prod_{i=1}^{n}(t - \zeta^i)$). Let also a, b be non-zero elements of the field F. We will consider the algebra generated by two elements x and y, which satisfy the (defining) relations

$$x^n = a, \quad y^n = b, \quad xy = \zeta yx \,. \tag{6}$$

We will denote this algebra by (a, b, n, F, ζ). It consists of polynomials in x and y with x- and y-degrees both less than n, whose addition and multiplication are defined by the associative and distributive laws together with equations (6).

Theorem 1. *The algebra (a, b, n, F, ζ) is central and simple, and every extension L of the field F in which b has an n-th root is its splitting field.*

We will now show that if $K \supset F$ is a cyclic extension with Galois group G and $\zeta \in F$, then every cyclic algebra (K, G, γ) has the form (a, b, n, F, ζ). Let $G = \{1, s, \ldots, s^{n-1}\}$. We will view s as a linear transformation of the space K over F. The characteristic polynomial of this mapping equals $\lambda^n - 1$ and, by hypothesis, it factors in F into linear binomials. This means that the transformation s has a non-zero eigenvector x corresponding to the root ζ: $x^s = \zeta x$. Now we have $x x^s x^{s^2} \ldots x^{s^{h-1}} \in F$, i.e. $\zeta^{n(n-1)/2} x^n \in F$ and hence $x^n = a \in F$. We will let $b = \gamma$ and $y = u_s$. Then it is easy to see that the algebra (K, G, γ) is generated over F by the elements x, y and that $x^n = a$, $y^n = b$ and $xy = xu_s = u_s x^s = \zeta u_s x = \zeta yx$. This implies that (K, G, γ) is a homomorphic image of the simple algebra (a, b, n, F, ζ), i.e. $(K, G, \gamma) \simeq (a, b, n, F, \zeta)$.

Element of the Brauer group determined by the algebra (a, b, n, F, ζ) is denoted by $[a, b, n, F, \zeta]$. When it is clear from the context what n, F and ζ are, it is simply denoted $[a, b]$. Algebra (a, b, n, F, ζ) obviously depends on the choice of the primitive root of 1. If ζ^t is another such root, then $[a, b, n, F, \zeta^t] = t[a, b, n, F, \zeta]$.

One of the most important accomplishments of the study of Brauer groups of fields is the following

Theorem 2 (A.S. Merkurjev, A.A. Suslin). *Let the field F contain a primitive n-th root of 1. Then every central simple algebra A of exponent n (i.e. $n[A] = 0$) is similar to a tensor product of algebras of the form (a, b, n, F, ζ), i.e. in the Brauer group of F every element of exponent n can be represented as a sum $[a_1, b_1] + [a_2, b_2] + \ldots + [a_k, b_k]$.*

Proof of this result heavily relies on achievements of the so-called algebraic K-theory. An important rôle is played in it by a functor K_2. The value of K_2 on a field F is defined to be the factor group of the tensor product of abelian groups $F^* \otimes_{\mathbb{Z}} F^*$ modulo the subgroup generated by all tensors $a \otimes (1 - a)$.

In order to explain links between the functor K_2 and the Brauer group, we will fix an integer n and a primitive n-th root of 1, ζ. Then elements of the Brauer group satisfy a number of relations concerning the bracket $[a, b]$ defined above: $[a, 1 - a] = 0$ (*Steinberg's identity*), $[ab, c] = [a, c] + [b, c]$, $[a, bc] = [a, b] + [a, c]$ (bi-multiplicative property) and $[a^n, b] = n[a, b] = 0$. These identities show that the mapping $a \otimes b \mapsto [a, b]$ extends to a homomorphism of abelian groups $a_\zeta : K_2(F) \to \mathrm{Br}\,(F)$, whose kernel contains the subgroup $nK_2(F)$ and whose image is contained in the subgroup ${}_n\mathrm{Br}\,(F)$ consisting of elements whose orders divide n. This means that there exists a natural homomorphism

$$R_{n,F,\zeta} : K_2(F)/nK_2(F) \to {}_n\mathrm{Br}\,(F) \,.$$

The Merkurjev-Suslin theorem stated in these terms says that $R_{n,F,\zeta}$ is an isomorphism.

2.12. Modules and Representation Type of Finite-Dimensional Algebras. Let A be a finite-dimensional algebra over a field F. A *representation* T of the algebra A is a homomorphism of A into the algebra $\mathrm{End}\, V$ of endomorphisms of a finite-dimensional vector space V over F. Dimension of V is called the *degree of the representation.* Fixing a basis of V we obtain a matrix representation $a \mapsto [T(a)]$ for $a \in A$, of the algebra A, i.e. a homomorphism of A into the algebra $M_n(F)$ (where n is the degree of T. Two representations $T : A \to \mathrm{End}\, V$ and $T_1 : A \to \mathrm{End}\, W$ are called *equivalent* if there is an isomorphism $\varphi : V \to W$ such that $\varphi T(a) = T_1(a)\varphi$ for all $a \in A$. In terms of matrices this means that $[T_1(a)] = C[T(a)]C^{-1}$ for some non-singular matrix C. With each representation $T : A \to \mathrm{End}\, V$ we can associate a left A-module structure on V, defined by $av = T(a)(v)$ for $a \in A$ and $v \in V$. It isn't difficult to show that equivalence of representations is synonymous with an isomorphism between the corresponding modules. The task of describing all representations of the algebra A is therefore equivalent to describing all A-modules which are finite-dimensional over F.

If A is a semisimple algebra, then by Wedderburn's theorem $A = \bigoplus_{i=1}^{k} M_{n_i}(D_i)$, where the D_i are division rings. It can be directly verified that, say, the modules $M_{n_i}(D_i)e_{1,1}$ (where $1 \leq i \leq k$), i.e. modules of first columns of matrices from $M_{n_i}(D_i)$ are simple, pairwise non-isomorphic A-modules (in other words – minimal left ideals of A). We will denote these modules by $M_1, \ldots, M_k$. Similarly, the modules of first rows $e_{1,1}M_{n_i}(D_i) = N_i$ (for $1 \leq i \leq k$) are simple, pairwise non-isomorphic right A-modules. The following theorem describes, in terms of these simple modules, all finite-dimensional A-modules.

Theorem 1. *Let A be a finite-dimensional semisimple algebra. Then every A-module is completely reducible, i.e. A is a classically semisimple ring. Furthermore, every simple left (right) A-module is isomorphic to one of the minimal left (resp. right) ideals M_i (resp. N_i) of A. In particular, the number of non-isomorphic simple left (right) A-modules equals the number of simple summands of the algebra A.*

In principle, this theorem describes all simple A-modules, even though such description doesn't allow to explicitly produce those modules unless a decomposition of A into simple summands is obtained first. In some situations, for example for a group algebra FS_n of the *symmetric group* S_n (i.e. the group of all permutations of the set $\{1, \ldots, n\}$) over a field of characteristic 0, simple modules can be described explicitly (in case of FS_n, by means of the so-called Young tableaux). For more information on this subject and its applications to the theory of identities, see Part II.

Let now A be an arbitrary (generally speaking, not semisimple) finite-dimensional algebra. In this case, as we have already mentioned, indecomposable (finite-dimensional) modules are the most important ones for the module-theoretic point of view. Depending on the number of such modules, algebras fall into two categories.

1) *Finite type.* In this case, by definition, we have only finitely many non-isomorphic indecomposable A-modules. For a long time the *first Brauer-Thrall conjecture* remained unproved: if dimensions of indecomposable A-modules are bounded then the algebra A is of finite type. This conjecture was established by A.V. Roiter (1968).

2) *Infinite type.* In this situation by definition there are infinitely many pairwise non-isomorphic indecomposable A-modules. The *second Brauer-Thrall conjecture* was also open for a long time. It was proved by L.A. Nazarova and A.V. Roiter (1972).

Another categorization of algebras by the type of their representations is the division into algebras of wild and tame types. Informally speaking, an algebra is of wild type when the classification problem for indecomposable

A-modules includes the classification problem for indecomposable modules over the free algebra on two generators (i.e. the "wild" classification problem for pairs of matrices by simultaneous conjugation). Algebras of tame type are those algebras which are not wild. Several characterizations of algebras of tame type have been obtained in terms of certain graphs – schemes of algebras (in the sense of Gabriel). Definition of a scheme of an algebra is given in the next section.

Morita equivalence (see Sections 1.20, 1.22 and 4.8) is an especially important concept in the classification of algebras according to the type of their representations. Since equivalent categories of modules are essentially indistinguishable, it is clear that Morita-equivalent algebras have identical respresentation types. Hence from the point of view of module theory (representations), such algebras are indistinguishable. The following theorem allows us to find, for every algebra, an algebra which is Morita-equivalent to the original one, and has the simplest possible structure. We will call a finite-dimensional algebra B *reducible* if the factor algebra $B/\operatorname{Rad} B$ is isomorphic to a direct sum of division rings.

Theorem 2. *Every finite-dimensional algebra A is Morita-equivalent to a uniquely determined reducible algebra B.*

This result follows from Morita's theorem (see Sect. 4.8). It turns out that reducible algebras can be studied by means of generators and defining relations. This will also be discussed in the next section.

2.13. Schemes and Reducible Algebras. By a *scheme* we will mean any finite set of vertices connected with each other by arrows. If vertices of the scheme are labeled by integers $1, 2, \ldots, s$ then the scheme S is described by its $s \times s$ matrix $(t_{i,j})$, where $t_{i,j}$ is the number of arrows leading from vertex i to vertex j. With each vertex i we will also associate an 'empty' arrow ε_i (which begins and ends at i). By $K(S)$ we will denote the F-algebra whose generators are all (empty and non-empty) arrows, and whose defining relations consist of all equalities $\alpha\beta = 0$ – whenever the endpoint of arrow α is not the same as the starting point of β, and $\varepsilon_i\alpha = \alpha$ ($\alpha\varepsilon_i = \alpha$) – when the starting point (endpoint) of α equals i. Elements of $K(S)$ are therefore linear combinations of paths over S, i.e. sequences of consecutive arrows in which the endpoint of the preceding arrow coincides with the starting point of the following one (including the empty arrows ε_i). The unity element of $K(S)$ is $1 = \varepsilon_1 + \ldots + \varepsilon_s$. The algebra $K(S)$ can be infinite-dimensional. However, if the scheme S has no cycles (closed paths), then all paths have finite length and hence $K(S)$ is a finite-dimensional algebra. There is a standard method of obtaining a finite-dimensional algebra out of any $K(S)$. It is based on equating to zero all paths whose lengths are greater than or equal to some given integer n. We will denote the resulting algebra by $K_n(S)$. If I is an ideal generated by all non-empty arrows, then $K_n(S) \simeq K(S)/I^n$. The radical of $K_n(S)$ equals I/I^n. It is easy to notice that the algebra $K_n(S)$ is reducible:

$K_n(S)/\mathrm{Rad}\, K_n(S) \simeq \underbrace{F \oplus F \oplus \ldots \oplus F}_{s \text{ times}}$. Algebras of the form $K_n(S)$ are in a certain sense universal models of reducible algebras. A reducible algebra A will be called *basic* if $A/\mathrm{Rad}\, A \simeq F \oplus \ldots \oplus F$. For example, if the field F is algebraically closed, then every reducible algebra over F is basic, since there are no finite-dimensional division algebras over F other than F itself. A universal property of this example is established in the next theorem.

Theorem 1. *Any basic algebra A is isomorphic to a factor algebra of $K_n(S)$ modulo an ideal contained in the square of the radical.*

The scheme S in the above theorem is uniquely determined by the algebra A (needless to say, up to a permutation of vertices) and is called the *scheme* of A. We will describe the process of constructing a scheme for any, not necessarily basic or even reducible, algebra. We will consider the (right) regular module A_A and decompose it into a direct sum of indecomposable submodules $A = P_1 \oplus \ldots \oplus P_n$. The summands P_i (and all modules isomorphic to them) are called principal (indecomposable) A-modules. Let $P_1, \ldots, P_s$ be all, pairwise non-isomorphic, principal modules. Their number equals the number of simple summands in the decomposition of the semisimple algebra $A/\mathrm{Rad}\, A$. We will write $R_i = P_i \mathrm{Rad}\, A$ and $V_i = R_i/R_i \mathrm{Rad}\, A$. Then V_i is a semisimple module, and hence

$$V_i \simeq \bigoplus_{j=1}^{s} U_j^{t_{i,j}} ,$$

where $U_j = P_j/R_j$ are indecomposable modules (it can be shown that the U_j's include all indecomposable A-modules and are pairwise non-isomorphic). To each module V_i we will assign a point in the plane, label it with the integer i, and connect vertex i with vertex j with $t_{i,j}$ arrows. The set of vertices and arrows obtained in this way turns out to be the scheme $S(A)$ of the algebra A.

Example 1. If the algebra A is semisimple, then $R_i = 0$ and $S(A)$ is simply a set of points with no arrows between them.

Example 2. Let $A = T_n(F)$ be the algebra of $n \times n$ upper-triangular matrices over F. The matrix units $e_{i,i}$ are its minimal idempotents. It is easy to see that $P_i = e_{i,i}A$ are pairwise non-isomorphic principal A-modules, with $R_i = e_{ii}A\mathrm{Rad}\, A = e_{i,i+1}F + \ldots + e_{i,n}F \simeq P_{i+1}$. The scheme of A then has the form

$$\underset{1}{\cdot} \longrightarrow \underset{2}{\cdot} \longrightarrow \underset{3}{\cdot} \longrightarrow \underset{4}{\cdot} \longrightarrow \ldots \longrightarrow \underset{n}{\cdot}$$

Example 3. It is also easy to prove that a scheme of an algebra $K_n(S)$, where $n \geq 2$, equals (modulo a permutation of vertices) the scheme S.

We will remark that by construction, the scheme of an algebra depends only on the factor algebra $A/(\operatorname{Rad} A)^2$, i.e. $S(A) = S(A/(\operatorname{Rad} A)^2)$. Statements expressed in terms of schemes will therefore carry over from $A/(\operatorname{Rad} A)^2$ to A. For example, the following fact holds.

Theorem 2. *Algebra A cannot be decomposed into a direct sum if and only if the scheme $S(A)$ is connected.*

Corollary. *Both algebras A and $A/(\operatorname{Rad} A)^2$ are either decomposable or indecomposable at the same time.*

2.14. Artinian Rings. Wedderburn's theorems on the structure of simple and semisimple finite-dimensional algebras have been generalized by E. Artin to left Artinian rings (see Sect. 1.8). The class of (left, right) Artinian rings is closed under homomorphic images, finite direct sums and forming of matrix rings. There exist left, but not right Artinian rings (for example the matrix ring $\begin{pmatrix} \mathbb{R} & \mathbb{C} \\ 0 & \mathbb{Q} \end{pmatrix}$). The ring $\mathbb{Z}$ is not Artinian, but its ring of fractions $\mathbb{Z}$ is a field, hence Artinian (in general, rings of fractions of certain – not necessarily Artinian – rings turn out to be simple, or semisimple, Artinian rings; see Sect. 4.6 for more information).

Theorem 1. *A ring R is simple left (right) Artinian if and only if R is isomorphic to a ring of matrices $M_n(D)$ over a division ring D. The integer n and the division ring D (up to isomorphism) are uniquely determined by the ring R.*

Theorem 2. *A ring R is semisimple Artinian if and only if R is isomorphic to a finite direct sum of matrix rings over division rings.*

The decomposition in the above theorem is unique, which follows from an easy general result.

Theorem 3. *If $S_1 \oplus \ldots \oplus S_n \simeq R_1 \oplus \ldots \oplus R_m$, where S_i and R_j are non-zero rings indecomposable into direct sums, then $m = n$ and $S_i \simeq R_{\sigma(i)}$, where σ is some permutation of subscripts.*

Theorems 1 and 2 are known in the literature as "Wedderburn-Artin theorems".

Representation theory of Artinian rings, as well as their structure theory, is largely parallel to the representation theory of finite-dimensional algebras. For example, the following theorem – which we already mentioned before – holds.

Theorem 4. *Every R-module is completely reducible (i.e. R is a classically semisimple ring) if and only if R is a semisimple Artinian ring.*

One fragment of the theory of Artinian rings – the so-called quasi-Frobenius rings – will be discussed in Sect. 3.

2.15. Bibliographical Notes. Material covered in Sections 2.3 through 2.9 is related to the classical achievements of the theory of finite-dimensional algebras, and is treated most exhaustively in the monographs by Albert [1961] and Pierce [1982]. First examples of non-crossed products were constructed by Amitsur [1972] (see also Jacobson [1975] and Rowen [1980]). For fundamentals of the algebraic K-theory one can refer to the concisely written book by Milnor [1971], as well as the more detailed monograph by Bass [1968], the latter being primarily devoted to the functor K_1. Recent developments in the algebraic K-theory can be found in surveys of A.A. Suslin [1984a, 1984b]. A proof of the theorem from Sect. 2.11 is given in Merkurjev and Suslin [1982] as well as Suslin [1984a]. The textbook by Drozd and Kirichenko [1980] contains the structure theory of finite-dimensional algebras, and includes material discussed in Sect. 2.13. Beginnings of the structure theory of finite-dimensional algebras and the representation theory of finite groups can be found in the classical algebra handbook by van der Waerden [1971, 1967]. The fullest and most detailed treatment of the representation theory of finite-dimensional algebras is presented in the monograph by Curtis and Reiner [1962]. Proof of Brauer-Thrall conjectures is contained in the original paper by Nazarova and Roiter [1973].

Details of the theory of Artinian rings can be found in the textbooks Kasch [1977] and Kertesz [1968].

§ 3. Modules and Some Classes of Rings

3.1. Introduction. Modules over commutative rings (more precisely, categories of modules) play an important part in mathematics in general, and in ring theory in particular. Categories of modules (and – more generally – abelian categories) are the main objects of study in homological algebra. As was said before, modules arise in a natural way in the theory of linear representations of groups and algebras, i.e. homomorphisms into groups and algebras of linear transformations of finite-dimensional vector spaces over a field. Several natural classes of rings (e.g. clasically semisimple, hereditary, perfect and semiperfect, quasi-Frobenius rings etc.) which are often encountered in various problems of ring theory, module theory and homological algebra, can be characterized in terms of properties of modules.

In the general case it is necessary to consider separately left and right modules over rings. There is, however, a standard method of passing from modules of one kind to the other, based on passing from a ring R to the *opposite (anti-isomorphic)* ring R^{op}, whose underlying set equals that of R, but whose multiplication differs from the one in R by the order of factors: $r_1 \circ r_2 = r_2 r_1$, where $\circ$ denotes multiplication in the ring R^{op}. In these terms, if M is a left R-module then it can be made into a right R^{op}-module by setting $mr = rm$, where $m \in M$ and $r \in R$. In the same way the concept of a bimodule can be related to that of a left module. Let $R \otimes_{\mathbb{Z}} S$ be the tensor product of rings R and S^{op} over $\mathbb{Z}$; then the study of (R, S)-bimodules M can be reduced to the study of left $R \otimes_{\mathbb{Z}} S$-modules, letting $(r \otimes s)m = rms$ (for $r \in R$, $s \in S$ and $m \in M$). Below, for sake of consistency, we will consider only left modules.

In a sense, the simplest modules over a ring R are the *cyclic* modules, i.e. modules M generated by a single element, $M = Rx$. It is easy to see that every cyclic module $M = Rx$ is isomorphic to a factor module R/I, where I is a left ideal of R, namely the annihilator $\{r \in R \mid rx = 0\}$ of the element x. Conversely, any module R/I (with I – a left ideal of R) is cyclic: it is generated by the element $\bar{1} = 1 + I$ $(1 \in R)$. A module M with a finite set of generators is called *finitely generated.* It follows from a fundamental theorem on finitely generated abelian groups that every finitely generated $\mathbb{Z}$-module is a direct sum of cyclic $\mathbb{Z}$-modules (for finite abelian groups, this assertion dates back to Gauss). The same statement holds for all finitely generated modules over (left and right) principal ideal domains. The above theorem is important in the task of reducing (by means of mappings similar to linear transformations) matrices over principal ideal domains to a diagonal form (the so-called problem of finding invariant factors of a matrix). A special case of this is the problem, well-known from linear algebra, of reducing λ-matrices (i.e. matrices over the algebra of polynomials $F[\lambda]$) to the canonical form.

3.2. Artinian and Noetherian Modules. A module M over R is called *Noetherian* (*Artinian*), if it satisfies the maximality (resp. minimality) condition on submodules. This is equivalent to the statement that every ascending chain $N_1 \subseteq N_2 \subseteq \ldots$ (respectively, descending chain $N_1 \supseteq N_2 \supseteq \ldots$) stabilizes. Let $R = \mathbb{Z}$. All finitely generated groups turn out to be Noetherian $\mathbb{Z}$-modules. Among them, finite groups are the only Artinian ones. There exist Artinian, but not Noetherian abelian groups – e.g. the quasicyclic group $\mathbb{Z}_{p^\infty}$ (the group of all complex roots of 1 of degrees $p, p^2, \ldots$). This group is a union of all its cyclic subgroups $\mathbb{Z}_{p^n}$ and each of its proper subgroups coincides with one of the subgroups $\mathbb{Z}_{p^n}$. Every Artinian abelian group is isomorphic to a subgroup of a finite direct sum of quasicyclic groups.

The class of Noetherian (Artinian) R-modules is closed under passing to submodules, homomorphic images and extensions. This last property means that if a module B contains a Noetherian (resp. Artinian) submodule A such that B/A is Noetherian (Artinian), then B itself is Noetherian (Artinian). In particular, if $N_1, \ldots, N_k$ are Noetherian (Artinian) submodules of a module M, then the submodule $N_1 + \ldots + N_k$ is also Noetherian (Artinian). Therefore every finitely generated left module over a left Noetherian (left Artinian) ring is left Noetherian (left Artinian). From this, using the Hilbert's basis theorem, we deduce that every finitely generated module over a commutative finitely generated ring is Noetherian. It also follows from the above that if R is a left Noetherian (left Artinian) ring, then $M_n(R)$ is of the same type (alternately, this follows from Morita equivalence of the rings R and $M_n(R)$).

3.3. Modules of Finite Length. *Length* $l(M)$ of a module M is the least upper bound of the integers n such that M contains a strictly ascending chain of submodules

$$0 = M_0 \subset M_1 \subset \ldots \subset M_n = M\,. \tag{1}$$

Modules of length 1 are the simple modules (and of length 0 – the zero modules). A chain of the form (1) is called a *composition series* of M, if all factors M_i/M_{i-1} are simple modules (or, equivalently, if the chain (1) cannot be refined).

It is easy to show that the following statements about a module M are equivalent: (a) M has finite length; (b) M is both Artinian and Noetherian; (c) M has a composition series.

We will introduce the notion of equivalence between composition series. We say that the chain (1) is *equivalent* to the chain of submodules

$$0 = N_0 \subset N_1 \subset \ldots \subset N_k = M\,, \tag{2}$$

if $k = n$ and there is a permutation σ of the integers $1, \ldots, n$ such that $N_i/N_{i-1} \simeq M_{\sigma(i)}/M_{\sigma(i)-1}$ for $i = 1, \ldots, n$), i.e. if the chains (1) and (2) have isomorphic factors modulo their order.

Theorem 1 (Jordan, Hölder). *If a module M has two composition series, then they are equivalent.*

It follows from the Jordan-Hölder theorem that length of a module equals the length of any of its composition series, or is infinite – if such series do not exist. We will remark that for any submodule N of a module M the equality $l(M) = l(N) + l(M/N)$ holds. In particular,

$$l(M_1 \oplus \ldots \oplus M_n) = \sum_{i=1}^{n} l(M_i)\,.$$

The classical theorem on the uniqueness of decomposition into a direct sum of indecomposable modules is also true for modules of finite length.

Theorem 2 (Krull-Schmidt). *Every module of finite length decomposes in a unique way, up to isomorphism and order of the summands, into a direct sum of indecomposable modules.*

The essence of the Krull-Schmidt theorem consists of the fact that isomorphism classes $[M]$ of modules M of finite length form a free abelian monoid under the operation $[M] + [N] = [M \oplus N]$. Both the Krull-Schmidt and Jordan-Hölder theorems are useful, for example, in determining uniqueness of invariant factors of matrices over principal ideal domains.

3.4. Projective Modules. Definition of a projective module was given in Sect. 1.21. The following criterion of projectivity is often used.

Theorem 1 (Dual basis lemma). *Module P over a ring R is projective if and only if there exist elements $x_i \in P$ $(i \in I)$ and homomorphisms $f_i : P \to R$ $(i \in I)$ such that for every element $y \in P$ we have $f_i(y) = 0$ for all but finitely many $i \in I$, and $y = \sum_{i \in I} f_i(y) x_i$.*

We will apply this criterion to establish the fact that invertible ideals of a commutative integral domain R are projective R-modules. A non-zero ideal I of R is called invertible, if there exists a non-zero ideal I' of R such that $II' = dR$ is a principal ideal (we will recall that $II' = \{\sum x_i x_i' \mid x_i \in I,\ x_i' \in I'\}$). We have $d = \sum_{i=1}^n x_i x_i'$, and if $y \in I$ then $dy = \sum_{i=1}^n x_i(x_i' y)$. However, $x_i' y \in dR$, i.e. $x_i' y = f_i(y) d$. Substituting and cancelling d on both sides we obtain the required equality.

In a new and independent area of algebra – the algebraic K-theory, which we already mentioned in Sect. 2.11 – situations in which projective modules turn out to be free are of particular importance. For example, a result of A.A. Suslin and D. Quillen on Serre's conjecture asserts that every projective module over the algebra of polynomials $F[x_1, \ldots, x_n]$ (F – a field) is free. We will now quote other results of this type. Let I be a two-sided ideal of the ring R, and let P be a projective R-module. We will consider the module $\bar{P} = P/IP$ over the ring $\bar{R} = R/I$. This module is projective; indeed, if $P \oplus Q = F$ is a free R-module, then $\bar{P} \oplus \bar{Q} = \bar{F}$ is a free $\bar{R}$-module. We will call the module $\bar{P}$ over $\bar{R}$ a *reduction of the module* P by the ideal I. The following theorem is due to Beck.

Theorem 2. *Let P be a projective R-module, $J(R)$ – the Jacobson radical of R and $\bar{P}$ – a reduction of P by the ideal $J(R)$. If $\bar{P}$ is a free module, then P is also free.*

Since for a local ring R (see Sect. 1.24) the factor ring $R/J(R)$ is a division ring, the preceding theorem (together with the fact that every module over a division ring is free) immediately implies a corollary, due to Kaplansky.

Corollary 3. *Every projective module over a local ring is free.*

Proof of Theorem 2 is very complicated. It relies on a result which is of independent interest.

Theorem 4. *Let P, P_1 are projective R-modules and $\bar{P}, \bar{P}_1$ – their reductions by $J(R)$. If $\bar{P} \simeq \bar{P}_1$ and one of the modules P, P_1 is finitely generated, then $P \simeq P_1$.*

Proof of this theorem is based on a lemma of Bass, saying that reduction of a non-zero projective module by the Jacobson radical is non-zero. It is interesting to note that, as was shown by V.N. Gerasimov and I.I. Sakhaev, there are examples of projective, not finitely generated R-modules P, whose reduction by the Jacobson radical is finitely generated (even though when, for instance, R is a commutative ring, the property of being finitely generated lifts from $\bar{P}$ to P).

The following theorem, due to Kaplansky, in a certain sense reduces the study of arbitrary projective modules to the case of countably generated projective modules.

Theorem 5. *Every projective module over any ring R is a direct sum of countably generated ones.*

Since every module M is a homomorphic image of a projective – even free – module, there exist exact sequences of the form $0 \to K \to P \to M \to 0$, where P is a projective module. We will call them *projective presentations* of the module M. Taking a projective representation for K etc., we can construct, for every n, an exact sequence (a *partial projective resolution*) of the form $0 \to K_n \to P_{n-1} \to \cdots \to P_0 \to M \to 0$, where the modules P_i are projective. If the module K_n is also projective, then the above sequence is called a *projective resolution of length n for M*. Not every module has such a resolution. In the general case there exists an infinite projective resolution $\cdots \to P_1 \to P_0 \to M \to 0$. The following theorem proves crucial for the study of projective resolutions.

Theorem 6 (Schanuel's lemma). *Let $0 \to K \to P \to M \to 0$ and $0 \to K' \to P' \to M \to 0$ be two projective representations of the module M. Then $K \oplus P' \simeq K' \oplus P$.*

From this, by induction on n, we have

Corollary 7. *Let $0 \to P_n \to P_{n-1} \to \cdots \to P_0 \to M \to 0$ and $0 \to P'_n \to P'_{n-1} \to \cdots \to P'_0 \to M \to 0$ be two partial projective resolutions of length n for the module M. Then there is an isomorphism*

$$P_0 \oplus P'_1 \oplus P_2 \oplus \ldots \simeq P'_0 \oplus P_1 \oplus P'_2 \oplus \ldots .$$

This directly implies

Corollary 8. *If M has a projective resolution of lenght n, then every partial projective resolution of length n for M is a projective resolution for M.*

We will remark that projective, as well as injective resolutions (see Sect. 3.5) have important applications in homological algebra, in defining the derived functors (e.g. $\mathrm{Ext}^n(A, C)$ and $\mathrm{Tor}_n(A, C)$), and in definitions of homological dimensions of modules and rings (see Sect. 3.6).

3.5. Injective Modules. Definition of an injective module was also given in Sect. 1.21. An effective instrument in verification of injectivity of a module is *Baer's criterion*:

Theorem 1. *For a (left) R-module Q to be injective, it is necessary and sufficient that for any left ideal I of R and any R-module homomorphism $f : I \to Q$ there exist an element $q \in Q$ such that $f(a) = aq$ for all $a \in I$.*

In particular, if R is a principal left ideal domain, then for an R-module Q to be injective it is necessary and sufficient that it be *divisible*, i.e. $rQ = Q$ for any $0 \neq r \in R$. Indeed, by Baer's criterion it is necessary that for any $0 \neq r \in R$ and any $f : Rr \to Q$ there exist an element $q \in Q$ such that $f(r) = rq$. But any element of Q can be chosen as $f(r)$. It follows, for example, that an abelian group (over $R = \mathbb{Z}$) is injective if and only if it is divisible (or, in other words, complete). It is known that every abelian group can be embedded in an injective (i.e. complete) abelian group. An analogous result holds for arbitrary modules as well.

Theorem 2. *Every module can be embedded in an injective one.*

A submodule N is called an *essential submodule* of a module M (and M – an *essential extension* of N), denoted by $N \leq_{\text{ess}} M$, if every non-zero submodule of M has a non-zero intersection with N. An essential extension E of a module M is called an *injective hull* of M if E is an injective module.

Theorem 3. *Every module M has an injective hull $E(M)$, and all injective hulls of M are isomorphic (as extensions of M).*

The last assertion means that if E_1 and E_2 are two injective hulls of M, then there is an isomorphism $f : E_1 \to E_2$ which is the identity on M.

We will outline the proof of Theorem 3. Let Q be an injective extension of the module M. By Zorn's lemma we can find a submodule E of Q such that $M \leq_{\text{ess}} E$, and E does not have proper essential extensions contained in Q. Then E will not have proper essential extensions at all. Indeed, if $E \leq_{\text{ess}} W$, then there is a homomorphism $f : W \to Q$ in the diagram

$$\begin{array}{ccccc} 0 & \to & E & \xrightarrow{1_E} & W \\ & & \downarrow 1_E & \swarrow f & \\ & & Q & & \end{array}$$

, where $\ker f \cap E = 0$, i.e. $\ker f = 0$; then $f(W)$ is an essential extension of E contained in Q, so that $f(W) = E$ and hence $W = E$. However, we have the following

Lemma 4. *A module E is injective if and only if it has no proper essential extensions.*

By this lemma, $M \leq_{\text{ess}} E$ is an injective hull of M. If E_1 is another injective hull of M, then there exists a monomorphism $f : E_1 \to E$ such that $f|_M = 1_M$. Moreover, $f(E_1) \leq_{\text{ess}} E$, and hence $f(E_1) = E$ (say, by lemma 4 applied to the module $f(E_1)$).

3.6. Homological Dimensions of Rings and Modules. In this section we will present some basic facts about various homological dimensions of modules and rings. These concepts are widely used in ring theory, module theory and homological algebra. They migrated into algebra from geometry and topology. We will first describe how the global dimension (see below) of a local ring is related to the dimension of an algebraic variety in an affine space. We begin with the notion of a *regular local commutative ring*. Let

R be a local Noetherian commutative ring, $\mathbf{m}$ – its maximal ideal (i.e. the Jacobson radical) and let n be the minimal number of generators of this ideal. R is called *regular*, if R contains a chain $\wp_0 \subset \wp_1 \subset \ldots \subset \wp_n = \mathbf{m}$ of prime ideals (by one of Krull's theorems there are no chains of prime ideals of greater length in R). Regular local rings have many noteworthy properties; for example, they turn out to be unique factorization domains. Moreover, they have finite global dimension (see definition below), and are characterized by this property among all local Noetherian (commutative) rings. Also, three dimension functions coincide for a regular local ring R: global dimension, Krull dimension (maximum of lengths of chains of prime ideals) and the minimal number of generators of the maximal ideal $\mathbf{m}$ (i.e. the dimension of the vector space $\mathbf{m}/\mathbf{m}^2$ over the residue field $R/\mathbf{m}$). If K is an algebraically closed field, $V \subseteq K^n$ – an irreducible algebraic variety and $K[V]$ – the ring of polynomial functions over it, then the dimension $\dim V$ of the variety V is defined as the transcendence degree of the domain $K[V]$ over K (or as the Krull dimension of the ring $K[V]$). If $v \in V$, then the local ring $\mathcal{O}_{v,V}$ of the point v is defined as the ring of fractions of $K[V]$ relative to the maximal ideal $\mathbf{m}_{v,V} = \{f \in K[V] \mid f(v) = 0\}$, that is

$$\mathcal{O}_{v,V} = \{f/g \mid f, g \in K[V],\, g(v) \neq 0\}\,.$$

Krull dimension of this ring is always equal to that of $K[V]$, i.e. is the same as the dimension of the variety V. If the ring $\mathcal{O}_{v,V}$ is regular, then the point v is called non-singular (or simple). There are 'fewer' singular points – they form a proper sub-variety of V. This way, for every non-singular point $v \in V$ the local ring $\mathcal{O}_{v,V}$ has global dimension equal to $\dim V$. If V is a smooth variety, i.e. it contains no singular points, then the global dimension of $K[V]$ is also equal to $\dim V$. Otherwise, the ring $K[V]$ has infinite global dimension.

Projective dimension $\operatorname{pd} M$ of a non-zero module M is the minimal length of its projective resolutions, and -1 if $M = 0$. In view of corollary 8 in Sect. 3.4 we see that the projective dimension of the module M does not exceed n (for $n \geq 0$) if and only if every partial resolution of M of length n is in fact projective.

Theorem 1. *Let $0 \to A' \to A \to A'' \to 0$ be an exact sequence of modules, d', d and d'' – projective dimensions of A', A and A'' respectively. Then $d \leq \max(d', d'')$, $d'' \leq \max(d, d' + 1)$ and $d' \leq \max(d, d'' - 1)$.*

In particular, if $d \neq \max(d', d'')$ then $d'' = d' + 1$. Hence when $A \simeq A' \oplus A''$, we necessarily have $d = \max(d', d'')$. Indeed, if $d < \max(d', d'')$ then $d'' = d' + 1$; however, there is also the exact sequence $0 \to A'' \to A'' \oplus A' \to A' \to 0$, meaning that $d' = d'' + 1$ – which is impossible.

By duality one defines the *injective dimension* $\operatorname{id}(M)$ of a module M. Namely, $\operatorname{id}(M) = -1$ for $M = 0$, while if $M \neq 0$ then $\operatorname{id}(M)$ is the minimal length of an injective resolution of M. Here an *injective resolution* of length n is understood to be an exact sequence of the form $0 \to M \to Q_0 \to$

$Q_1 \to \ldots \to Q_n \to 0$, where the modules $Q_0, \ldots, Q_n$ are injective. If only $Q_0, \ldots, Q_{n-1}$ are required to be injective, the above sequence is called a *partial injective resolution* of M. A dual version of Theorem 1, in which d' and d'' are interchanged, holds in this situation (so that, for example, $\mathrm{id}\,(A') \le \max(\mathrm{id}\,(A), \mathrm{id}\,(A'') + 1)$). Among the injective resolutions of a module M one can find a certain canonical one, namely – a minimal resolution. An injective resolution $0 \to M \overset{\alpha_0}{\to} Q_0 \overset{\alpha_1}{\to} Q_1 \overset{\alpha_2}{\to} \ldots \to$ (whether finite or infinite) is *minimal*, if $\mathrm{Im}\,\alpha_i \le_{\mathrm{ess}} Q_i$, i.e. when Q_i is an injective hull of $\mathrm{Im}\,\alpha_i$ for all $i \ge 0$. It follows from Theorem 3 in Sect. 3.5 that every module has a minimal injective resolution $0 \to M \overset{\alpha_0}{\to} E_0 \overset{\alpha_1}{\to} E_1 \overset{\alpha_2}{\to} \ldots$, where $E_0 = E(M)$ is an injective hull of M, $E_1 = E(E_0/M)$ and $E_{i+1} = E(E_i/\alpha_i(E_{i-1}))$. Moreover, a minimal injective resolution is unique up to isomorphism. If $\mathrm{id}\,(M) = \infty$, then $E_i \ne 0$ for all $i \ge 0$. If, however, $\mathrm{id}\,(M) = n < \infty$, then $E_i = 0$ for $i > n$ and $E_i \ne 0$ when $i \le n$.

We will note that the injective dimension of modules is applied in an essential way in homology theory of commutative rings (see I. Kaplansky, *Commutative rings*, Boston, Allyn and Bacon 1970). In particular, Gorenstein rings which give rise to a rich theory, are defined as commutative Noetherian rings of finite injective dimension (as modules over themselves).

Left global dimension of a ring R (l.gl.dim R) is the supremum of projective dimensions of left R-modules. Considering right R-modules instead, we obtain the definition of the right global dimension (r.gl.dim R).

The following theorem is fundamental in this theory.

Theorem 2. *For any ring R and an integer $n \ge 0$, the following statements are equivalent:*
(a) $\mathrm{pd}_R(A) \le n$ *for all* ${}_RA$*;*
(b) $\mathrm{id}_R(A) \le n$ *for all* ${}_RA$*;*
(c) for any A and B, $\mathrm{Ext}^i_R(A, B) = 0$ *whenever $i > n$;*
(d) for any A and B, $\mathrm{Ext}^{n+1}_R(A, B) = 0$.

Corollary 3. l.gl.dim $R = \sup \mathrm{id}_R(M)$, *i.e. for any ring R the supremum of projective dimensions of R-modules coincides with the supremum of their injective dimensions.*

3.7. Flat Modules. A left R-module B is called *flat* if the functor $-\otimes_R B$ is exact, i.e. it transforms every exact sequence of right R-modules $0 \to A' \to A \to A'' \to 0$ into an exact sequence of groups $0 \to A' \otimes_R B \to A \otimes_R B \to A'' \otimes_R B \to 0$. Clearly, it is enough to require that if A' is a submodule of a right R-module A, then $A' \otimes B \to A \otimes B$ is an injection (so that even though in general $A' \otimes B$ should not be identified with a subgroup of $A \otimes B$, it can be done when B is a flat module). All projective modules are flat; indeed, for any family B_i $(i \in I)$ of left R-modules, we have $A \otimes_R (\bigoplus_{i \in I} B_i) \simeq \bigoplus_{i \in I} (A \otimes_R B_i)$ (i.e. the tensor product commutes with direct sums of modules). It follows that a direct sum of flat modules is a flat module, and that a direct summand of a flat module is also flat. Since $A \otimes_R R \simeq A$, this means that R – and so every free R-module as well – is flat, as are all projective modules (as summands of free modules). Other examples of flat modules are provided by some useful constructions known from commutative algebra. For example, if A is a Noetherian commutative ring and I – an ideal of A, then the so-called I-adic completion $\hat{A}$ of the ring A is a flat A-module. Furthermore, if A is a commutative ring and S – its multiplicative subset (i.e. $1 \in S$, $0 \notin S$ and S is closed under multiplication), then $S^{-1}A$ is a flat A-module.

The next theorem gives a criterion for a module B to be flat.

Theorem 1. *For a (left) R-module B to be flat it is necessary and sufficient that for any finitely generated ideal I of R the homomorphism $I \otimes_R B \to B$, $i \otimes b \mapsto ib$ ($i \in I$, $b \in B$) be injective.*

This theorem implies several interesting corollaries. One of them concerns (von Neumann) regular rings. In such a ring R every finitely generated right ideal I is generated by an idempotent, $I = eR$ with $e^2 = e$, and hence it splits as a direct summand of R ($R = I \oplus (1-e)R$). The mapping $I \otimes_R B \to R \otimes_R B \simeq B$ is therefore an injection. In view of Theorem 1 we thus conclude that every module over a regular ring is flat. This property characterizes regular rings. Another consequence has to do with principal right ideal domains. Using the criterion proved above it isn't difficult to show that if R is a principal right ideal domain, then an R-module B is flat if and only if it is *torsion-free*, i.e. $rb = 0 \Rightarrow r = 0$ or $b = 0$ (where $r \in R$ and $b \in B$). In particular, an abelian group is a flat $\mathbb{Z}$-module if and only if it is torsion-free.

Another criterion of flatness is given by the following

Theorem 2. *For an R-module B to be flat it is necessary and sufficient that for any elements $a_i \in R$ ($1 \le i \le n$) such that $\sum_{i=1}^n a_i b_i = 0$, there exist elements $r_{i,j} \in R$ and $v_j \in B$ such that $b_i = \sum_j r_{i,j} v_j$ and $\sum_i a_i r_{i,j} = 0$.*

This criterion is applied, in particular, to analytical problems in Malgrange's book [1966].

3.8. Classically Semisimple Rings. We will recall that a rings R is called classically semisimple, if every module over R is completely reducible. Examples of classically semisimple rings were presented in Sect. 1 – they were: finite-dimensional semisimple algebras, group algebras of finite groups over a 'good' field, and Clifford algebras of non-degenerate quadratic forms. This class of rings has many important characterizations and properties.

Theorem. *The following conditions are equivalent for a ring R:*
(a) R is a classically semisimple ring;
(b) all R-modules are projective;
(c) all R-modules are injective;
(d) R is a semisimple Artinian ring.

As was proved by B. Osofsky, for a ring R to be classically semisimple it suffices that all cyclic R-modules be injective. This, however, is a far more difficult and subtle result than the preceding theorem.

3.9. Hereditary and Semihereditary Rings. Ring R is called *left hereditary* if each of its left ideals is a projective R-module. For example, a commutative domain R is hereditary if and only if it is a Dedekind domain (i.e. if every non-zero ideal of R is invertible, see Sect. 3.4). A special case of left hereditary rings is the class of *left free ideal rings*, i.e. rings in which every left ideal is a free module of a fixed rank. Examples of left free ideal rings include principal left ideal domains, free algebras over fields, as well as group algebras of free groups. Left hereditary rings can be characterized by a number of equivalent statements.

Theorem 1. *The following properties of a ring R are equivalent:*
(a) R is left hereditary;
(b) $\mathrm{l.gl.dim}\, R \leq 1$;
(c) all submodules of every projective R-module are projective;
(d) all factor modules of every injective R-module are injective.

We will remark that a left hereditary ring may not only fail to be right hereditary – it can even have right global dimension equal to any integer $n > 0$ or infinity. Relevant examples were first constructed by Jategaonkar.

A ring R is called *left semihereditary* if each of its finitely generated left ideals is a projective R-module. All valuation rings of fields or, more generally, all (commutative) *Prüfer rings* are semihereditary. The latter can be defined as commutative domains, in which every finitely generated ideal is invertible. Every (von Neumann) regular ring is also (left and right) semihereditary, since each of its finitely generated (left or right) ideals is generated by an idempotent. A ring, whose left (right) finitely generated ideals are free R-modules of some given fixed rank, also turns out to be left (right) semihereditary. Submodules (finitely generated submodules) of free modules over a hereditary (resp. semihereditary) ring are descibed by the following theorem.

Theorem 2. *Let R be a left (semi-) hereditary ring, F – a free left R-module and P – its (finitely generated) submodule. Then the module P is isomorphic to a direct sum of (finitely generated) left ideals of R.*

In case of a projective module P and a semihereditary ring R, the hypothesis of P being finitely generated can be removed (Kaplansky, Bergman):

Theorem 3. *Every projective left module over a left semihereditary ring R is isomorphic to a direct sum of finitely generated left ideals of R.*

3.10. Local Rings. Local rings cam be defined by any of the following equivalent properties: (a) non-invertible elements of R form an ideal; (b) R

contains a unique maximal right (left) ideal; (c) a sum of two non-invertible elements of R is non-invertible; and (d) the factor ring $R/J(R)$ of R modulo its Jacobson radical is a division ring.

An example of a non-commutative local ring was given in Sect. 1; it was the algebra of non-commutative formal power series $F\langle\langle X\rangle\rangle$. Other examples are obtained as rings of endomorphisms of certain modules.

Theorem 1. *Let M be an indecomposable R-module of finite length (or an indecomposable injective module). Then* $\mathrm{End}_R M$ *is a local ring.*

We will note that indecomposability of a module M is equivalent to the fact, that its endomorphism ring does not contain non-trivial (i.e. distinct from 0 and 1) idempotents. Every non-zero module with local endomorphism ring is therefore indecomposable.

A module P is called a *generator* in a category R-Mod, if every R-module M is a homomorphic image of a direct sum of copies of P. Since a free module is a direct sum of copies of ${}_RR$, the regular module ${}_RR$ is a generator. Similarly, if ${}_RR$ is a direct summand of a module P then P is a generator. It turns out that for a local ring R the converse of this statement is also true.

Theorem 2. *Let R be a local ring. A left R-module P is a generator if and only if ${}_RR$ is a direct summand of P.*

3.11. Perfect and Semiperfect Rings. Ring R is called *semilocal* if $R/J(R)$ is a classicaly semisimple ring. The Wedderburn-Artin theorem implies that every left (right) Artinian ring is semilocal. In this section we will consider more general classes of semilocal rings. We say that in a ring R *idempotents can be lifted modulo the radical,* if for any idempotent $\bar{e} \in R/J(R)$ there exists an idempotent $f \in R$ such that $\bar{e} = f + J(R)$. It can be shown that if $J(R)$ is a nil ideal (i.e. each of its elements is nilpotent), then R has the above property. In every Banach algebra idempotents also lift modulo the radical. An R-module is said to have a (finite) Azumaya diagram if $M = \bigoplus_{i=1}^n M_i$, and $\mathrm{End}\, M_i$ are local rings. Ring R is said to *have an Azumaya diagram* if the left regular module ${}_RR$ has an Azumaya diagram. A submodule A of a module M is called *superfluous* (or *small*) if, for any submodule $B \subseteq M$, the equality $A + B = M$ implies $B = M$. This notion is dual to that of an essential extension. *Radical* $J(M)$ of a module M is the intersection of all kernels of homomorphisms from M into simple modules. We clearly have $J(R)M \subseteq J(M)$ for any left R-module M. It turns out, that $J(M)$ is a sum of all superfluous submodules of M. Module M is said to have a *projective cover*, if there exists a projective module P and an epimorphism $\varphi : P \to M$, whose kernel is a superfluous submodule of P. Projective cover is a notion dual to that of the injective hull of a module. In contrast with the injective hull, which always exists, a projective cover often does not. We will now state the main definition of this section. A ring is called *(left) semiperfect* if every cyclic left R-module has a projective cover. If every left R-module has a projective cover, then the ring is *left perfect*. The next theorem, due to H. Bass, characterizes semiperfect rings and, at the same time, provides a source of examples of such rings.

Theorem 1. *For a ring R, the following conditions are equivalent:*
(a) R is semiperfect;
(b) R is a semilocal ring in which idempotents lift modulo the radical;
(c) every finitely generated R-module has a projective cover;
(d) R is the endomorphism ring of a module which has an Azumaya diagram;
(e) R has an Azumaya diagram.

It is clear from this theorem (item (b)) that the classes of right and left semiperfect rings coincide.

We will now give a characterization of left perfect rings. Ideal I of a ring R is called *left T-nilpotent* if for any infinite sequence $a_1, a_2, \ldots$ of elements of I, there is an integer n for which $a_1 a_2 \cdots a_n = 0$. Ring R is called *left socular* if every non-zero sub-factor (i.e. a homomomorphic image of a submodule) of the module ${}_RR$ has a simple submodule. In general, the *socle* $\operatorname{Soc} M$ of a module M is the sum of its simple submodules. This way, a ring R is left socular if every factor module of ${}_RR$ has non-zero socle. Perfect rings have been characterized by Bass and Björk as follows.

Theorem 2. *For any ring R the following conditions are equivalent:*
(a) R is right perfect;
(b) $R/J(R)$ is classically semisimple and the Jacobson radical $J(R)$ is left T-nilpotent;
(c) R is a left socular ring with finitely many pairwise orthogonal idempotents;
(d) R satisfies the minimal condition on principal right ideals;
(e) right R-modules satisfy the minimal condition on finitely generated submodules;
(b) every flat left R-module is projective.

This theorem shows that every finite-dimensional algebra (even more, every left or right Artinian ring) is a perfect ring.

3.12. Quasi-Frobenius Rings. Quasi-Frobenius rings arise as a natural generalization of Frobenius algebras, with which we begin.

Theorem 1. *Let A be a finite-dimensional algebra over a field F, and let $f \in A^* = \operatorname{Hom}_F(A, F)$ be a linear form on A. Then the following conditions are equivalent:*
(a) $\operatorname{Ker} f$ *does not contain non-zero left ideals of A;*
(b) $\operatorname{Ker} f$ *does not contain non-zero right ideals of A;*
(c) the bilinear form $\varphi(x, y) = f(xy)$ is non-degenerate.

A finite-dimensional algebra with such a linear form f is called a *Frobenius algebra*. For example, if $A = FG$ is a group algebra of a finite group G then A is a Frobenius algebra, because the form $\operatorname{tr} : A \to F$ defined by

$\operatorname{tr}(\sum \alpha_g g) = \alpha_1$ obviously satisfies condition (a). The form φ constructed in condition (c) is associative, i.e. $\varphi(ab, c) = \varphi(a, bc)$. Conversely, every associative bilinear form $\gamma : A \times A \to F$ arises as $\gamma(x, y) = f(xy)$ for some linear form $f \in A^*$ (namely, $f(a) = \gamma(1, a)$). This implies the following

Theorem 2. *A finite-dimensional F-algebra A is Frobenius if and only if it has a non-degenerate associative bilinear form.*

There are several other characterizations of Frobenius algebras.

Theorem 3. *For a finite-dimensional F-algebra A to be Frobenius it is necessary and sufficient that the following conditions be satisfied for any left ideal L and any right ideal R: (a) $l(r(L)) = L$ and $r(l(R)) = R$; (b) $\dim r(L) = \dim A - \dim L$ and $\dim l(R) = \dim A - \dim R$.*

We will note that condition (b) is stronger than condition (a), since it implies equalities $\dim l(r(L)) = \dim L$ and $\dim r(l(R)) = \dim R$, from which in turn (a) follows – because $l(r(L)) \supseteq L$ and $r(l(R)) \supseteq R$. We will recall that $l(R) = \{x \in A \mid xR = 0\}$ and $r(L) = \{x \in A \mid Lx = 0\}$.

A finite-dimensional algebra which satisfies condition (a) is called a *quasi-Frobenius* algebra. The difference between Frobenius and quasi-Frobenius algebras is underscored by Theorem 4 below. If M is a right module over an F-algebra A, then the space $M^* = \operatorname{Hom}_F(M, F)$ has a structure of a left A-module. Namely, for $a \in A$ and $f \in M^*$ the element $af \in M^*$ is defined by the formula $(af)(m) = f(ma)$ (where $m \in M$). In particular, considering A to be a right A-module, we obtain a left A-module A^*.

Theorem 4. *A finite-dimensional algebra A over a field F is Frobenius if and only if the left A-modules A and A^* are isomorphic. Algebra A is quasi-Frobenius if and only if these modules have identical (up to isomorphism) indecomposable components (possibly with different multiplicities).*

For a Frobenius algebra an isomorphism $A \simeq A^*$ can be obtained by means of the non-degenerate associative bilinear form $\varphi : A \times A \to F$ by assigning to each element $a \in A$ the functional $a^* = \varphi(a, -)$.

We will remark that the class of Frobenius algebras is closed under direct sums and tensor products (it is enough to consider direct sums and tensor products of the respective bilinear forms). Matrix algebra $M_n(F)$ is Frobenius, when one defines the form f from Theorem 1 to be the ordinary trace. If A is a division algebra, then A is Frobenius – with any non-zero form $f \in A^*$ satisfying the conditions in Theorem 1. It follows from these observations that every semisimple finite-dimensional algebra is Frobenius. The first example of a quasi-Frobenius algebra which is not Frobenius was constructed by Nakayama (1939). An Artinian (left and right) ring A is *quasi-Frobenius* (or a *QF*-ring), if it satisfies condition (a) in Theorem 3. Quasi-Frobenius rings have many different characterizations. We recall (Sect. 1.21) that a ring A is *left self-injective* if the left regular module ${}_AA$ is injective.

Theorem 5. *Every quasi-Frobenius ring is both left and right self-injective.*

Theorem 6. *If a ring A is left or right self-injective and satisfies the maximal condition on left or right annihilators of subsets, then A is a quasi-Frobenius ring.*

These theorems imply

Theorem 7. *For a ring A, the following conditions are equivalent:*
(a) A is a quasi-Frobenius ring;
(b) A is left or right Artinian and self-injective on either side;
(c) A is left and right Noetherian and self-injective on both sides.

From the point of view of module theory, quasi-Frobenius rings are interesting because the classes of projective and injective modules over them coincide (Faith, Walker):

Theorem 8. *Each of the following conditions on a ring R is equivalent to R being quasi-Frobenius:*
(a) every projective (left) A-module is injective;
(b) every injective (left) A-module is projective.

A left A-module M is called a *cogenerator* if every left A-module can be embedded in a direct sum of copies of the module M. It follows from the last theorem that if A is a quasi-Frobenius ring, then the left regular module ${}_AA$ is an injective cogenerator. The following theorem and its proof clarify the nature of the definition of a quasi-Frobenius ring.

Theorem 9. *Let A be a ring for which the module ${}_AA$ is a cogenerator. Then $l(r(L)) = L$ for every left ideal L of A.*

Indeed, let $f : A/L \to \bigoplus_{i\in I} A$ be an embedding. If $(a_i)_{i\in I}$ is the image of $1 + L$ under f, then $La_i = 0$, i.e. $a_i \in r(L)$. When $x \in l(r(L))$, we have $xa_i = 0$, and so $f(x + L) = xf(1 + L) = x(a_i)_{i\in I} = 0$ – meaning that $x \in L$.

Theorem 10. *If A is a left or right Artinian ring and ${}_AA$ is a cogenerator, then A is a quasi-Frobenius ring.*

Theorem 11. *In every quasi-Frobenius ring left and right socles coincide.*

Theorem 12. *Let A be a finite-dimensional algebra over a field F,* Soc (A) – *its left socle. If the left A-modules* Soc (A) *and $A/J(A)$ are isomorphic, then A is a Frobenius algebra.*

Indeed, since $A/J(A)$ is a semisimple algebra, it is Frobenius and there is a linear form on it which does not annihilate any non-zero left ideal. It follows that there is a linear form on Soc (A), which does not annihilate any non-zero submodule. Let f be any extension of this form to A. Then Ker f does not contain a minimal left ideal of A, and hence condition (a) of Theorem 1 is satisfied.

Theorem 13. *If A is a finite-dimensional Frobenius algebra and* Soc (A) – *its socle, then the left A-modules* Soc (A) *and $A/J(A)$ are isomorphic.*

Because $A/J(A)$ is a Frobenius algebra, $A/J(A) \simeq (A/J(A))^*$. The second module is isomorphic to $J(A)^\perp = \{x \in A \mid \varphi(J(A), x) = 0\}$, where φ is a non-degenerate associative form on A (every element $x \in J(A)^\perp$ determines a form $\gamma_x \in (A/J(A))^*$, $\gamma_x(a + J(A)) = \varphi(a, x)$). But we also have $J(A)^\perp = r(J(A)) = \mathrm{Soc}\,(A)$, and the theorem follows.

The last results motivate the following definition. A quasi-Frobenius ring A is called a *Frobenius ring*, if the left A-modules $A/J(A)$ and Soc (A) are isomorphic.

Quasi-Frobenius rings are also interesting in the commutative case.

We will remark that any commutative Artinian ring A uniquely decomposes into a direct sum of local Artinian rings $A_1, \ldots, A_n$. Moreover, A is a quasi-Frobenius ring if and only if all rings A_i are quasi-Frobenius, i.e. self-injective. The discussion thus reduces to the case of local rings. Let A be a commutative local Artinian ring, $\mathbf{m}$ – its maximal ideal and $F = A/\mathbf{m}$ – the residue field. Then the socle $S = \mathrm{Soc}\,(A) = l(\mathbf{m})$ is a vector space over F. It is known from commutative algebra that A is self-injective if and only if S is one-dimensional over F (this property defines the class of Artinian Gorenstein rings), i.e. when A has a unique minimal ideal or, equivalently, when it is subdirectly irreducible (see Sect. 4.2). In this way we obtain a theorem.

Theorem 14. *For a commutative Artinian ring A, the following conditions are equivalent:*
(a) A is quasi-Frobenius;
(b) A is Frobenius;
(c) $A = A_1 \oplus \ldots \oplus A_n$, where A_i are subdirectly irreducible rings.

We will now consider some additional examples of (quasi-) Frobenius rings and algebras.

1) If R is a Dedekind domain and I – its non-zero ideal, then R/I is a Frobenius ring.

2) If G is a finite group then the group ring RG is a QF-ring if and only if R is a QF-ring.

3) Let M, N be one-dimensional vector spaces over a field F. We formally define $(M, N) = (N, M) = 0$, obtaining a Morita context (F, M, N, F). The corresponding ring $R = \begin{pmatrix} F & N \\ M & F \end{pmatrix} = \bigoplus_{i,j=1}^2 Fe_{i,j}$, where $e_{1,2}e_{2,1} = e_{2,1}e_{1,2} = 0$, is a Frobenius algebra (with remaining products of elements $e_{i,j}$ similar to those in ordinary matrices). The ordinary trace is a form f satisfying conditions of Theorem 1.

Interesting examples of commutative Frobenius algebras arise in the theory of singularities of smooth manifolds (see the book by Arnol'd, Varchenko and Husein-Zade [1982]). We will consider the ring $\mathbb{C}\{X_1, \ldots, X_n\}$ consisting of all formal series $f = \sum_{i_\nu \geq 0} a_{i_1,,,i_n} X_1^{i_1} \ldots X_n^{i_n} \in \mathbb{C}[[X_1, \ldots, X_n]]$ convergent in a neighborhood of zero, i.e. such that there exist real numbers $M, r_1, \ldots, r_n > 0$ with the property that $|a_{i_1,,,i_n}| \leq \frac{M}{r_1^{i_1} \ldots r_n^{i_n}}$ for all $i_1, \ldots, i_n \geq 0$. This ring can be identified with the ring $\mathcal{O}_n$ of germs of holomorphic (complex-valued) functions at zero, defined in a neighborhood of 0 in $\mathbb{C}^n$. $\mathcal{O}_n$ is a Noetherian local ring, with maximal ideal $\mathbf{m}_n$ which consists of the series with no constant term. This ideal is generated by $X_1, \ldots, X_n$. The residue ring of $\mathcal{O}_n$ modulo $\mathbf{m}_n$ equals $\mathbb{C}$, the map $f \mapsto f(0)$ from $\mathcal{O}_n$ to $\mathbb{C}$ being a homomorphism with kernel $\mathbf{m}_n$. The ring $\mathcal{O}_n$ is a regular local ring and so, in particular, a unique factorization domain. Germs at zero of holomorphic functions $\mathbb{C}_n \to \mathbb{C}_n$ which map zero to zero, turn out to be n-tuples $P = (f_1, \ldots, f_n) \in \mathbf{m}_n \times \ldots \times \mathbf{m}_n$ of series from $\mathbb{C}\{X_1, \ldots, X_n\}$ with no constant term. A singularity (at zero) of such a germ is determined by the local ring $Q_P = \mathcal{O}_n / I_P$, where $I_P = (f_1, \ldots, f_n)$. Dimension of Q_P as a vector space over $\mathbb{C}$ is called the germ's multiplicity. It turns out that if the algebra Q_P is finite-dimensional (i.e. the germ P has finite multiplicity), then it is automatically subdirectly irreducible, and so it is a Frobenius algebra. Moreover, the Jacobian $J_P = \det\left(\frac{\partial f_i}{\partial x_j}\right)_{i,j}$ does not belong to I_P, and its image $\bar{J}_P$ in the algebra Q_P generates a unique minimal (i.e. one-dimensional over $\mathbb{C}$) ideal. In other words, $\mathbb{C}\bar{J}_P$ is the socle of the algebra Q_P. We will note that the proof of the Frobenius property of the algebra Q_P is highly non-trivial. The text by Arnol'd, Varchenko and Husein-Zade [1982] mentioned above gives an explicit construction of a form $Q_P \to \mathbb{C}$ such that the corresponding bilinear form $Q_F \times Q_F \to \mathbb{C}$ (see Theorem 1) is non-degenerate.

3.13. Bibliographical Notes. The most classical source on theory of modules is the monograph by Cartan and Eilenberg [1956]. It contains, aside from purely homological problems, definitions and properties of projective, injective and flat modules, as well as Prüfer, hereditary and semihereditary rings. Definitions of the projective and injective dimensions of a module and of the global dimension of a ring can also be found here. For a deeper study of the current state of module theory one can recommend Faith [1973, 1976], Stenström [1975] as well as Anderson and Fuller [1974]. Elements of the theory of modules are also contained in the handbook by Lang [1965] and in numerous books on the theory of associative rings (see e.g. Jacobson [1964], Bokut' [1977, 1981], Kasch [1977], Cohn [1985] and Lambek [1966]). Artinian modules are studied in Curtis and Reiner [1962]. Perfect and semiperfect rings are found in many sources: Lambek [1966], Stenström [1975], Anderson and Fuller [1974], Kasch [1977] and Faith [1973, 1976]. Faith [1976] contains a good exposition of Azumaya diagrams. Frobenius algebras are studied in Curtis and Reiner [1962]. Quasi-Frobenius rings are considered in Curtis and Reiner [1962], Kasch [1977] and Stenström [1975]. The result of Osofsky, mentioned at the end of Sect. 3.9, is in her monograph [1973]. Free ideal rings are studied in Cohn [1985]. Here one can also find supplementary material on hereditary and semihereditary rings, as well as proofs of some of the theorems on projective modules from Sect. 3.4. For Beck's theorem, see his original paper [1972]. The result of Gerasimov and Sakhaev is in their article [1984], which also contains a survey of results on reduction of projective modules. Flat modules are well presented in Bourbaki [1961]. Dimensions of rings and modules are considered in monographs by MacLane [1963], Faith [1973, 1976] and Osofsky [1973]. Dimensions of commutative local rings and modules over them are studied in Serre [1964].

§ 4. Structure of Rings

4.1. Introduction. The general structure theory of rings was initiated by Jacobson in the mid-40's. This involved the construction of Jacobson's radical of an arbitrary ring (see definition in Sect. 1.24), which allowed to describe, to some extent, semisimple rings (i.e. with their Jacobson radical equal to 0) as subdirect products of primitive rings (see definitions and statements below). Finally, the Jacobson-Chevalley density theorem describes primitive rings in terms of rings of linear transformations of modules over division rings. From that point on, the general structure theory of rings became an integral part of ring theory, and is often applied in its widely varying aspects.

On the other hand, the development of ring theory over the past decade brought to life a deep structure theory based on Baer's radical. This can be explained by the fact that, first of all, significant progress was made in the investigation of prime rings (which included study of prime rings with generalized identities, development of Galois theory of prime rings, investigation of derivations and involutions of prime rings etc.); secondly, a method of orthogonal completion (see Sect. 4.3) developed recently by K.I. Beidar and A.V. Mikhalev, enables to transfer theorems from prime to semiprime rings almost automatically (with unavoidable changes in their formulation). The essence of this method lies in the fact that, from the point of view of elementary language of logic, a semiprime orthogonally complete ring behaves like a direct sum of prime rings.

Apart from Jacobson and Baer radicals, modern ring theory widely employs other radicals – primarily the Andrunakievich radical, the Levitzki radical and the upper nil-radical. The so-called general theory of radicals, whose foundations were laid by A.G. Kurosh, S.A. Amitsur and V.A. Andrunakievich, is being developed as well.

Over the past 25 years, theory of Noetherian rings came to occupy one of the central places in non-commutative ring theory. There were several reasons for that. First of all, the Noetherian condition allows to develop a fine and deep technique for investigating such rings. Secondly, the already classical theory of commutative Noetherian rings exerts a stimulating influence.

Thirdly, there exist examples of Noetherian rings which are important for analysis and geometry (universal enveloping algebras of finite-dimensional Lie algebras, Weyl algebras), study of which has direct consequences, for example, in the theory of differential equations and in representation theory. Noetherian ring theory also produced classes of rings with conditions weaker than even a one-sided Noetherian property (Goldie rings, rings with Krull dimension; see below). Developments in these new directions is stimulated by the theory of Noetherian rings.

Structure theory of PI-rings (i.e. rings satisfying polynomial identities) is another central parts of modern ring theory. It has numerous links with the theory of identities of rings and algebras; however, we will make almost no mention of those connections.

4.2. Jacobson Structure Theory. Definition and some discussion of the Jacobson radical can be found in Sect. 1.24. Before stating a theorem which describes semisimple rings (i.e. rings with zero Jacobson radical), we will define subdirect product of rings. Let $\{R_i\}_{i\in I}$ be an indexed family of rings, and let $\prod_{i\in I} R_i$ be their direct product, i.e. the $\mathbb{Z}$-module $\prod R_i$ (see Sect. 1.21) with multiplication defined by $(a_i)(b_i) = (a_i b_i)$. Ring R is called a *subdirect product* of the family $\{R_i\}_{i\in I}$, denoted by $R = \prod^S_{i\in I} R_i$, if R is isomorphic to a subring of the direct product $\prod R_i$, with all the induced projection homomorphisms $R \to R_i$ $(i \in I)$ being surjective. Denoting kernels of these homomorphisms by $\mathbf{n}_i$, we can give another, equivalent definition of a subdirect product. Namely, R is a subdirect product of rings R_i $(i \in I)$, if one can find in R a family of ideals $\{\mathbf{n}_i\}_{i\in I}$ such that $R/\mathbf{n}_i \simeq R_i$ (for $i \in I$) and the intersection $\bigcap_{i\in I} \mathbf{n}_i$ equals zero. It follows directly from the definition that the only rings which cannot be non-trivially identified with a direct product, are those which have a smallest non-zero ideal. Such rings are called *subdirectly irreducible*. It can be shown that every ring is a subdirect product of subdirectly irreducible rings.

Ring R is called *(left) primitive* if it has at least one faithful simple left module M, i.e. a simple module whose annihilator in R is zero: $rM = 0 \Rightarrow r = 0$ for any $r \in R$. If M is any simple left R-module and $\operatorname{Ann} M = \{r \in R \mid rM = 0\}$ – its annihilator, then the ring $R/\operatorname{Ann} M$ is primitive. The rôle of primitive rings is revealed by the following theorem.

Theorem 1. *Every semisimple ring is a subdirect product of primitive rings.*

This theorem shows that the study of primitive rings reduces, in a certain sense, to the description of primitive rings. Moreover, since $R/J(R)$ is a semisimple ring, it can be said that investigation of arbitrary rings is reduced to that of radical rings (i.e. those which are equal to their radicals) and semisimple rings. We will list some non-trivial examples of primitive and semisimple rings: the free algebra $F\langle X \rangle$ on any set of free generators is primitive. Group algebra FG of any group G over a non-countable field of characteristic 0 is semisimple. The question of semisimplicity of group algebras over an arbitrary field of characteristic zero remains open. Let R be a ring with no non-zero nil ideals; then the polynomial ring $R[x]$ is semisimple. Group algebra FG of a so-called polycyclic (see Sect. 5) torsion-free group is primitive.

There exist left primitive rings, which are not right primitive. The first such example was constructed by Bergman in 1965. An important (and, as it turns out, universal – see next theorem) example of primitive rings is provided by dense rings of linear transformations of a vector space over a division ring. Namely, let D be a division ring, V_D – a right vector space over D (i.e. a right D-module), and let $\operatorname{End} V$ be its endomorphism ring (its elements written on the left of their arguments, and multiplying in a corresponding way: $\varphi\psi(x) = \varphi(\psi(x))$). A subring $R \subseteq \operatorname{End} V$ is called *dense* if, for any linearly independent elements $v_1, \ldots, v_n$ and any $w_1, \ldots, w_n \in V$, there exists an element $\varphi \in R$ for which $\varphi(v_i) = w_i$ $(1 \le i \le n)$. This situation can be also described as density of R in $\operatorname{End} V$ in topological sense (i.e. the closure of R is the entire $\operatorname{End} V$), by considering the finite topology in which a basis of neighborhoods of a point $\varphi \in \operatorname{End} V$ consists of all subsets $U(\varphi; v_1, \ldots, v_n) = \{\psi \in \operatorname{End} V \mid \psi(v_i) = \varphi(v_i)\,(1 \le i \le n)\}$, where $\{v_1, \ldots, v_n\}$ runs over all finite subsets of V.

Primitive rings are described by the following *Jacobson–Chevalley density theorem.*

Theorem 2. *A left primitive ring R is isomorphic to a dense subring of a ring of linear transformations of a right vector space over a division ring. Furthermore, if V is a faithful simple left R-module and D is its endomorphism ring then the mapping $r \mapsto \hat{r}$, $\hat{r}(m) = rm$ (where $m \in V, r \in R$) is an isomorphism of the ring R onto a dense subring of* $\operatorname{End}_D V$.

It is possible to obtain a more detailed description of primitive rings with non-zero socle. The *socle* of a ring R is the socle of the regular left module $_RR$, i.e. the sum of all minimal left ideals of the ring R. It can be shown that the socle of a ring is a two-sided ideal, which coincides with its right analogue. We will recall that the rank of a linear transformation is the dimension of its image.

Theorem 3. *The following properties of a ring R are equivalent:*
(a) R is a primitive ring with non-zero socle;
(b) R is isomorphic to a dense subring of a ring of linear transformations of a vector space V over a division ring D, containing a non-zero transformation of finite rank.

We will mention that the division ring D in Theorem 3 and the faithful simple module V are determined by the ring R up to isomorphism. If, namely, $e \in R$ is an idempotent such that Re is a minimal left ideal of R, then eRe is a division ring isomorphic to D, and Re is a unique faithful simple left R-module. On the other hand, in Theorem 2 the division ring D and the vector space V are not unique for a ring R. For example, let D be a division ring with center C, for which there is an embedding $\varphi : D \to D$ – identical on C – such that the centralizer $\Delta = \{d \in D \,|\, da = ad \quad \forall a \in \varphi(D)\}$ of the division subring $\varphi(D)$ in D is not commutative. This is the case when D is the ring of fractions of an infinitely-generated Weyl algebra $A_\infty(F) = F\langle x_1, y_1, x_2, y_2, \dots ; [x_i, y_i] = \delta_{i,j},\ [x_i, x_j] = [y_i, y_j] = 0\rangle$, so that $C = F$ and φ is defined by $\varphi(x_i) = x_{i+1}$ and $\varphi(y_i) = y_{i+1}$. Then $x_1, y_1 \in \Delta$. The division ring D can be made into a (D, D)-bimodule in two ways, which yield two different structures of D as a simple left module over the ring $R = D \oplus_C D^{\mathrm{op}}$:

$$(i) \quad (d \oplus d')v = dvd' , \qquad (ii) \quad (d \oplus d')v = dv\varphi(d') \qquad (d, d', v \in D) .$$

Endomorphism rings of the respective modules are C and Δ, and they are not isomorphic (this example was communicated to the authors by K.I. Beidar).

4.3. A New Structure Theory. In case of the Baer radical, the rôle of semisimple rings is played by semiprime ones, and primitive rings correspond to prime ones. We recall (Sect. 1.8) that a ring R is called *prime* if a product of any two non-zero ideals is non-zero, while a ring is *semiprime* if the square of any non-zero ideal is non-zero.

Theorem 1. *Every semiprime ring is a subdirect product of prime rings.*

We will now define the Martindale ring of quotients, the extended centroid and the central closure of a semiprime ring. Let R be a semiprime ring and $\mathcal{F}$ – the collection of all of its two-sided essential ideals (i.e. ideals whose right annihilators in R are zero: $I \in \mathcal{F} \Leftrightarrow (\forall r\ Ir = 0 \Rightarrow r = 0)$). We will consider the set $V = \bigcup_{I \in \mathcal{F}} \mathrm{Hom}(I_R, R_R)$, i.e. the union of abelian groups of of right R-module homomorphisms. We will define an equivalence relation on it: $\varphi_1 \sim \varphi_2$ whenever there exists an ideal $I \in \mathcal{F}$, contained in the intersection of domains of φ_1 and φ_2, on which φ_1 and φ_2 coincide. On the factor space $V/\sim$ we will define operations of addition and multiplication in the following way: if $\varphi_1 \in \mathrm{Hom}(I_1, R)$ and $\varphi_2 \in \mathrm{Hom}(I_2, R)$ then we will regard $\varphi_1 \pm \varphi_2$ and $\varphi_1\varphi_2$ as elements of $\mathrm{Hom}(I_1 I_2, R)$, with $(\varphi_1 \pm \varphi_2)(a) = \varphi_1(a) \pm \varphi_2(a)$ and $(\varphi_1\varphi_2)(a) = \varphi_1(\varphi_2(a))$. The set $V/\sim$ becomes a ring, denoted by $R_\mathcal{F}$. We will denote elements of $R_\mathcal{F}$ simply by φ, keeping in mind the equivalence $\sim$. If $r \in R$ then the mapping $\bar{r} : x \mapsto rx$ $(x \in R)$ is a right module homomorphism, $\bar{r} \in \mathrm{Hom}(R_R, R_R)$, and so it determines an element $\bar{r} \in R_\mathcal{F}$. Identifying r with $\bar{r}$ one can assume that $R \subseteq R_\mathcal{F}$.

The ring $R_\mathcal{F}$ – by definition – has the property that, for any $\varphi \in R_\mathcal{F}$, there exists an ideal $I \in \mathcal{F}$ such that $\varphi(I) \subseteq R$. It can be said that I is the ideal of 'right denominators' of the element φ. The ring $R_\mathcal{F}$ contains a subset called the *Martindale quotient ring* $Q(R)$, which consists of those elements whose ideals of left denominators are also essential: $Q(R) = \{q \in R_\mathcal{F} \,|\, \exists I \in \mathcal{F}$ for which $Iq \subseteq R\}$. It is clear that $R \subseteq Q(R)$. The ring $Q(R)$ is semiprime (as is $R_\mathcal{F}$), and is closely related to the original ring R. For example, if R contains no zero divisors, then neither does $Q(R)$ (in contrast with $R_{\mathcal{F}}$); moreover, every automorphism and every derivation of R uniquely extends to $Q(R)$. It can be shown that when $|X| \geq 2$, $Q(F\langle X \rangle) = F\langle X \rangle$. The *extended centroid* of the ring R is the center C of $Q(R)$ (see Sect. 1.6).

Theorem 2. *The extended centroid of a semiprime ring is a commutative von Neumann regular (see Sect. 5.9) self-injective ring.*

It is easy to show that C is a field if and only if R is a prime ring. The *central closure* of R is the ring RC, generated in $Q(R)$ by the subrings R and C.

Let $C\langle X \rangle$ be a free algebra over C, and let $Q = Q(R)$. Elements of the free product $Q *_C C\langle X \rangle$ (see Sect. 1.13) are called *generalized polynomials.* In other words, generalized polynomials are linear combinations of words of the form $q_1 x_1 q_2 x_2 \dots q_n x_n q_{n+1}$, where $q_i \in Q$ and $x_i \in X$. A non-zero element $f \in Q *_C C\langle X \rangle$ is a *generalized identity* of the ring R if, for any substitution $x_i = r_i$ ($x_i \in X, r_i \in R$), it becomes equal to zero as an element of the ring Q.

Theorem 3 (Martindale). *If a prime ring R satisfies a generalized identity, then its central closure RC is primitive, has non-zero socle and acts densely as endomorphisms on a vector space over a division ring. The converse is also true.*

This theorem also provides an instrument for studying prime rings without generalized identities, since for any non-zero generalized identity $f(x_1, \dots, x_n)$ it allows to find elements $r_1, \dots, r_n \in R$ such that $f(r_1, \dots, r_n) \neq 0$.

We will present a general outline of the method of orthogonal completions, which enables to reduce the study of semiprime rings to that of prime ones. Theorem 2 shows that the extended centroid C of a semiprime ring R has many idempotents: for any $c \in C$ there exists an element $c' \in C$ such that $c^2 c' = c$, whence $c(cc') = c$ and $(cc')^2 = cc'$, i.e. $cc' = e$ is an idempotent and $ce = c$. We will recall that idempotents e_1 and e_2 are called *orthogonal* if $e_1 e_2 = e_2 e_1 = 0$. Let $\{e_\alpha \,|\, \alpha \in A\}$ be a family of pairwise orthogonal idempotents in C, and let $\{q_\alpha \,|\, \alpha \in A\}$ be a collection of elements of the ring Q. It can be shown that there is an element $q \in Q$ such that $e_\alpha q = q_\alpha e_\alpha$ for all $\alpha \in A$. Moreover, if $\{e_\alpha\}$ is a dense set, i.e. if its annihilator in C (or, equivalently, in R or $R_{\mathcal{F}}$) is zero, then that element is unique and is called the *orthogonal sum* of elements q_α (relative to the system of idempotents $\{e_\alpha\}$). A subset $S \subseteq Q$ is called *orthogonally complete* if it is closed under orthogonal sums of its elements (relative to arbitrary systems of idempotents). If S is any set, then by $\mathcal{O}(S)$ we denote the *orthogonal completion* of S, i.e. the smallest orthogonally complete subset containing S. It can be shown that $\mathcal{O}(R)$ is a semiprime subring of R. Furthermore, R turns out to be everywhere dense in $\mathcal{O}(R)$ with respect to a certain natural topology, so that the connection between R and $\mathcal{O}(R)$ is very strong.

Let $\wp$ be a maximal ideal of the ring C. A *fiber* at the point $\wp$ (or over the point $\wp$) is the factor ring $R_\wp = \mathcal{O}(R)/\wp\mathcal{O}(R)$. Maximal ideals of the ring C form a topological space $\operatorname{Spec} C$, in which closed sets are of the form $\{\wp \mid \wp \supseteq M\}$, where M is any ideal of C (this is well known from commutative algebra). In this context, maximal ideals are referred to as 'points'. We will remark that in view of Theorem 2, all simple ideals (in the sense of commutative algebra) of C are maximal, so that the meaning of $\operatorname{Spec} C$ agrees with the generally accepted one.

It can be proved that all fibers $R_\wp$ are prime rings. In relation to them, the ring $\mathcal{O}(R)$ behaves largely like their subdirect product.

In the theory of algebraic systems, among the class of elementary formulæ one can distinguish a subclass of the so-called Horn formulæ, whose principal property lies in the fact that validity of such a formula for the components implies its validity for their subdirect product. In the class of Horn formulæ one finds the statements

$$(P_1 \& P_2 \& \ldots \& P_n) \to P_0, \quad \neg P_1 \vee \ldots \vee \neg P_n, \quad P_0,$$

as well as conjunctions of such formulæ, and statements constructed from these by an arbitrary combination of quantifiers $\forall, \exists$ (here P_i are statements of the form $f(x_1, \ldots, x_n) = 0$, with f – a polynomial).

Meta-Theorem. *If a theorem can be formulated in terms of Horn formulæ, and if the set of those points $\wp \in \operatorname{Spec} C$ fibers over which satisfy that statement, is everywhere dense in $\operatorname{Spec} C$, then the theorem is valid in $\mathcal{O}(R)$.*

Importance of this result is reinforced by the fact that in the class of prime rings, every elementary formula is equivalent to a Horn formula. Therefore any elementary theorem (i.e. theorem about elements) can be reformulated in such a way that the meta-theorem will apply to it. Clearly, there are many such reformulations, each of them yielding a generalization of the theorem from the class of prime rings to the class of semiprime ones. The question which one of them should be chosen is the remaining creative element. Naturally, the sphere of applications of the meta-theorem has its limits. For example, elementary statements do not allow phrases such as "there exists an ideal I for which...", or "for any ideal I...". Nevertheless, such obstacles can sometimes be circumvented by means of extension of the signature of an algebra (see Part II).

4.4. Radicals of Rings. We will define the Levitzki (or locally nilpotent) radical $\mathcal{L}(R)$, the nilradical $\mathcal{N}(R)$ and the Andrunakievich (or completely prime) radical $\mathcal{A}(R)$; we will also state the basic structure theorems for these radicals. In this section it will be convenient to regard ideals of rings as being rings themselves. This makes us temporarily revoke the assumption that all rings considered here have a unity element.

We begin with the general concept of a radical. Let r be a mapping of the class of all rings (or algebras) into itself, assigning to each ring R its ideal $r(R)$ (we assume that r is an abstract mapping, i.e. it commutes with isomorphisms). Ring R is called *r-radical* if $R = r(R)$, and *r-semisimple* if $r(R) = 0$. The mapping r is a *radical* (in the sense of Amitsur-Kurosh) if: the class of r-radical rings is homomorphically closed, for every R the ring $R/r(R)$ is r-semisimple, and that $r(R)$ is the largest r-radical ideal of R. Let K be a class of rings (or algebras), not necessarily with 1. Ideals of a ring (or an algebra) R which belong to K will be called K-ideals. A non-empty class K will be called a *radical class* if it is closed under homomorphic images, extensions and unions of ascending chains of K-ideals. If K is a radical class then every algebra R contains the largest K-ideal $r_K(R)$. The mapping r_K is a radical. This establishes a one-to-one correspondence between radicals and radical classes of rings (or algebras).

Locally nilpotent rings – in which every finitely generated subring is nilpotent, as well as nil rings, form a radical class. The corresponding radical is called the *Levitzki radical* $\mathcal{L}(R)$ (resp. the *nilradical* $\mathcal{N}(R)$) of the ring R. $\mathcal{L}(R)$ also turns out to be the largest one-sided locally nilpotent ideal of R, while $\mathcal{N}(R)$ is the largest nil ideal of R. The question of whether $\mathcal{N}(R)$ is the largest left nil ideal is equivalent to the following open *Köthe problem*: is the sum of any two left nil ideals of a ring R a left nil ideal? The answer to this is affirmative for algebras over an uncountable field (Amitsur). The study of rings semisimple with respect to the Levitzki radical (resp. the nilradical) reduces to the study of prime rings of the same type (A.M. Babich):

Theorem 1. *Every ring with no non-zero locally nilpotent ideals (nil ideals) is isomorphic to a subdirect product of prime rings with no non-zero locally nilpotent ideals (resp. nil ideals).*

The class of those rings whose homomorphic images all contain zero divisors, is a radical class. The corresponding radical is called the *Andrunakievich radical* $\mathcal{A}(R)$ of the ring R. For any integer $n = 0, 1, 2, \dots$ we will define an ideal $\mathcal{A}_n(R)$ as follows: $\mathcal{A}_0(R) = 0$, and $\mathcal{A}_{n+1}(R)$ is the ideal of R generated by all those elements of R which are nilpotent modulo $\mathcal{A}_n(R)$. We will also let $\mathcal{A}_\omega(R) = \bigcup \mathcal{A}_n(R)$. An ideal I of a ring R is called *completely semiprime (completely prime)* if R/I contains no non-zero nilpotent elements (resp. zero divisors). The ideal $\mathcal{A}(R)$ happens to be the smallest completely semiprime ideal of R, and it equals the intersection of all completely prime ideals of R. Moreover, $\mathcal{A}(R) = \mathcal{A}_\omega(R)$. Rings semisimple with respect to the Andrunakievich radical are the rings with no non-zero nilpotent elements. They are described by the following theorem (V.A. Andrunakievich, Yu.M. Ryabukhin).

Theorem 2. *Every ring with no non-zero nilpotent elements is isomorphic to a subdirect product of rings without zero divisors.*

We will remark that for every ring R the following inclusions hold:

$$\operatorname{Rad}(R) \subseteq \mathcal{L}(R) \subseteq \mathcal{N}(R) \begin{matrix} \subseteq J(R) \\ \subseteq \mathcal{A}(R) \end{matrix} .$$

General theory of radicals specifically studies these and other radicals of rings (and classifies various radicals according to their properties).

4.5. Examples of Noetherian Rings. We will recall that by a Noetherian ring we mean a ring which is right and left Noetherian at the same time (see Sect. 1.10). First of all, we will remark that – in contrast with the commutative case – not every finitely generated non-commutative ring is Noetherian (or even one-sided Noetherian). For example, the free algebra $F\langle x_1, \dots, x_n \rangle$ for $n \geq 2$ is neither left nor right Noetherian. To verify this, it is convenient to note that, say, the left Noetherian condition is equivalent to the fact that every left ideal of the ring is finitely generated (as a left ideal, i.e. a left module over the ring). For instance, the left ideal generated by the elements $x_1 x_2^k x_1$ (for $k = 1, 2, \dots$) is not generated by any finite set. It is easiest to see this as follows. If an element, say, $x_1 x_2^{n+1} x_1$ were in the left ideal generated by the $x_1 x_2^i x^1$ (with $i \leq n$), then it would certainly belong to the two-sided ideal generated by these elements. It follows from the composition lemma (see Sect. 1.18) however, that in the algebra $\langle x_1, x_2; x_1 x_2^i x_1 = 0 \ (i = 1, 2, \dots, n) \rangle$ the element $x_1 x_2^{n+1} x_1$ is distinct from 0, which contradicts our hypothesis.

The following result, due to Hopkins, yields a broad (though not the most typical) subclass of left-Noetherian rings. We emphasize that only rings with 1 are being considered here.

Theorem 1. *Every right (left) Artinian ring (see Sect. 1.8) is left (right) Noetherian.*

A proof of the fact that a given ring is Noetherian requires a certain technique. One of the possible methods is as follows. Let R be a ring. A (positive) *filtration* of R is a system R_i $(i \geq 0)$ of its additive subgroups such that $R_i \subseteq R_j$ for $i \leq j$, $\bigcup_i R_i = R$, $R_i R_j \subseteq R_{i+j}$ and $1 \in R_0$ (in this situation we say that R is a *filtered ring*). We will define the abelian groups $\operatorname{gr}_i R = R_i / R_{i-1}$ (assuming that $R_{-1} = 0$), and $\operatorname{gr} R = \bigoplus_{i \geq 0} \operatorname{gr}_i R$. We now introduce multiplication in $\operatorname{gr} R$. Let $\bar{x} \in \operatorname{gr}_i R$ and $\bar{y} \in \operatorname{gr}_j R$; we will set $\bar{x}\bar{y} = xy + R_{i+j-1}$. Since $xy \in R_{i+j}$, the element $\bar{x}\bar{y}$ is in $\operatorname{gr}_{i+j} R$. It is easy to see that $\bar{x}\bar{y}$ does not depend on the choice of elements x and y. Extending this operation to all elements of $\operatorname{gr} R$ by means of the distributive law, we obtain a graded ring (see Section 1.13), corresponding to the filtered ring R. As a rule, the ring $\operatorname{gr} R$ has a simpler structure than R itself; nevertheless, we have the following

Theorem 2. *Let R be a filtered ring, for which the graded ring* $\operatorname{gr} R$ *is left (right) Noetherian. Then R itself is left (right) Noetherian. Furthermore, if* $\operatorname{gr} R$ *has no zero divisors then neither does R.*

From this we obtain the well-known Tamari's theorem on universal enveloping algebras:

Theorem 3. *Let L be a finite-dimensional Lie algebra over a field F, and let $\mathcal{U}(L)$ be its enveloping algebra. Then $\mathcal{U}(L)$ is a Noetherian algebra with no zero divisors.*

It follows from the Poincaré-Birkhoff-Witt theorem that the associated graded algebra of $\mathcal{U}(L)$ relative to the filtration $U_0 = F$, $U_m = U_{m-1} + L^m$ $(m \geq 1)$, is isomorphic to an algebra of polynomials in finitely many indeterminates (their number being equal to the dimension of L). The assertion of Theorem 3 is therefore a consequence of Hilbert's Basis Theorem and Theorem 2 above.

Theorem 2 can also be applied to the Weyl algebra (see Sect. 1.14) $W = A_n(F)$. If L is the subspace spanned by the generators x_i, y_i $(1 \leq i \leq n)$ of W, then L defines filtration $W_0 = F$, $W_m = W_{m-1} + L^m$. The associated graded ring corresponding to this filtration again turns out to be a polynomial algebra (in $2n$ indeterminates). This yields a theorem.

Theorem 4. *Weyl algebra $A_n(F)$ over any field F is a Noetherian integral domain.*

If F is a field of characteristic 0, then it can be shown that $A_n(F)$ is a simple algebra. This is done as follows: first one proves that every non-zero ideal I of $A_n(F)$ contains a non-zero polynomial $f \in F[x_1, \ldots, x_n]$; since the characteristic of F is 0, using $I \ni [f, y_i] = \partial f / \partial x_i$ one clearly obtains a non-zero element of F contained in I, i.e. $I = A_n(F)$.

We will show that a left Noetherian ring R with no zero divisors satisfies the left Ore condition (see Sect. 1.19). Let s, r be non-zero elements of R. We will consider the sequence of elements $s, sr, sr^2, \ldots, sr^n, \ldots$. The left ideal generated by these elements, thanks to the Noetherian condition, is finitely generated. Hence one of the elements of this sequence can be expressed as a linear combination (over R) of the preceding ones: $sr^n = r_{n-1}sr^{n-1} + \ldots + r_k sr^k$, where $r_k \neq 0$ and $k \leq n-1$. Since there are no zero divisors in R we can cancel r^k – thus obtaining $s_1 r = r_k s$, with $s_1 = sr^{n-k-1} - \ldots - r_{k+1}s$, as required. It follows from Ore's theorem (see Sect. 1.19) that the algebras $A_n(F)$ and, for a finite-dimensional Lie algebra L, $\mathcal{U}(L)$ have classical rings of fractions.

We now describe an example of a left Noetherian ring, which is not right Noetherian. Let $K = F(t_1, \ldots, t_n, \ldots)$ be the field of rational functions in countably many variables over F. We will consider a homomorphism $\alpha : K \to K$, mapping t_i into t_{i+1} $(i \geq 1)$ and acting as identity on elements of the field F. Let $A = K[x, \alpha]$ be the ring of skew polynomials (see Sect. 1.12). Then A is a left Noetherian domain (even a principal left ideal domain – which is proved, just as for polynomial rings, by means of a left analogue of Euclid's division algorithm). At the same time, A is not right Noetherian: otherwise A would be a right Ore domain (see Sect. 1.19), but it is easy to notice that $t_1 xA \cap xA = 0$.

4.6. Goldie Rings. Let us begin with the following problem. Suppose that R is a ring which has the classical ring of fractions $Q_{\text{cl}}(R)$ (see Sect. 1.19). What can be said about R if $Q_{\text{cl}}(R)$ is a semisimple (resp. simple) Artinian ring? It is easy to show that R contains no infinite direct sums of left ideals (i.e. any sequence of left ideals $I_1, I_2, \ldots, I_n, \ldots$ such that $I_n \cap \sum_{k<n} I_k = 0$ $(n \geq 2)$ has finitely many non-zero elements), that it satisfies the maximal condition on left annihilators (i.e. left ideals of the form $l(S) = \{x \in R \,|\, xS = 0\}$, where S is a subset of R) and that it is a semiprime (resp. prime) ring. It turns out that this statement can be reversed, which is the substance of Goldie's theorem.

First, a definition. A ring R is called a *(left) Goldie ring* if it contains no infinite direct sums of non-zero left ideals and satisfies the maximal condition on left annihilators. It is clear that every left Noetherian ring is left Goldie. The latter class, however, is larger than the class of left Noetherian rings. For example, every left Ore domain is a left Goldie domain (the converse being valid as well). In particular, every commutative integral domain is a Goldie ring. The following *Goldie's theorem* is a central result in the theory of Goldie rings.

Theorem 1. *A ring R has the classical left ring of fractions, isomorphic to a semisimple (simple) Artinian ring if and only if R is a semiprime (resp. prime) left Goldie ring.*

Various proofs of this theorem employ many properties of semiprime Goldie rings, as well as certain important concepts. We will mention some of them:

1. An *essential left ideal* is a left ideal L such that $L \cap V \neq 0$ for every non-zero left ideal V (see also Sect. 3.5). It turns out that every essential left ideal of a semiprime left Goldie ring contains a regular element (i.e. a non-zero divisor, see Sect. 1.1), and vice versa – every left ideal of such a ring containing a regular element is essential.

2. The *left singular ideal* $Z_l(R)$ is the set of those elements whose left annihilators are essential. In a semiprime left Goldie ring, $Z_l(R) = 0$. This follows from a result of Mewborn and Winton: if a ring R satisfies the maximal condition on left annihilators then $Z_l(R)$ is nilpotent.

3. A semiprime left Goldie ring contains no non-zero one-sided nil ideals. We will recall that a (one- or two-sided) nil ideal is an ideal all of whose elements are nilpotent.

One of the consequences of Goldie's theorem stated above is the following result, also due to Goldie, concerning principal left ideal rings.

Theorem 2. *Every semiprime principal left ideal ring is isomorphic to a finite direct sum of prime principal left ideal rings. Every prime principal left ideal ring is isomorphic to a ring of $(n \times n)$ matrices over a left Ore domain (in which every left ideal is generated by n elements).*

4.7. Krull Dimension. We will recall that, in commutative algebra, the *Krull dimension* of a ring is the maximum of integers k such that R contains a strictly ascending chain of prime ideals $P_0 \subset \ldots \subset P_k$ of length $k+1$ (recall that an ideal P of a commutative ring is prime if R/P has no zero divisors). It turns out that Krull dimension can be defined for any ring R (this was done by Gabriel, Krause and Gordon), so that it will coincide with the above definition for commutative Noetherian rings (provided that one doesn't distinguish between infinite ordinals). This new Krull dimension of a ring can be any ordinal number. In certain important cases – for example, for Weyl algebras – it is finite. We will define classes of (left) R-modules $\mathcal{A}_\alpha$ (for $\alpha = 0, 1, \ldots$ – ordinal numbers). For convenience, we let $\mathcal{A}_{-1}$ be the class consisting of the zero module. Assume that $\mathcal{A}_\beta$ has been defined for $\beta < \alpha$ (and $\alpha \geq 0$). Then $\mathcal{A}_\alpha$ is the class of all modules M such that for any chain $M_0 \supseteq M_1 \supseteq \ldots$ of submodules of M, almost all (i.e. all but finitely many) factor modules M_i/M_{i+1} are in $\mathcal{A}_{\beta_i}$ for some $\beta_i < \alpha$. We will say that module M has Krull dimension, if M belongs to some class $\mathcal{A}_\alpha$. The minimal such ordinal α is called the *Krull dimension* $\mathrm{K.dim}\, M$ of the module M. In particular, $\mathrm{K.dim}\, M = 0$ if and only if M is a non-zero Artinian module. One of the important properties of the Krull dimension lies in the fact that every Noetherian module has Krull dimension (which may be infinite). Modules with Krull dimension can be regarded as a generalization of Noetherian modules, since many arguments in the theory of Noetherian modules can be carried over to this class. The *left Krull dimension* $\mathrm{l.K.dim}\, R$ of a ring R is the Krull dimension of its left regular module ${}_R R$. The following theorem, due to Gordon, often allows to argue by induction on the Krull dimension and thus has many applications.

Theorem 1. *Let M be a module with Krull dimension. Then, for any ordinal α, the module M contains the largest submodule with Krull dimension $\leq \alpha$.*

In particular, this result is used to establish that if $\mathrm{l.K.dim}\, R = \alpha$ then, for any R-module M with Krull dimension, $\mathrm{K.dim}\, M \leq \alpha$.

If R is a commutative Noetherian ring with $\mathrm{K.dim}\, R = n < \infty$, then the Krull dimension of R in the sense of commutative ring theory also equals n. For any ordinal α there exists a commutative Noetherian domain with Krull dimension α (i.e. in the commutative case, $\mathrm{K.dim}\, R$ is a finer invariant than the classical Krull dimension – which does not distinguish between infinite ordinals). Finally, we have the following

Theorem 2. *The left (right) Krull dimension of a Weyl algebra $A_n(F)$ equals n.*

Rings with right Krull dimension have a rich structure theory, similar to the theory of left Noetherian rings. For example, every such ring R has the largest nilpotent ideal N such that R/N is a semiprime left Goldie ring. Moreover, such a ring R satisfies the maximal condition on semiprime ideals (i.e. ideals $P \triangleleft R$ for which the ring R/P is semiprime).

4.8. Simple Noetherian Rings. As we already mentioned, simple Artinian rings have a very precise description – they are matrix rings over division rings. Is it possible to find an analogous characterization of simple Noetherian rings? Since there exist simple Noetherian domains (for example, the Weyl algebra $A_n(F)$ in characteristic zero), it is clear that one can at most hope for a description of simple Noetherian rings as matrix rings over domains. Furthermore, examples (first constructed by A.E. Zalesskij and O.M. Neroslavskij) show that it is also necessary to abandon the matrix ring structure and substitute for it, in the very least, Morita equivalence (see Sect. 1.22), since the Noetherian condition and simplicity are Morita-invariant properties. As a result, we arrive at the following question: is it true that every simple Noetherian ring is Morita-equivalent to a (simple Noetherian) domain? It turns out that, in general, this is not the case. Suitable counterexamples have also been constructed by A.E. Zalesskij and O.M. Neroslavskij. The question, however, still stands if one additionally assumes that the ring has finite global dimension (for Noetherian rings, the left and right global dimensions agree). So far it is only known that when the global dimension of a simple Noetherian ring is ≤ 2, then the ring is Morita-equivalent to a domain (a result of Faith and Michler).

We begin by discussing the Morita equivalence of rings. The following *Morita's theorem* clarifies this notion.

Theorem 1. *Let R and S be rings. Then the following statements are equivalent:*
(a) the rings R and S are Morita-equivalent;
(b) $S \simeq \operatorname{End}_R Q$, where Q_R is a right finitely generated projective generator (i.e. R_R is a direct summand of some finite direct sum of copies of Q_R);
(c) there exists a surjective Morita context (R, P, Q, S) (see Sect. 1.20).

In particular, it follows that Morita equivalence is a left-right symmetric property (that is, the categories of right R- and S-modules are also Morita-equivalent, because $(S^{\mathrm{op}}, P^{\mathrm{op}}, Q^{\mathrm{op}}, R^{\mathrm{op}})$ is a surjective Morita context as well). Moreover, in the cases when finitely generated projective R-modules are free (e.g. when R is a local ring), every ring Morita-equivalent to R has the form $M_n(R)$. This in particular implies that for finite-dimensional central simple algebras the notions of similarity (in the Brauer group sense) and Morita equivalence coincide. We will also note that Morita's theorem allows to prove invariance of properties of rings under Morita equivalence. Namely, if (R, P, Q, S) is a surjective Morita context then, using set mappings $R_R \supseteq I \mapsto IP \subseteq P_S$, ${}_RR_R \supseteq I \mapsto QIP \subseteq {}_SS_S$ etc., one can easily show that simplicity, primeness, semiprimeness, right and left Noetherian property, right and left Artinian property, right and left Krull dimensions as well as right and left global dimensions are Morita invariants of a ring. In particular, classical semisimplicity and the right and left hereditary property are preserved by Morita equivalence. Finally, regularity (in the sense of von Neumann; see Sect. 5.9) is also a Morita invariant.

Returning to simple Noetherian rings, we will begin with positive results.

Theorem 2. *If R is a simple Noetherian ring with global dimension ≤ 2, then R is Morita-equivalent to a simple Noetherian domain.*

In proving this theorem, Faith and Michler made use of a corollary to a theorem of Bass stated below. We will recall that the dual of a right R-module M is the left module $M^* = \mathrm{Hom}_R(M, R)$.

Theorem 3. *The global dimension of a Noetherian ring R does not exceed 2 if and only if the dual module of any finitely generated right R-module is projective.*

Corollary. *Let R be a Noetherian ring with global dimension ≤ 2. Then every left annihilator ideal in R is a projective R-module.*

Statement of the following theorem of Stafford requires the notion of a *uniform* (or *Goldie*) *dimension* u.dim M of a module M, which is defined as the supremum of integers n such that M contains a direct sum of n non-zero submodules.

Theorem 4. *Let R be a simple Noetherian ring with finite global dimension. If the left Krull dimension of R does not exceed an integer n, then the ring R is Morita-equivalent to a simple Noetherian ring with uniform dimension $\leq n$.*

In particular, if $n = 1$ then R is similar to a simple Noetherian domain.

We now describe an example of a simple Noetherian ring, which is not Morita-equivalent to a domain. Let G be a group of automorphisms of a ring R (we will write $g : x \mapsto x^g$ for $g \in G$). Define the ring $B = R\langle G \rangle$ as the free right R-module with basis G, in which multiplication is given by the rule $xg = gx^g$ (where $x \in R, g \in G$). This construction is a special case of the crossed product of a ring and a group. Let now F be a field of characteristic 2, $K = F(t)$ – the field of rational functions, $R_1 = K[x, x^{-1}]$ – localization of the polynomial ring $K[x]$ at the element x, and g – the automorphism of the K-algebra R_1 mapping x to tx. Let $R_2 = R_1[y, y^{-1}; g]$ (i.e. $ya = a^g y$ and $y^{-1}a = a^{g^{-1}} y^{-1}$ for $a \in R_1$). Then R_1 is a principal ideal domain and so, in particular, a Noetherian hereditary commutative domain. The following result can be applied to R_2:

Theorem 5. *Let R be a commutative Noetherian domain which is not a field, and let α be an automorphism of R such that $I^\alpha \neq I$ for any proper ideal I of the ring R. Then $A = R[x, x^{-1}; \alpha]$ is a simple Noetherian domain, and the global dimensions of A and R are equal.*

By this theorem, R_2 is a simple Noetherian domain. Let now h be the automorphism of the K-algebra R_2, which maps x to x^{-1} and y to y^{-1}. This automorphism generates a subgroup H of order 2 in the group of automorphisms of R_2. We will consider the ring $R = R_2\langle H \rangle$. It turns out to be the required one.

Theorem 6. *R is a simple Noetherian ring with infinite global dimension, which is not Morita-equivalent to a domain.*

4.9. Structure of PI-Rings. A natural generalization of commutative rings is provided by rings satisfying a polynomial identity

$$x_1 \dots x_n \sum_{\sigma \neq \mathrm{id}} \lambda_\sigma x_{\sigma(1)} \dots x_{\sigma(n)} , \tag{1}$$

where λ_σ are integers and the σ's are permutations of integers $1, \dots, n$. A ring R satisfying such an identity is called a *PI-ring*. It can be shown that if R is an algebra over a commutative ring Φ, and if R satisfies some polynomial identity $f(x_1, \dots, x_n) = 0$ (in the sense described in Sect. 1.13) then, under very weak restrictions the coefficients of f, an identity of the form (1) is satisfied in R. For example, this is the case if some linear combination of the coefficients of f equals 1 (e.g. when Φ is a field, or when one of the coefficients of f is 1). A typical example of a PI-ring is a ring of matrices over a commutative ring. Indeed, every $n \times n$ matrix over a ring K can be written as a linear combination $X = \sum k_{i,j} e_{i,j}$, where $e_{i,j}$ are the matrix units; hence the skew-symmetric mulitilinear mapping $S(x_1, \dots, x_{n^2+1}) = \sum (-1)^\sigma x_{\sigma(1)} \dots x_{\sigma(n^2+1)}$ is identically 0 on $M_n(K)$. In fact, the lowest degree of a polynomial identity satisfied by $M_n(K)$ equals $2n$.

One of the first results concerning the structure of PI-rings was the following *Kaplansky's theorem.*

Theorem 1. *Ring R is a primitive PI-ring if and only if R is isomorphic to a ring of matrices over a division ring, the latter being finite-dimensional over its center C, i.e. R is a finite-dimensional simple algebra over its center C. Moreover,* $\dim_C R \leq [d/2]^2$, *where d is the degree of any identity on R.*

The structure of prime PI-rings became much more clear after the solution of Kaplansky's problem about *central polynomials* of the matrix algebra $M_n(F)$ over a field F (i.e. non-constant polynomials whose evaluations all lie in the field F). A proof of the existence of such polynomials was given independently by E. Formanek and Yu.P. Razmyslov. We will describe the simplest Razmyslov's central polynomial. Let $m = n^2$, and let

$$p_n = \sum_{k=1}^{m} y_{k+1} x_{k+1} \dots y_m x_m y_1 x_1 \dots y_k x_k .$$

Then

$$c_n = \sum_{\sigma \in \mathrm{Sym}_m} (-1)^{\mathrm{sgn}\, \sigma} p_n(x_{\sigma(1)}, \dots, x_{\sigma(m)}, y_1, \dots, y_m)$$

is a central polynomial of the matrix algebra $M_n(F)$.

The original variant of the next theorem was obtained by Posner. In the modern version, its proof exploits the existence of a central polynomial.

Theorem 2. *Let R be a prime PI-ring, C – its center (an integral domain). Let $S = C \setminus \{0\}$ and $K = S^{-1}C$ – the ring of fractions of C. Then $K \otimes_C R \simeq S^{-1}R$ is a finite-dimensional central simple algebra over K, and $R \subseteq S^{-1}R$ (i.e. R is a torsion-free C-module; see Section 3.7). Furthermore, the rings R and $S^{-1}R$ satisfy the same identities.*

It follows from Theorem 2 that for any prime PI-ring R, there exists a field F and an integer n such that R and $M_n(F)$ have the same identities. The integer n is called the *PI-degree* of the prime ring R ($n = \text{pi.d}\,(R)$). For an arbitrary PI-ring R, its *PI-degree* is defined as the maximum of PI-degrees of its prime homomorphic images. It can be shown that the PI-degree of a semiprime ring R equals $[d/2]$, where d is the minimal degree of an identity of the form (1) satisfied in R (as a matter of fact, d turns out to be even).

Semiprime PI-rings also submit to investigation. Let R be a semiprime PI-ring with PI-degree n. Localizing R relative to certain central elements, we will obtain rings with desirable properties. Namely, let λ be a non-zero evaluation of the polynomial c_n in the ring R. Consider the subring $R_\lambda = \lambda^{-1}R$. If R is prime then R is a subring of R_λ. In the semiprime case, the natural homomorphism $R \to R_\lambda$ may have non-zero kernel, but R_λ itself turns out to be a PI-ring with PI-degree n, on which values of c_n are central – one of them being 1. Because of that, the next theorem can be applied to the rings R_λ.

Theorem 3. *Let A be a ring in which the following conditions hold for some n: (1) the 'Capelli identity' $d_{n^2+1} = 0$, where $d_m = \sum_{\sigma \in \mathrm{Sym}_m} (-1)^{\mathrm{sgn}\,\sigma} y_0 x_{\sigma(1)} y_1 \ldots x_{\sigma(m)} y_m$, is satisfied in A; (2) the polynomial c_n admits central values on A, one of them being 1. Then A is a free module of rank n^2 over its center C, and there is a one-to-one correspondence between ideals of A and those of C (in which an ideal of A corresponds to its intersection with C), preserving primeness of the ideal.*

This method of studying semiprime PI-rings was applied not long ago in the proof of the following theorem (Yu.P. Razmyslov, A.R. Kemer, A. Braun):

Theorem 4. *Let R be a finitely generated PI-algebra over a Noetherian ring Φ. Then the nil radical $\mathcal{N}(R)$ of R is nilpotent.*

If Φ is a field or the ring of integers, then the nil radical of a finitely generated PI-algebra coincides with its Jacobson radical. This was proved by Amitsur and Procesi.

Theorem 3 is important not only in the study of semiprime PI-rings, but also as a convenient for the theory of PI-rings variant of an Artin-Procesi theorem on Azumaya algebras. We will recall that a ring A is called an *Azumaya algebra* over its center C, if A is a projective left module over the ring $A \otimes_C A^{\text{op}}$ (in which case A is a finitely generated projective C-module). The class of rings in Theorem 3 turns out to be a large enough (for the PI-ring theory) subclass of Azumaya algebras of constant rank (a projective module P over a commutative ring C is called a *module of constant rank* r, if $\dim_{C/\mathbf{m}} P/\mathbf{m}P = r$ for all maximal ideals $\mathbf{m} \subset C$).

In view of the fact that algebras of matrices over commutative algebras are typical examples of PI-algebras, it is natural to seek conditions for a PI-algebra to be embeddable in such a matrix algebra. A F-algebra A (not necessarily with 1) will be called *presentable* (more precisely, n-presentable), if it embeds in a matrix algebra $M_n(C)$ over some commutative algebra C. This concept was first considered by A.I. Mal'tsev in 1943. He noticed that presentability of finitely generated algebras is closely linked with the property of being approximable by finite-dimensional algebras. We say that an algebra A is approximated by algebras A_i if there is a system of surjective homomorphisms $\varphi_i : A \to A_i$, with intersection of their kernels equal to zero (in other words, A is isomorphic to a subdirect product of the algebras A_i – see Sect. 4.2). An algebra which can be approximated by finite-dimensional algebras is called *finitely approximable.* Theorems 5 and 6 below were obtained by A.I. Mal'tsev.

Theorem 5. *Every finitely generated presentable algebra A over a field F is finitely approximable. If the field F is infinite then A can be approximated by algebras of bounded dimension. Conversely, every F-algebra (where F is any field) which can be approximated by algebras of bounded dimension, is presentable.*

A.I. Mal'tsev also showed that every presentable finitely generated F-algebra embeds in some algebra of matrices over the field of rational functions $F(t_1, \ldots, t_s)$ in indeterminates t_i.

An algebra A is a *Hopfian algebra* if every surjective homomorphism φ of A into itself is an isomorphism. It is clear that if an algebra A satisfies the maximal condition on two-sided ideals, it is a Hopfian algebra (if $\operatorname{Ker}\varphi \neq 0$ then $\operatorname{Ker}\varphi^i$ is a strictly increasing chain of ideals in A).

Theorem 6. *Let A be a finitely generated F-algebra, n – a fixed positive integer. A then satisfies the maximal condition on ideals $I \triangleleft A$ such that A/I is n-presentable.*

Mal'tsev concludes from this that every presentable finitely generated algebra is a Hopfian algebra. His results imply an even stronger theorem, whose group-theoretic analogue is well-known.

Theorem 7. *Every finitely generated, finitely approximable algebra A is a Hopfian algebra.*

Namely, if A_n is the intersection of all ideals of A of codimension $\leq n$, then $\bigcap_n A_n = 0$, with the A_n being fully characteristic ideals of A. The algebras A/A_n are presentable (Theorem 5), and hence they are Hopfian algebras.

Consequently, φ induces automorphisms on the algebras A/A_n – and so φ itself is an automorphism (this argument was communicated to the authors by V.T. Markov).

The research of A.I. Mal'tsev has been continued by many authors. Small constructed examples of finitely generated PI-algebras which satisfy all the identities of a matrix algebra, but are not presentable. On the other hand, V.N. Latyshev described varieties of F-algebras (F – a field of characteristic zero) in which every finitely generated algebra is left Noetherian. Furthermore, several authors (I.V. L'vov, A.Z. Anan'in, V.T. Markov, Yu.N. Mal'tsev) have noted that an analogue of Latyshev's theorem for two-sided ideals describes a very interesting class of varieties of algebras, which have a number of different characterizations.

Theorem 8. *Let $\mathcal{M}$ be a variety of algebras over an infinite field F. The following conditions are equivalent:*
(a) all finitely generated algebras from $\mathcal{M}$ are presentable;
(b) all finitely generated algebras from $\mathcal{M}$ are finitely approximable;
(c) all finitely generated algebras from $\mathcal{M}$ are Hopfian algebras;
(d) all finitely generated algebras from $\mathcal{M}$ satisfy the maximal condition on two-sided ideals;
(e) every finitely generated algebra from $\mathcal{M}$ satisfies an identity of the form

$$[x_1, \dots, x_n]y_1 \dots y_n[x_1, \dots, x_n] = 0 ,$$

where $[x_1, \dots, x_n] = [[x_1, \dots, x_{n-1}], x_n]$;
(f) algebras from $\mathcal{M}$ satisfy an identity of the form

$$[x, y, y, \dots, y]y^n[x, y, \dots, y] = 0 ;$$

(g) algebras from $\mathcal{M}$ satisfy an identity of the form

$$xy^n x = \sum_{i+j>0} \alpha_{i,j} y^i xy^{n-i-j} xy^j \quad (\alpha_{i,j} \in F) .$$

(h) $\mathcal{M}$ does not contain a subvariety of F-algebras defined by the identity $x[y, z]t = 0$.

Recently, A.Z. Anan'in described those varieties of algebras over a field F of characteristic zero in which all algebras are presentable.

Theorem 9. *If algebras from a variety $\mathcal{M}$ over a field F satisfy identities of the form*

$$[x_1, y_1] \dots [x_n, y_n] = 0 ,$$

$$[x_1, \dots, x_n]y_1 \dots y_n[z_1, \dots, z_n] = 0 ,$$

then all algebras from $\mathcal{M}$ are presentable. If the field F has characteristic 0, *then the converse holds as well.*

An important invariant of any algebra over a field, and of a PI-algebra in particular, is its Gelfand-Kirillov dimension. Let R be a finitely generated algebra over a field F, with generators $a_1, \ldots, a_r$. Let A be the subspace of R spanned by $a_1, \ldots, a_r$ and 1_R. We will consider in R the filtration $A^0 = F \cdot 1_R \subseteq A \subseteq A^2 \subseteq \ldots$. Growth of the dimensions of these subspaces is measured by the number $\mathrm{GK}\, R = \limsup_{n>0} \log_n \dim A^n$, which does not depend on the choice of generators, and is called the *Gelfand-Kirillov dimension* of the algebra R. In the general case, $\mathrm{GK}\, R$ is defined as the supremum of $\mathrm{GK}\, S$, with S running over all finitely generated subalgebras of R. If R is a commutative F-algebra then $\mathrm{GK}\, R$ is the supremum of integers n such that R contains n elements algebraically independent over F, i.e. the Gelfand-Kirillov dimension is a non-commutative variant of the notion of transcendence degree. Shirshov's height theorem (see Part II) implies that the Gelfand-Kirillov dimension of any finitely generated PI-algebra R is finite (even though it may not be an integer). If R is a finitely generated prime PI-algebra, then $\mathrm{GK}\, R$ coincides with the transcendence degree (over F) of the center C of the algebra R, and is also equal to the classical Krull dimension of R (the supremum of lengths of chains of prime ideals in R). The following result, due to Procesi, establishes a link between the PI-degree, the number of generators and the Gelfand-Kirillov dimension.

Theorem 10. *Let R be a finitely generated prime PI-algebra over a field F. Let n be the PI-degree of R, r – the minimal number of generators and d – the Gelfand-Kirillov dimension of R. Then $d \leq rn^2 - (n^2 - 1)$, with equality holding if and only if R is isomorphic to the algebra of generic matrices* (see Sect. 2.10) $F[X_1, \ldots, X_r]$ *(where X_i are $n \times n$ matrices).*

4.10. Bibliographical Notes. The Jacobson structure theory is presented in most detail in Jacobson [1964]. A more concise account of the main results of this theory can be found in books Bokut' [1977, 1981] and in Herstein [1968].

The monograph by Andrunakievich and Ryabukhin [1979] is devoted to the theory of radicals. Here, apart from the general radical theory, basic results related to the Baer, Levitzki, Andrunakievich and other radicals are discussed. A concise proof of Martindale's theorem is given in his original article [1969]. Theory of generalized identities is presented in detail in Rowen [1980]. The method of orthogonal completion was employed, in one way or another, in a number of papers: Beidar [1977, 1978a, 1978b], Beidar and Slavova [1980] and Kharchenko [1979]. As a general technique, however, it has been recognized and isolated in Beidar and Mikhalev [1985].

Goldie's theorems are contained in many modern handbooks on associative rings (Bokut' [1977, 1981], Faith [1973, 1976], Herstein [1968], Chatters and Harajnavis [1980], Stenström [1975]). Goldie rings (not only commutative ones) are studied in Chatters and Hajarnavis [1980]. Here one finds an extensive account of the theory of Noetherian rings as well. We will also mention the volume by Jategaonkar [1970], which includes not only the theory of non-commutative principal ideal rings, but also material on Goldie rings, Noetherian rings as well as interesting examples of Noetherian rings. The Krull dimension of non-commutative rings is thoroughly studied in the monograph by Gordon and Robson [1973]. This dimension was introduced in the article by Rentschler and Gabriel [1967], which contains a computation of the Krull dimension of Weyl algebras.

Material on simple Noetherian rings can be found in Faith [1973, 1976] and in Chatters and Hajarnavis [1980]. Specifically devoted to them is the monograph by Cozzens and Faith [1975], which contains a large number of examples of simple Noetherian rings. For examples of A.E. Zalesskij and O.M. Neroslavskij, see their papers [1977a, 1977b].

Foundations of the structure theory of PI-rings are found in the textbooks Jacobson [1964], Bokut' [1977, 1981] and Herstein [1968]. PI-rings are specifically discussed in the monographs Rowen [1980], Jacobson [1975] and Procesi [1973]. Theorem 4 (Sect. 4.9) on the radical of finitely generated algebras is proved in Braun [1984]. Results of A.I. Mal'tsev on presentability and approximability can be found in his article [1943], which also appears as a part of his collected works [1976]. Theorem 6 (Sect. 4.9) on local properties of varieties of algebras is included in L'vov [1969], Anan'in [1976] and Yu.N. Mal'tsev [1976]. The Gelfand-Kirillov dimension was introduced in the article by Gelfand and Kirillov [1966]. Its basic properties are well presented in Borho and Kraft [1976]. Theorem 2 from Sect. 4.6 is in the book by Procesi [1973].

§ 5. Miscellaneous

This section differs from the preceding ones essentially in that it touches on several various topics. In it, we present a number of fragments of modern ring theory. All of them are in one way or another related to previous material (in particular, to the examples in Sect. 1). Overall, the material of this section – in our opinion – complements the picture, presented in other sections, of non-commutative ring theory and its place in mathematics. Naturally, the selection of topics reflects the authors' tastes and interests.

5.1. Group Algebras of Finite Groups. This section is devoted mainly to the representation theory of finite groups. In what follows, G stands for a finite group, F – a field, and all modules are left modules. We know from the theory of finite-dimensional algebras that the group algebra FG contains the largest nilpotent ideal $\operatorname{Rad} FG$, and that the factor algebra $FG/\operatorname{Rad} FG$ can be written as a direct sum of algebras $A_1, \ldots, A_s$, where $A_i = M_{n_i}(D_i)$ and D_i are finite-dimensional division algebras over F. It also follows from general theory that s equals the number of non-equivalent irreducible representations (i.e. simple modules) of G over the field F. We say that F is a *splitting field of the group* G, if $D_i \simeq F$ for all i, that is, if $FG/\operatorname{Rad} FG \simeq \bigoplus_{i=1}^{s} M_{n_i}(F)$; then the n_i's ($1 \leq i \leq s$) are the dimensions of the irreducible representations of G. It is clear that if F is algebraically closed then F is a splitting field for any group G. Since every field embeds in its algebraic closure, for every group G and a field F there is a field extension of F which is a splitting field of G. The following questions are fundamental for the representation theory of groups:

1) what is the number s of non-equivalent irreducible representations of a group G over F?
2) what can be said about dimensions n_i of irreducible representations of G over F?
3) how to find a 'smallest' splitting field F (among fields of given characteristic)?

Initially (in the classical works of Frobenius, F.E. Molin, Burnside, Schur) representation theory of finite groups was being developed in the case when the base field has characteristic zero. In this situation, a solution is provided by Maschke's theorem, which was already mentioned before.

Theorem 1. *Let G be a finite group, and let F be a field whose characteristic does not divide the order of G (e.g. is zero). Then the group algebra FG is semisimple.*

The case when characteristic of F is as in Maschke's theorem, is called *regular* (or that it deals with *regular representations* of G). If this is not so, then we talk about a *modular* case (or *modular representations*). Maschke's theorem and the description of the center of a group algebra (see Sect. 1.9) yield an answer to question (1) in the case of regular representations:

Theorem 2. *Let G be a finite group, F – a splitting field of G whose characteristic does not divide the order of G. Then the number of non-equivalent irreducible representations of G over F equals the number of conjugacy classes of elements of G.*

Let now G and F represent the modular case. Then the algebra FG is not semisimple. Namely, the one-dimensional subspace spanned by the element $\sum_{g \in G} g$ is an ideal of FG, whose square equals 0 (so that $\operatorname{Rad} FG \neq 0$). An analog of Theorem 2, however, can be proved in this case as well. An element $g \in G$ is called *p-regular*, if its order (i.e. the smallest $m > 0$ such that $g^m = 1$) is not divisible by the integer p.

Theorem 3. *Let G be a finite group and F – a field of characteristic $p > 0$. If F is a splitting field of G, then the number of non-equivalent irreducible representations of the group G over F is equal to the number of conjugacy classes of p-regular elements of G.*

An answer to question (1) can also be given when F is not a splitting field of G. In this case, conjugacy of elements in G is replaced with the notion of F-conjugacy, which will not be described here. Aside from this, formulations of the above two theorems remain unchanged. We will only remark that when $F = \mathbb{Q}$, then the number of $\mathbb{Q}$-conjugacy classes of elements of the group G is the number of conjugacy classes of cyclic (i.e. generated by a single element) subgroups of G.

If $\varphi : G \to \mathrm{GL}_n(F)$ is some matrix representation of the group G, then the function $\chi_\varphi : G \to F$ defined by $\chi_\varphi(g) = \mathrm{tr}\,\varphi(g)$ (the trace of $\varphi(g)$) is called the (F-)*character* of the representation φ. If φ is irreducible then χ_φ is called an *irreducible* (F-)*character* of G. We note that every character is constant on any conjugacy class in G. Moreover, equivalent representations have equal characters. The following theorem gives an answer to the third question in the case of finite characteristic (Brauer).

Theorem 4. *Let L be an algebraically closed field of characteristic $p > 0$, $\chi_1, \ldots, \chi_s$ – all irreducible L-characters of a finite group G, and F – the subfield of L generated by the values of those characters. Then F is the smallest among splitting fields of G contained in L.*

Using this theorem one shows, for example, that if G is a *group of exponent* m (i.e. $x^m = 1$ for all $x \in G$ and m is the smallest integer with this property) and p is a prime, then the field $GF_p(\sqrt[m]{1})$ (i.e. the field obtained from the finite field GF_p of p elements by adjoining all m-th roots of 1) is a splitting field (in characterictic p) of the group G.

The following theorem, also due to Brauer, is proved using an altogether different reasoning, and produces a natural and not too large splitting field of characteristic 0.

Theorem 5. *Let G be a group of exponent m. Then $\mathbb{Q}(\sqrt[m]{1})$ is a splitting field of G.*

Naturally, this splitting field for a given group G may not be minimal. For example, $\mathbb{Q}$ is a splitting field of any symmetric group Sym_n.

As to the second of our questions, the following theorem of Ito gives an answer to it in the case of characteristic 0.

Theorem 6. *Let A be an abelian normal subgroup of a finite group G (e.g. $A = \{1\}$), F – a splitting field of characteristic 0 for G. Then the dimension of every irreducible representation of G over F divides the index $[G : A]$ of the subgroup A in G (i.e. the number of cosets of A in G).*

The case $A = \{1\}$ (in which $[G : (1)] = |G|$ is the order of G) was known already in the classical theory. The above result remains valid in characteristic $p > 0$, if the group G is *solvable*, i.e. satisfies an identity of the form $\{x_1, \ldots, x_{2^n}\} = 1$, where $\{x_1, x_2\} = [x_1, x_2] = x_1^{-1}x_2^{-1}x_1x_2$ is the commutator of x_1 and x_2, and

$$\{x_1, \ldots, x_{2^n}\} = [\{x_1, \ldots, x_{2^n}\}, \{x_{2^{n-1}+1}, \ldots, x_{2^n}\}] .$$

A major part in the theory of group representations is played by the so-called induced representations (modules). Let H be a subgroup of a group G and let V be some module over FH. Then the FG-module $V^G = FG \otimes_{FH} V$ is called the module *induced* from the module V. Induced representations are used, for example, in the proof of the following theorem of Higman. We will recall that a *Sylow p-subgroup* (p – a prime integer) of a finite group is a subgroup of order p^m which is not contained in any larger subgroup whose order is a power of p.

Theorem 7. *Let F be a field of characteristic $p > 0$, and let G be a finite group whose Sylow p-subgroups are not cyclic. Then G has irreducible representations over F of arbitrarily large dimensions.*

If the Sylow p-subgroups of G are cyclic, then G has only finitely many non-equivalent irreducible representations over a field F of characteristic p (this was also shown by Higman).

The study of characters of representations of a group G constitutes the so-called character theory of finite groups. One of the fundamental results of this theory is the following *Burnside's theorem.*

Theorem 8. *Let $\varphi_i : G \to \mathrm{GL}_{n_i}(F)$ (where $1 \le i \le s$) be non-equivalent irreducible representations of the group G over a field F of characteristic 0. Then the characters χ_i ($1 \le i \le s$) of those representations are linearly independent over F.*

This immediately yields a

Corollary 9. *Let G be a finite group, F – a field of characteristic 0. If two representations of G over F have equal characters, then they are equivalent.*

In case of the field complex numbers, orthogonality relations for irreducible characters of a finite group are well known. We will present similar relations for an arbitrary field F, whose characteristic does not divide the order of G (i.e. for regular characters).

Theorem 10. *Let $FG = A_1 \oplus \ldots \oplus A_s$ and $A_i \simeq M_{n_i}(D_i)$, where D_i are division algebras. Let χ_i be the character of the irreducible representation of the group G corresponding to the summand A_i, and let ε_i be the unity element of the algebra A_i. Then $\varepsilon_i = \frac{n_i}{|G|} \sum_{g \in G} \chi_i(g^{-1})g$.*

Comparing coefficients of the elements of G in the equality $\varepsilon_i \varepsilon_j = \delta_{i,j} \varepsilon_i$, we obtain from this theorem the desired orthogonality relations

$$|G|^{-1} \sum_{g \in G} \chi_i(g) \chi_j(g^{-1}) = \delta_{i,j} \dim_F D_i .$$

5.2. Group Algebras of Infinite Groups. Group algebras were for a long time regarded as formal objects, suited mainly to the problems of representation theory of finite groups. Intensive research into group algebras of infinite groups as an independent object of study began only in the 1950s. Connections with other branches of mathematics – algebraic topology, coding theory, not to mention the theory of infinite groups – were discovered at that time. We will focus here on a few directions of these investigations.

Zero divisors and embeddings in fields. If g is an element of finite order $m > 1$ in a group G, and if $x = 1+g+\ldots+g^{m-1}$, then $(g-1)x = 0$ – i.e. the algebra FG has zero divisors. To this day, the following *Kaplansky's problem* remains open. Let G be a *torsion-free* group (i.e. containing no elements of finite order, other than the neutral element), F – a field. Is it true that the algebra FG has no zero divisors? Group algebra of a *free group* (i.e. the group of all formal words in some alphabet X, identified only according to the relation $xx^{-1} = x^{-1}x = 1$) does not contain zero divisors. This follows from the fact that a free group is orderable (see below). One of the deepest results on Kaplansky's problem is the following fact, obtained by Farkas and Snider. We recall that a group G is called *polycyclic* if G contains a finite normal series with cyclic factors. We will call a group G a *finite extension* of a group H, if H is a normal subgroup of G and $|G/H| < \infty$.

Theorem 1. *If G is a torsion-free group which is a finite extension of a polycyclic group then the group algebra FG has no zero divisors.*

A group G is *ordered*, if G is a linearly ordered set, with order relation $\leq$ compatible with the operation in G (i.e. $a \leq b$ implies that $xay \leq xby$ for all $x, y \in G$). For example, free groups and torsion-free *nilpotent* groups (i.e. groups satisfying an identity $[x_1, \ldots, x_n] = 1$, where $[x_1, \ldots, x_n] = [[x_1, \ldots, x_{n-1}]x_n]$) turn out to be ordered. Generalizing the construction of formal power series, A.I. Mal'tsev and B.H. Neumann defined, for any ordered group G and a field F, a certain division ring $H(FG)$ which contains FG. Namely, a formal (infinite in general) sum $x = \sum_{g\in G} \alpha_g g$ (where $\alpha_g \in F$) is called an l-series if the subset $\operatorname{Supp} x = \{g \in G \mid \alpha_g \neq 0\}$ is totally ordered by the ordering it inherits from G. Addition and multiplication of l-series is defined in the usual manner, using commutation of coefficients with the group elements, as well as the group multiplication. The set $H(FG)$ of all l-series is a ring containing FG.

Theorem 2. *The ring $H(FG)$ is a division ring, containing the group algebra FG.*

Since every free associative algebra embeds in a group algebra of a free group, the Mal'tsev-Neumann theorem implies, in particular, that a free associative algebra can be embedded in a division ring.

In connection with the problem of embeddability in division rings, existence of the classical ring of fractions $Q_{\mathrm{cl}}(FG)$ of the group algebra FG presents an interesting question. If G is a torsion-free nilpotent group then the division ring $Q_{\mathrm{cl}}(FG)$ exists, and the following result holds.

Theorem 3. *If G and G_1 are finitely generated torsion-free nilpotent groups and the division rings $Q_{\mathrm{cl}}(FG)$ and $Q_{\mathrm{cl}}(FG_1)$ are isomorphic, then the groups G and G_1 are isomorphic.*

Algebraic elements of group algebras. If $x = \sum_{g\in G}\alpha_g g$ is an element of a group algebra FG, then the coefficient α_1 of the neutral element of G is called the *trace $T(x)$ of the element x* (if G is a finite group and $[x]$ – the matrix corresponding to x under the left regular representation, then $\operatorname{tr}[x] = |G|T(x)$). Trace of an element x plays an important rôle when, for instance, x is nilpotent or idempotent; namely:

Theorem 4. *If x is a nilpotent element of the algebra FG and the characteristic of F is 0 or does not divide the orders of those elements of G which appear in x with non-zero coefficients, then $T(x) = 0$.*

Theorem 5 (A.E. Zalesskij, Kaplansky)**.** *If e is an idempotent of an algebra FG over a field F of characteristic 0, then the trace $T(e)$ is a rational number. Moreover, if e is distinct from zero or one, then $0 < T(e) < 1$.*

Let $x = \sum_{g\in G}\alpha_g g$. *Support of the element x* is the set $\operatorname{Supp} x = \{g \in G \mid \alpha_g \neq 0\}$. If x is an algebraic element, then the subgroup $\langle \operatorname{Supp} x \rangle$ generated by this set has many useful properties. For example:

Theorem 6. *If an idempotent e of FG is central (i.e. belonging to the center), then the subgroup $\langle \operatorname{Supp} x \rangle$ is finite and normal in G.*

Group algebras with identities. The group algebra FG of a finite group G, as every finite-dimensional algebra, satisfies (non-central) identities. The task of describing those groups G for which FG satisfies an identity has been fully accomplished by now (in works of Passman, Isaacs and others).

Theorem 7. *Let F be a field of characteristic 0. Then the group algebra FG satisfies an identity if and only if G contains an abelian subgroup of finite index (i.e. when G is almost abelian).*

In the case of characteristic $p > 0$ in the above theorem, 'abelian subgroup' should be replaced with the phrase 'p-abelian subgroup', i.e. a subgroup whose commutant is a finite p-group (we will recall that the *commutant* $[G, G]$ of a group G is the normal subgroup of G generated by all products of commutators $[g, h] = g^{-1}h^{-1}gh$).

Modules over group algebras. Study of modules over a group algebra FG is equivalent to the investigation of linear representations (generally speaking, infinite-dimensional) of the group G over F. The question when all irreducible representations are finite-dimensional is an interesting one. This problem has been solved for finite extensions of polycyclic groups. Field F is called *absolute* if its multiplicative group is periodic, i.e. if F is a union of finite fields. P.

Hall showed that if a field F is not absolute and G is a finite extension of a polycyclic group, then all simple FG-modules are finite-dimensional over F if and only if G is almost abelian. The case of an absolute field F was resolved by Roseblade.

Theorem 8. *If F is an absolute field and G – a finite extension of a polycyclic group, then all simple FG-modules are finite-dimensional over F.*

5.3. Localization of Rings and Embeddings in Division Rings. Construction of fractions used to obtain rational numbers directly generalizes to commutative rings, where it plays a fundamental part. As regards non-commutative rings, as we saw in Sect. 1, such construction applies to a suitably narrow class of rings – namely those satisfying the (left) Ore condition. Nevertheless, one can formally define the ring $S^{-1}R$ for any subset S of a ring R (see Sect. 1.19). Until recently the prevailing opinion was that the ring $S^{-1}R$ in the non-commutative case does not submit to investigation (in particular, questions about the kernel of the natural homomorphism $R \to S^{-1}R$ presented grave difficulties) and that, consequently, it cannot be applied in an effective way. This proved to be false thanks to a matrix construction of the ring $S^{-1}R$, discovered by V.N. Gerasimov (and, a little later, by Malcolmson). Ideas involved in this construction can be traced back to Cohn. We will now describe it.

Let $\mathcal{M}$ be the set of those square matrices over a ring R which have the form $a = \begin{pmatrix} a' & \tilde{a} \\ a^0 & 'a \end{pmatrix}$, where a' is a row vector, $'a$ – a column vector, $\tilde{a} \in R$ and a^0 is an upper triangular matrix with elements from S on the diagonal (the case $a = (\tilde{a})$ is not excluded). We will introduce on $\mathcal{M}$ an equivalence relation $\sim$, induced by the elementary operations of the following types:

1) addition of a row (column) with a lower (higher) index, multiplied on the left (right) by an element of the ring, to a row (column) with a higher (lower) index;
2) deletion of a row (column) passing through the matrix a^0, all of whose elements are zero – except possibly the element on the diagonal of a^0, with simultaneous deletion of the column (row) containing that diagonal element of a^0.

Further, we will define operations $\oplus$ and $\odot$ in $\mathcal{M}$ by

$$a \oplus b = \begin{pmatrix} a' & b' & \tilde{a} + \tilde{b} \\ a^0 & 0 & 'a \\ 0 & b^0 & 'b \end{pmatrix}, \quad a \odot b = \begin{pmatrix} a' & \tilde{a}b' & \tilde{a}\tilde{b} \\ a^0 & 'ab' & 'a\tilde{b} \\ 0 & b^0 & 'b \end{pmatrix}.$$

It turns out that if $a \sim c$ and $b \sim d$, then $a \oplus b \sim c \oplus d$ and $a \odot b \sim c \odot d$, i.e. our operations induce operations in the factor space $M = \mathcal{M}/\sim$, and that the ring M is isomorphic to $S^{-1}R$.

Theorem 1. *Let R be a ring, S – a subset of R. Then $\langle M, \oplus, \odot \rangle$ is a ring isomorphic to $S^{-1}R$ (via the mappings $r \mapsto (r)$ and $s^{-1} \mapsto \left(\begin{pmatrix} 1 & 0 \\ s & -1 \end{pmatrix}\right)$. Kernel of the mapping $R \to M$ ($r \mapsto (r)$) consists of those elements $r \in R$ for which there exist matrices $a = \begin{pmatrix} a' & a'' \\ a^0 & a_1 \end{pmatrix}$, $b = \begin{pmatrix} b_1 & 'b \\ b^0 & ''b \end{pmatrix}$ (where a^0, b^0 are as above, a', a'' are row vectors and $'b,''b$ – column vectors over R) such that block multiplication yields $ab = \begin{pmatrix} 0 & r \\ 0 & 0 \end{pmatrix}$ (with the 0's standing for zero matrices of suitable sizes).*

An earlier construction due to Cohn was also based on matrices. In order to comprehend the idea, we will consider the set $\mathcal{P}$ of all singular square matrices over some field (or division ring) F. It is easy to see that $\mathcal{P}$ has the following properties:

1) all matrices of the form AB, where A is $n \times m$ and B – $m \times n$ with $n > m$ (such matrices are called *non-full*) lie in $\mathcal{P}$;
2) if two matrices differ only by a single i-th, row (column), then the result of adding corresponding elements of this row (column) (without changing the others – such summation is called *determinantal summation*) is again an element of $\mathcal{P}$;
3) if $A \in \mathcal{P}$ and X is any square matrix, then the diagonal sum $A \dot{+} X$ lies in $\mathcal{P}$;
4) if $1 \dot{+} A \in \mathcal{P}$ then $A \in \mathcal{P}$;
5) if $A \dot{+} B \in \mathcal{P}$ then $A \in \mathcal{P}$ or $B \in \mathcal{P}$;
6) $1 \notin \mathcal{P}$.

Let now R be any ring, and $\mathcal{P}$ – a set of square matrices over R. We will call the set $\mathcal{P}$ a *prime matrix ideal* if $\mathcal{P}$ satisfies conditions (1) through (6).

We have the following

Theorem 2 (Cohn). *Let $\mathcal{P}$ be a prime matrix ideal over a ring R. Then there exists a homomorphism $R \to K$ of R into a division ring K such that $\mathcal{P}$ coincides with the set of square matrices over R which map to singular matrices over K.*

Using Theorem 2, Cohn showed that every free left ideal ring (see Sect. 3.9) embeds in a division ring. Another application of Theorem 2 is the following criterion of embeddability of a ring in a division ring, which was also obtained by Cohn.

Theorem 3. *Ring R embeds in a division ring if and only if R has no zero divisors and no scalar matrix $a \cdot 1$ ($a \neq 0$) can be expressed as a determinantal sum of non-full matrices.*

The problem of embeddability of rings in division rings has had a long history. In the first edition of van der Waerden's book "Modern Algebra" [1931] a question is posed: is every ring without zero divisors embeddable in a division ring? In 1937, A.I. Mal'tsev constructed the first example of a ring without zero divisors, which cannot be embedded in a division ring – namely, the algebra $R = \langle x, y, z, t, a, b, c, d;\ ax = by, cx = dy, az = bt \rangle$ over any field F (R does not embed in a division ring because in every division ring containing R, we would have $cz = dt$ – this, however, is false in R). In connection with this example A.I. Mal'tsev asked another question (see Mal'tsev [1976, p.6]): is there a ring R which is not embeddable in a division ring, but whose multiplicative semigroup embeds in a group? It turned out that such examples do exist. This was shown by L.A. Bokut', A. Bowtell and A.A. Klein in 1966.

The notion of an invertible element makes sense not only in the context of rings, but also for monoids and categories. For a monoid R, localization $S^{-1}R$ (where S is a subset of R) was first considered by A.I. Mal'tsev. This led him to the search for necessary and sufficient conditions for embeddability of a monoid in a group. Localizations of categories first appeared in topology, where certain mappings which are not isomorphisms in the original category, become isomorphisms in the localized category. This is the way in which, for example, the so-called homotopic category is studied in algebraic topology.

5.4. Identities and Rational Identities Over a Field of Characteristic Zero. The theory of identities of algebras over a field of characteristic 0 is, formally speaking, a part of the theory of identities of rings and algebras over fields of arbitrary characteristic. The principal method used in characteristic 0 does not apply to other cases, however. Here we have in mind the representation theory of symmetric groups Sym_n over a field of characteristic zero. The main question in the theory of identities of rings and algebras is, as we already mentioned, Specht's problem (see Sect. 1.13).[1]

We will restate Specht's problem in a different language. A *variety of rings (algebras)* is a class of all rings (algebras) satisfying a fixed set of identities (by an identity of a ring we mean an identity with integer coefficients). A variety generated by a given ring (or by a collection of rings) is, by definition, determined by all identities of that ring (resp. all identities common to the rings in the given collection). A subvariety of a given variety is defined in an obvious way. Specht's problem, in terms of varieties, is equivalent to the following question: does every variety of algebras over a field of characteristic 0 satisfy the minimal condition on subvarieties? It turns out that every such variety does have certain properties; they are, however, far removed from those that resemble the 'Artinian' one. In order to formulate a result, we will introduce a few definitions. If $\mathfrak{M}, \mathfrak{N}$ are two varieties of algebras over a field F, then their sum $\mathfrak{M} \cup \mathfrak{N}$ is defined as the smallest variety which contains both $\mathfrak{M}$ and $\mathfrak{N}$ (it is determined by the identities common to $\mathfrak{M}$ and $\mathfrak{N}$). Next, if $\mathfrak{U}, \mathfrak{B}$ are varieties, then the *product of varieties* $\mathfrak{U} \circ_{\mathfrak{M}} \mathfrak{B}$ within a variety $\mathfrak{M}$ is the class of those algebras from $\mathfrak{M}$ which contain an ideal from

[1] See footnote on page 14.

$\mathfrak{U}$ such that the factor algebra modulo that ideal belongs to $\mathfrak{B}$ (i.e. those algebras from $\mathfrak{M}$ which are extensions of algebras from $\mathfrak{U}$ by algebras from $\mathfrak{B}$). It can be verified that $\mathfrak{U} \circ_{\mathfrak{M}} \mathfrak{B}$ is a variety itself (a subvariety of $\mathfrak{M}$). Let G be the Grassman algebra of countable rank over F (see Sect. 1.15), G_0 – its even component (the subspace spanned by all words of even length in the generators) and G_1 – the odd component (spanned by odd-length words). Let $M_{n,k}(G)$ $(n, k \geq 1)$ be the algebra of block matrices $\begin{pmatrix} A & B \\ C & D \end{pmatrix}$, where $A \in M_n(G_0)$, $D \in M_k(G_0)$ and B, C are rectangular matrices of suitable size, with entries from G_1. The following theorem was proved by A.R. Kemer.

Theorem 1. *Every proper (i.e. distinct from the variety of all algebras) variety $\mathfrak{M}$ of algebras over a field F of characteristic zero can be presented in the form $\mathfrak{M} = \mathfrak{N}_k \circ_{\mathfrak{M}} (\mathcal{P}_1 \cup \ldots \cup \mathcal{P}_s)$, where $\mathfrak{N}_k$ is the variety of nilpotent algebras with nilpotence index $\leq k$ and $\mathcal{P}_i$ is a variety generated by one of the algebras $M_{n,k}(G)$, $M_n(F)$ and $M_n(G)$ for $n, k \geq 1$.*

This result implies, in particular, that in every proper variety of algebras over a field of characteristic zero, all identities of the algebra $M_n(G)$ (for some n) are satisfied. This theorem is, at least in appearance, an analogue of the Wedderburn-Artin structure theorem for Artinian rings.

We will now turn to rational identities. Let X be a countable set of non-commuting indeterminates. Any expression $r(x_1, \ldots, x_n)$, obtained from the elements $x_1, \ldots, x_n$ of X by means of finitely many operations of addition, subtraction, multiplication and taking formal inverses, is called a *rational expression.* We say that a division ring D satisfies a *rational identity* $r(x_1, \ldots, x_n) = 0$ if, after substituting arbitrary elements $d_1, \ldots, d_n \in D$ for the indeterminates, either one of the inverses becomes undefined (i.e. the operation $(\ )^{-1}$ is applied to the zero element), or the resulting value equals zero. A rational identity is *non-trivial* if there exists a division ring in which it is not satisfied. Rational identities first arose in projective geometry. For example, in the proof of the so-called fundamental theorem of projective geometry, an important rôle is played by the (trivial) rational identity $x - (x^{-1} - (y^{-1} - x)^{-1})^{-1} - xyx = 0$. Division rings which satisfy polynomial identities (i.e. PI division rings) are finite-dimensional over their centers. A similar result for division rings which satisfy a rational identity and whose centers are infinite (e.g. in characteristic zero), was obtained by Amitsur.

Theorem 2. *Let $r(x_1, \ldots, x_n)$ be a non-trivial rational identity. There exists a positive integer N, depending on r, such that every division ring whose center is infinite, and which satisfies the identity r, is of dimension at most N over the center.*

The above theorem has applications in projective geometry. It is known that every Desarguian projective plane P is coordinatized by some division ring D in the sense that points (resp. straight lines) in P can be identified with triples (p_1, p_2, p_3) (resp. (l_1, l_2, l_3)), where p_i (resp. l_i) for $1 \leq i \leq 3$

are elements of D, not all equal to zero. Two points (lines) are considered identical if one of them is obtained from the other by multiplying it on the right (resp. left) by a non-zero element of D. A line $l = (l_1, l_2, l_3)$ is incident with (i.e. passes through) a point $p = (p_1, p_2, p_3)$ if $l_1p_1 + l_2p_2 + l_3p_3 = 0$. A *configuration* K in the plane P is any set of points and lines. We say that, for a configuration K in P and three of its points p, q, r, the intersection theorem holds, if for any configuration K' in P isomorphic to K, the points p', q', r' of K' which correspond to p, q and r are colinear in P. Examples of configurations include the Pappian and Fano's configurations (see Figure 1).

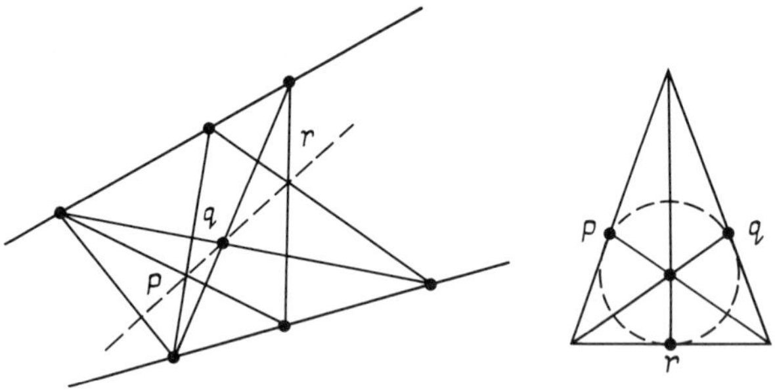

Fig. 1

Intersection theorems for these configurations and the selected points p, q and r describe the fact that P contains straight lines (marked with dotted pattern) on which the three points lie. For a division ring D, Pappian and Fano configurations mean that the polynomial identities $xy - yx = 0$ and $x + x = 0$ respectively are satisfied. In the general case of an arbitrary configuration K which satisfies the so-called constructibility condition (Pappian and Fano configurations both having this property), one can effectively associate with it a certain rational expression $r(x_1, \ldots, x_n)$ such that validity of the intersection theorem for K in P will be equivalent to the validity of the identity $r(x_1, \ldots, x_n) = 0$ in the division ring D. This is the basis upon which application of Amitsur's theorem to Desarguian projective planes relies.

5.5. Topological Rings. In this section we assume familiarity with the notion of a topological space. By a *topological ring* we mean a ring R endowed with a Hausdorff topology such that the ring operations are continuous in that topology. Important examples of topological rings include normed (and pseudo-normed) rings, i.e. rings with a real-valued, non-negative norm $\|x\|$, for which $\|x\| = 0 \Leftrightarrow x = 0$, $\|x + y\| \le \|x\| + \|y\|$, $\|-x\| = \|x\|$ and $\|xy\| = \|x\| \cdot \|y\|$ ($\|xy\| \le \|x\| \cdot \|y\|$ for pseudo-normed rings). For example, rational numbers, complex numbers and quaternion rings are all normed rings (a norm on $\mathbb{Q}$ can be defined in several different ways). If X is a bicompact topological space, then the ring $C(X)$ of continuous real-valued functions on X is a pseudo-normed ring, with a pseudo-norm given by $\|f\| = \max\{|f(x)|, x \in X\}$. Yet another example of a topological ring was already discussed before. It is the ring of endomorphisms $\operatorname{End} V$ of a vector space V_D over a division ring D, with finite topology (in which the basis of neighborhoods of zero in $\operatorname{End} V$ consists of the ideals $\operatorname{Ann}\{v_1, \dots, v_n\} = \{f \in \operatorname{End} V \mid f(v_i) = 0 \text{ for } 1 \leq i \leq n\}$ – annihilators of finite subsets of V). The presence of a topological structure compatible with the ring operations allows to apply both algebraic and topological methods in studying that ring. In the 1930s, this enabled L.S. Pontryagin to obtain a complete description of connected, locally bicompact division rings (i.e. connected topological division rings which contain a neighborhood of zero whose closure is a bicompact subspace).

Theorem 1 (L.S. Pontryagin). *Every connected, locally bicompact topological division ring is topologically isomorphic to $\mathbb{R}$, $\mathbb{C}$ or $\mathbb{H}$ (where $\mathbb{H}$ stands for the ring of quaternions).*

Another example of application of algebraic and topological methods in the investigation of topological rings is the description, due to Kaplansky, of the structure of semisimple (in the sense of the Jacobson radical) bicompact rings.

Theorem 2. *An associative semisimple bicompact ring is topologically isomorphic to a direct product of matrix rings over finite fields, relative to the product topology induced by finite topologies on the components.*

There exist several works which study the question whether it is possibile to define a topology on a given ring by means of a norm (or pseudo-norm). This direction was originated in 1943 by I.R. Shafarevich, who obtained a criterion of normalizability of a topological field. This criterion was later generalized by V.I. Arnautov to topological division rings. An element x of a topological ring R is called *topologically nilpotent* if for any neighborhood U of zero there exists an integer m such that $x^n \in U$ for all $n \geq m$. A subset S of R is called *bounded* if, for any neighborhood of zero U, there exists a neighborhood of zero V such that $SV \cup VS \subseteq U$.

Theorem 3. *Let K be a topological division ring. The topology of this ring is induced by a norm if and only if the following conditions hold:*

1) *the set N of all topologically nilpotent elements of K is open and bounded, and*
2) *if $a \in N$ and $b \notin N$, then $b^{-1}a \in N$.*

Another question considered in the theory of topological rings concerns the possibility of introducing a non-discrete topology in a given ring. The following result was proved by V.I. Arnautov.

Theorem 4. *Every countable (infinite) commutative ring possesses a non-discrete topology.*

In fact, every not necessarily associative countable ring admits a non-discrete topology. On the other hand, there exists a non-associative ring whose cardinality is continuum, which has no topology other than the discrete one.

5.6. Multiplicative Structure of Finite-Dimensional Simple Algebras. Research devoted to the multiplicative structure (i.e. the semigroup of non-zero elements and the group of invertible elements) of non-commutative rings lies at the junction of ring theory, group theory and the theory of semigroups. During the past decade, significant progress was made in the investigation of groups of invertible elements of finite-dimensional simple algebras. This resulted in the solution of certain important problems, which had arisen in algebraic K-theory, algebraic geometry and group theory.

Let F be a field, A – a central simple finite-dimensional algebra over F and A^* – the group of invertible elements of A. We know that $A = M_n(D)$, where D is a non-commutative central division ring over F. Let $\bar{F}$ be the algebraic closure of the field F; we have also noted that $\bar{F}$ is a splitting field of A, i.e. $A \otimes_F \bar{F} \simeq M_t(\bar{F})$. We will denote by φ the natural injection $A \to A \otimes_F \bar{F} = M_t(\bar{F})$, under which $a \mapsto a \otimes 1$. The *reduced norm* $\mathrm{N.rd}_A(a)$ for $a \in A$ is defined as $\mathrm{N.rd}_A(a) = \det \varphi(a)$. Formally speaking, $\mathrm{N.rd}_A(a) \in \bar{F}$, but it is easy to show that the reduced norm is in fact a mapping with values in the field F. This mapping is multiplicative, i.e. preserves the operation of multiplication (which follows from the well-known theorem about determinants of products). The kernel of $\mathrm{N.rd}_A$ restricted to the group A^* will be denoted by $A^{(1)}$. The group $A^{(1)} = \{a \in A^* \mid \mathrm{N.rd}_A(a) = 1\}$ contains, as the kernel of a homomorphism of the group A^* into the abelian group F^*, the commutant $[A^*, A^*]$ of A^*.

The *reduced Whitehead group* of an algebra A is the group $SK_1(A) = A^{(1)}/[A^*, A^*]$. It follows from Dieudonné's theorem on non-commutative determinants that calculation of $SK_1(A)$ can be reduced to the case when A is a division ring. The study of reduced Whitehead groups has its origins in the following question.

Tanaka-Artin problem. Do the groups $A^{(1)}$ and $[A^*, A^*]$ coincide, i.e. is $SK_1(A)$ trivial?

Immediately after the Tanaka-Artin problem was formulated, attempts were made at establishing affirmative answers in many important special cases. Nakayama and Matsushima showed that for a central simple algebra A over the field of p-adic numbers, the problem has an affirmative solution, i.e. the group $SK_1(A)$ is indeed trivial. Another important result of this nature was obtained by Wang, who proved triviality of $SK_1(A)$ for central simple algebras over algebraic number fields. Wang also obtained the following fact: if A is a division ring with center F, and $\dim_F A = p_1^2 \dots p_r^2$ for distinct prime integers $p_1, \ldots, p_r$, then $SK_1(A) = \{1\}$. A situation in which the Tanaka-Artin problem has negative answer was obtained by V.P. Platonov in 1975. Following Platonov, we will describe some examples of division rings A with non-trivial group $SK_1(A)$.

Let $k(x, y)$ be the field of rational functions in two variables x and y over a field k. For the time being, we will denote the field of formal Laurent series $\sum_{i \geq n} a_i x^i$ (where $n \in \mathbb{Z}$ and $a_i \in k$) in x over k by $k\langle x \rangle$. Let $K = k\langle x \rangle \langle y \rangle$ be the field of formal Laurent series in y over $k\langle x \rangle$.

If R is a cyclic extension of the field k with Galois group $\mathrm{Gal}(R/k) = \langle \sigma \rangle$, then R induces cyclic extensions of the fields $k(x, y)$ and K. To avoid cumbersome notation, we will use the same letter R, instead of $R(x, y)$ and $R\langle x \rangle \langle y \rangle$, to denote them. Let R_1 and R_2 be two cyclic extensions of the field k with Galois groups $\mathrm{Gal}(R_1/k) = \langle \sigma_1 \rangle$ and $\mathrm{Gal}(R_2/k) = \langle \sigma_2 \rangle$. We will use $A(x, R_1)$ to denote the cyclic algebra (R, σ_1, x) over the field $k(x, y)$ (i.e. we in fact consider the algebra $(R_1(x, y), \sigma, x)$, see Sect. 2). The analogous algebra over K, i.e. $(R_1\langle x \rangle \langle y \rangle, \sigma_1, x)$, will be denoted by $A\langle x, R_1 \rangle$. Carrying out similar constructions for the extension R_2 and the element y, we obtain algebras $A(y, R_2)$ over $k(x, y)$ and $A\langle y, R_2 \rangle$ over K. We set

$$A(R_1, R_2) = A(x, R_1) \otimes_{k(x,y)} A(y, R_2) \, ,$$

$$A\langle R_1, R_2 \rangle = A\langle x, R_1 \rangle \otimes_K A\langle y, R_2 \rangle \, .$$

A necessary and sufficient condition for the algebras $A(R_1, R_2)$ and $A\langle R_1, R_2 \rangle$ to be division rings is the condition that the fields R_1 and R_2 be *linearly separated* over k (i.e. that the homomorphism $R_1 \otimes_k R_2 \to R_1 R_2 \subseteq \bar{k}$, $r_1 \otimes r_2 \mapsto r_1 r_2$ be injective). Division rings of the type $A\langle R_1, R_2 \rangle$ turned out to be the first counterexamples to the Artin-Tanaka conjecture. In order to demonstrate this, it is convenient to interpret the group $SK_1\langle R_1, R_2 \rangle$ in terms of relative Brauer groups $\mathrm{Br}(T/k)$ (see Sect. 2.8).

We have the following isomorphism formula:

$$SK_1(A\langle R_1, R_2 \rangle) \simeq \mathrm{Br}(R_1 \otimes_k R_2/k)/\mathrm{Br}(R_1/k)\mathrm{Br}(R_2/k) \, .$$

If k is a locally compact or a *global* field (the latter meaning a finite field extension of $\mathbb{Q}$ or $GF(p)(x)$) then, by means of the so-called class field theory, the above formula yields explicit description of the group $SK_1(A\langle R_1, R_2 \rangle)$.

Theorem 1. *If k is a locally compact field then $SK_1(A\langle R_1, R_2 \rangle) \simeq \mathbb{Z}_m$, where $m = \gcd(\dim R_1, \dim R_2)$.*

If k is a global field and $\dim R_1 = \dim R_2 = p$ is a prime integer, then $SK_1(A\langle R_1, R_2 \rangle) \simeq \mathbb{Z}_p \times \ldots \times \mathbb{Z}_p$ (the number of factors in the last direct products can also be computed).

A problem which naturally arises at this point is that of characterizing those fields, for which there exist finite-dimensional central simple algebras with non-trivial reduced Whitehead group. In this direction, V.P. Platonov obtained the following results.

Existence theorem. Let K be a finitely generated field. If the transcendence degree of K over its prime subfield is greater than 1 in case of characteristic zero, and greater than 2 in case of positive characteristic, then for every positive integer m there exists a division ring A with center K such that $|SK_1(A)| > m$.

Stability theorem. For a purely transcendental extension F of any field K there is an isomorphism $SK_1(A \otimes_k F) \simeq SK_1(A)$.

In connection with the settling of the Tanaka-Artin problem, a converse problem is posed in reduced K-theory – namely, to describe those abelian groups which arise as reduced Whitehead groups. It is clear that $SK_1(A)$ is a group of finite exponent. Theorem 1 implies that a reduced Whitehead group can be finite with arbitrarily large cardinality. For infinite groups, this question is answered by

Realization theorem. For any countable abelian group M of finite exponent there exists an algebraic number field k and cyclic extensions R_1, R_2 such that $SK_1(A(R_1, R_2)) \simeq M$.

This theorem is deduced by means of class field theory from the following result:

Infiniteness theorem. Let R_1 and R_2 be cyclic extensions of degree n of a global field k, such that $[(R_1, R_2)_v : k_v] = n^2$ for some norm v on the field k. Then, for every Galois extension F of k, contained in k_v, the inequality $|SK_1(A(R_1, R_2) \otimes F\langle x, y\rangle)| \geq n^{[F:k]-1}$ holds; in particular, if F is an infinite extension of k, then $SK_1(A(R_1, R_2) \otimes F\langle x, y\rangle)$ is an infinite abelian group of exponent n.

5.7. The Brauer Group of a Commutative Ring. Let A be an algebra over a commutative ring Φ. Let also A^{op} be the opposite algebra of A, and $A^e = A \otimes_\Phi A^{\mathrm{op}}$. We will consider A as a left A^e-module (via $(a \otimes b)c = acb$). We will recall that the algebra A is called *separable*, if this module is projective. A is *central* if the mapping $\lambda \mapsto \lambda 1_A$ (for $\lambda \in \Phi$) is an isomorphism of Φ onto the center of A (i.e. A can be naturally regarded as an algebra over its center). A central separable algebra is called an *Azumaya algebra.* We remark that if A and B are Azumaya algebras over Φ, then A^{op} and $A \otimes_\Phi B$ are also Azumaya over Φ.

Examples of Azumaya algebras:

1) If Φ is a field, then Azumaya Φ-algebras include the central simple algebras.
2) If P is a faithful, finitely generated projective Φ-module, then $\mathrm{End}_\Phi P$ is an Azumaya Φ-algebra.
3) *The generalized quaternion algebra.* Let Φ be a commutative ring in which 2 is invertible. For any invertible $a, b \in \Phi$ we consider the free module of rank 4 over Φ: $A = \Phi \cdot 1 \oplus \Phi\mathbf{i} \oplus \Phi\mathbf{j} \oplus \Phi\mathbf{k}$. We introduce multiplication on

A by: $\mathbf{ij} = \mathbf{k}$, $\mathbf{ji} = -\mathbf{ij}$, $\mathbf{i}^2 = a$ and $\mathbf{j}^2 = b$. A turns out to be an Azumaya algebra over Φ.

Let A be any Φ-algebra. We will consider a homomorphism $h : A \otimes A^{\text{op}} \to \text{End}_\Phi A$, given by the equality $h(x \otimes y)(z) = xzy$. We have the following characterization of Azumaya algebras.

Theorem 1. *Let A be a Φ-algebra, finitely generated as a Φ-module. Then the following conditions are equivalent:*
1) A is an Azumaya algebra;
2) the Φ-module A is projective, and the homomorphism h is an isomorphism;
3) for every maximal ideal $\mathbf{m}$ of Φ, $A/\mathbf{m}A$ is a central simple algebra over the field $\Phi/\mathbf{m}$.

Two Azumaya algebras A and B over Φ are called *equivalent* if for some faithful, finitely generated projective Φ-modules E and F, the Φ-algebras $A \otimes_\Phi \text{End}_\Phi(E)$ and $B \otimes_\Phi \text{End}_\Phi(F)$ are isomorphic. The set of equivalence classes, denoted $\text{Br}(\Phi)$, can be equipped with addition operation $[A]+[B] = [A \otimes_\Phi B]$ under which $\text{Br}(\Phi)$ becomes an abelian group, called the Brauer group of the commutative ring Φ. When Φ is a field, $Br(\Phi)$ is nothing else but the Brauer group of a field. The fact that h is injective guarantees the equality $-[A] = [A^{\text{op}}]$. A homomorphism of commutative rings $\Phi \to \Psi$ induces a group homomorphism $\text{Br}(\Phi) \to \text{Br}(\Psi)$, $[A] \mapsto [A \otimes_\Phi \Psi]$, which turns $\text{Br}(-)$ into a functor from the category of commutative rings into the category of abelian groups.

Some properties of Brauer groups:

1) The Brauer group is periodic.
2) Let Φ be a ring regular in the commutative sense (i.e. Φ is a Noetherian commutative ring in which for every maximal ideal $\mathbf{m} \lhd \Phi$, the local ring $\Phi_\mathbf{m} = S^{-1}\Phi$, $S = \Phi \setminus \mathbf{m}$, is a regular local ring; see Section 3.6), and let K be its field of fractions. Then the natural homomorphism $\text{Br}(\Phi) \to \text{Br}(K)$ is injective.
3) For a regular algebra Φ over a field in the group $\text{Br}(K)$, we have the equality $\text{Br}(\Phi) = \bigcap_{\wp \in H} \text{Br}(\Phi_\wp)$, where H is the set of prime ideals of height 1 in Φ.
4) Let Φ be a regular ring of characteristic 0; then the induced homomorphism $\text{Br}(\Phi) \to \text{Br}(\Phi[x])$ is an isomorphism.
5) Let Φ be a complete local ring with maximal ideal $\mathbf{m}$; then the homomorphism $\text{Br}(\Phi) \to \text{Br}(\Phi/\mathbf{m})$ is an isomorphism.

Examples of Brauer groups of commutative rings:

1) If Φ is a finite ring then $\text{Br}(\Phi) = 0$.
2) Let Φ be the ring of entire elements of an algebraic number field K. If r is the number of real norms on the field K then

$$\text{Br}(\Phi) = \begin{cases} 0 & \text{if } r = 0, \\ (\mathbb{Z}_2)^{r-1} & \text{if } r \geq 1. \end{cases}$$

In particular, $\text{Br}(\mathbb{Z}) = 0$.

5.8. Non-Commutative Galois Theory. Let G be a finite group acting as isomorphisms on a ring R. By R^G we will denote the *fixed subring of R under G* (*fixed ring of G*, for short), i.e. the set $\{r \in R \mid \forall_{g \in G}\ r^g = r\}$. If S is an intermediate subring, i.e. $R^G \subseteq S \subseteq R$, then $A(S)$ will denote the subgroup of all those elements $g \in G$ which fix all elements of S: $A(S) = \{g \in G \mid \forall_{s \in S}\ s^g = s\}$.

The fundamental theorem of classical Galois theory of fields asserts that if R is a field, then the mapping $H \mapsto R^H$ defines a one-to-one correspondence between all subgroups of the group G and all subfields of R containing R^G. The inverse transformation is, naturally, the mapping $S \mapsto A(S)$. We are interested in analogous statements for non-commutative rings R. We will limit ourselves to the sufficiently general case of prime rings. An example of a result on Galois correspondence in the non-commutative case is the following

Theorem 1. *Let $F\langle X \rangle$ be a free algebra, G – a finite group of automorphisms acting linearly on the generators. Then the mapping $H \mapsto F\langle X \rangle^H$ determines a one-to-one correspondence between all subgroups of the group G and all intermediate free subalgebras of $F\langle X \rangle$.*

In the general case, the so-called inner automorphisms of R are an obstacle in investigating the correspondence $H \mapsto R^H$. Moreover, the notion of an inner automorphism should be generalized, passing from the ring R to the Martindale ring of quotients $Q(R)$ (see Sect. 4.3). Every automorphism of a prime ring R uniquely extends to an automorphism of the ring $Q(R)$. It is therefore possible to identify in the group $\operatorname{Aut} R$ of all automorphisms of R those automorphisms, which become inner automorphisms in $Q(R)$. In the literature such automorphisms are called *X-inner*. We can now formulate the fundamental theorem in the case when R has no X-inner automorphisms other than the identity. A subring $S \subseteq R$ will be called *rationally complete* if, for any $r \in R$ and a non-zero ideal I of S, the inclusion $Ir \subseteq S$ implies $r \in S$.

Theorem 2. *Let G be a finite group of automorphisms of a prime ring R, containing no X-inner automorphisms other than the identity. Then the mapping $H \mapsto R^H$ defines a one-to-one correspondence between all subgroups of G and all intermediate rationally complete subrings of the ring R.*

For example, the algebra of generic $n \times n$ matrices $F\langle X_1, \ldots, X_m \rangle$ (see Sect. 2.10) in general doesn't have non-identity X-inner automorphisms, and hence Theorem 2 applies to it.

Difficulties connected with X-inner automorphisms in the case of prime (and even semiprime) rings can be successfully overcome, but at the price of the formulation of Galois correspondence theorem becoming significantly more complicated.

The results quoted here are due to V.K. Kharchenko; they extend the research into the Galois theory of fields (Jacobson, Cartan, Hochschild) and full rings of linear transformations (Nakayama and Azumaya, Dieudonné, Rosenberg and Zelinsky).

In recent years there appeared a large number of articles devoted to the study of automorphisms and derivations of non-commutative rings. In particular, questions about the relationship between a ring R and its fixed subring are being intensively researched. An important result in this area is the following theorem of Bergman and Isaacs, which laid the foundation of modern theory of fixed subrings.

Theorem 3. *Let G be a group of order n consisting of automorphisms of a ring R. If R has no additive n-torsion and $R^G = 0$, then R is nilpotent.*

5.9. Regular Rings. A partially ordered set $(M, \leq)$ is called a *lattice* if every two elements $a, b \in M$ have a least upper bound $a \vee b$ and a greatest lower bound $a \wedge b$. A lattice M is called *complete* if every subset $\{a_i \mid i \in I\}$ of M has a least upper bound $\bigvee_{i \in I} a_i$ and a greatest lower bound $\bigwedge_{i \in I} a_i$. Recall that a ring R is called *von Neumann regular* if for every element $a \in R$ there exists an element $x \in R$ such that $a = axa$. For example, the ring of all linear transformations of a vector space over a field is regular, as is the direct product of any collection of regular rings. Consequently, every classically semisimple ring is regular. If a ring R is a directed union of regular rings (i.e. $\forall_{i,j} \exists_k R_i \cup R_j \subseteq R_k$ and $\bigcup_i R_i = R$), then R is regular. Any ring which can be realized as a directed union of classically semisimple rings is therefore regular. In particular, *locally-matrix* rings (i.e. rings in which every finite subset is contained in a subring isomorphic to $M_n(\Delta)$ for some division ring Δ) are regular. Every *algebraic algebra* (i.e. an algebra whose elements are all algebraic over the base field) with no nilpotent elements is regular. Regular rings are Jacobson semisimple, since every non-zero ideal of a regular ring contains a non-zero idempotent ($a = axa \Rightarrow (ax)^2 = ax$), while the Jacobson radical of any ring cannot contain non-zero idempotents. If N is an injective module over some ring A and R – its endomorphism ring, then $R/J(R)$ is a regular ring.

Regular rings were introduced for the purpose of coordinatizing so-called continuous geometries. We will remark that every finitely generated left ideal in a regular ring R is generated by an idempotent. Principal left ideals of R therefore form a *complemented modular lattice* $\mathcal{L}_R$. A lattice M with 0 and 1 (0 and 1 being, respectively, the unique smallest and largest elements of M, i.e. $\forall_{a \in M} a \wedge 0 = 0$ and $a \vee 1 = 1$) is said to be complemented if for every element $a \in M$ there exists an element $a' \in M$ (called the complement of a) for which $a \wedge a' = 0$ and $a \vee a' = 1$. If there is an antiautomorphism $a \mapsto a^{\perp}$ of M such that $a^{\perp}$ is the complement of $a \in M$, then M is called *ortho-complemented.* A lattice M is called *modular* if $a \leq b \Rightarrow (a \vee c) \wedge b = a \vee (c \wedge b)$ holds in it. If the lattice $\mathcal{L}_R$ is a *continuous geometry* (i.e. is complete and continuous), then the regular ring R is called *continuous.* A lattice M is *continuous* if

$$a \wedge \Big(\bigvee_{b \in B} b\Big) = \bigvee_{b \in B} (a \wedge b)\,,$$

and

$$a \vee \Big(\bigwedge_{b \in B} b\Big) = \bigwedge_{b \in B} (a \vee b)$$

for any chain B in M (i.e. a subset in which every two elements are comparable). A complete, continuous and complemented modular lattice is called a *continuous geometry.* A result of Kaplansky, unexpected in its time, asserts that every complete modular ortho-complemented lattice is continuous, i.e. is in fact a continuous geometry.

Continuous regular rings possess interesting properties. For example, if such a ring R is indecomposable then it is *self-injective* (i.e. the modules R_R and ${}_RR$ are both injective) and simple. Conversely, every self-injective regular ring is continuous. Classically semisimple rings provide simplest examples of continuous regular rings. A direct product of a family of continuous regular rings is again a continuous regular ring. Continuous regular rings have the following elementary characterization in the class of regular rings: every left (right) ideal of such a ring is contained, as an essential submodule, in some left (right) principal ideal.

It is well-known that if a ring R is regular, then the matrix rings $M_n(R)$ are also regular. Conversely, regularity of $M_n(R)$ for some n implies that R is regular (more generally, if e is an idempotent of a regular ring A, then eAe is a regular ring as well). The converse assertion above remains true for continuous regular rings, but the original statement fails in this case: let $n > 1$ be an integer, and let R be a regular ring. Then the regular ring $M_n(R)$ is continuous if and only if R is self-injective. We will now describe an example of a commutative regular ring which is continuous, but not self-injective. Let $K \subset F$ be fields, $K_n \subset F_n$ $(n = 1, 2, \ldots)$ – their copies, and let $Q = \prod_n F_n$. Then $R = \{x \in Q \,|\, x_n \in K_n$ for sufficiently large $n\}$ is a ring with the desired properties. $M_2(R)$ is therefore not continuous.

J. von Neumann did not restrict himself to the question of coordinatization of continuous geometries; he also considered the more general problem of coordinatizing complemented modular lattices. Not every such lattice, however, is isomorphic to the lattice of principal left ideals of a regular ring: non-Desarguian projective planes provide one example of this obstacle. In order to exclude pathological cases, the following restriction is imposed on the lattice M: there exist $n \geq 4$ *independent* elements $a_1, \ldots, a_n$ (independence means that $a_i \wedge \bigvee_{j \neq i} a_j = 0$ for all $i = 1, 2, \ldots, n$) such that $a_1 \vee a_2 \vee \ldots \vee a_n = 1$ and $a_i \sim a_j$ for all i, j ($a \sim b$ meaning that a and b have a common complement). The system $\{a_1, \ldots, a_n\}$ is called a homogeneous basis (of rank n) of the lattice M. Existence of such basis has a natural algebraic meaning: the lattice $\mathcal{L}_R$ of principal left ideals of a regular ring R has a homogeneous basis of rank n if and only if R contains n^2 matrix units, i.e. $R \simeq M_n(A)$ for some (regular) ring A. Namely, if $e_{i,j}$ are matrix units in R (that is, $e_{i,j}e_{k,l} = \delta_{j,k}e_{i,l}$

and $1 = e_{1,1} + \ldots + e_{n,n}$), then $a_i = Re_{i,i}$ form a homogeneous basis of the lattice $\mathcal{L}_R$. We will also note that $\mathcal{L}_R$ is then isomorphic to the lattice of finitely generated submodules of the free A-module A^n (this is, by the way, a special case of Morita-equivalence). The coordinatization theorem now takes the following form.

Theorem 1. *Let M be a complemented modular lattice, containing a homogeneous basis of rank $n \geq 4$. Then M is isomorphic to the lattice $\mathcal{L}_R$ of principal left ideals of some regular ring R. In that case we also have $R \simeq M_n(A)$, where A is a regular ring, and M is isomorphic to the lattice of finitely generated submodules of the left A-module A^n.*

We will remark that the center C of a regular ring R is also a regular ring, and the lattice $\mathcal{L}_C$ is isomorphic to the center of $\mathcal{L}_R$ (i.e. $Ca \in \mathcal{L}_C \Rightarrow Ra \in \mathcal{L}_R$), where by the center of a complemented modular lattice M we mean the set of those of its elements which have uniquely determined complements.

The coordinatization problem for ortho-complemented modular lattices leads to the notion of a $*$-regular ring. This name is used for regular rings with an involution $*$ such that $a^*a \neq 0$ whenever $a \neq 0$. In a $*$-regular ring R every left principal ideal is generated by a uniquely determined *projector*, i.e. a self-adjoint idempotent. In this situation the lattice $\mathcal{L}_R$ turns out to be isomorphic to the lattice $\mathcal{E}(R)$ of projectors of the ring R, and so it is in fact an ortho-complemented lattice. The following coordinatization theorem is also due to J. von Neumann.

Theorem 2. *Let M be an ortho-complemented modular lattice containing a homogeneous basis of rank $n \geq 4$. Then M is isomorphic to the lattice $\mathcal{L}_R$ of principal left ideals of some $*$-regular ring R (and also to the lattice $\mathcal{E}(R)$ of its projectors).*

For example, if $A \subseteq \mathcal{B}(H)$ is a so-called factor of type II_1 (here $\mathcal{B}(H)$ denotes the algebra of bounded operators on the Hilbert space H), then the preceding theorem shows that there exists a $*$-regular ring R such that the lattices of projectors $\mathcal{E}(A)$ and $\mathcal{E}(R)$ are isomorphic. The ring A, however, is not regular itself. Moreover, as was shown by Kaplansky, if a Banach algebra is regular in the sense of von Neumann (in the theory of Banach algebras regularity usually has different meaning), then it is finite-dimensional. In his proof, Kaplansky relied on the following theorem (obtained by him as well).

Theorem 3. *If a regular ring does not contain an infinite set of orthogonal idempotents, then it is clasically semisimple.*

The next result, due to Armendariz and Fisher, generalizes a classical theorem of Kaplansky dealing with characterization of commutative regular rings.

Theorem 4. *For a PI-ring R, the following statements are equivalent:*
1) R is a regular ring;

2) $I^2 = I$ *for every ideal* $I \lhd R$*;*
3) every simple left R*-module is injective;*
4) every left ideal of R *is an intersection of maximal left ideals.*

We will conclude by considering a question, which for a long time remained open, until it was settled by Goodearl. Let R and S be regular rings such that the matrix rings $M_n(R)$ and $M_n(S)$ are isomorphic for some positive integer n; does it follow that the rings R and S are isomorphic? If R and S are division rings, then this is indeed the case. We will now describe Goodearl's example showing that the answer is, in general, negative. Let $F \subseteq K$ be a field extension of degree n, V – an infinite-dimensional vector space over K and $E = \mathrm{End}_F(V)$; let also E_0 be the ideal of the F-algebra E, consisting of transformations of finite rank, $E_1 = \mathrm{End}_K(V)$ – its subalgebra, and $R = E_0 + E_1$; $e \in E_0$ will denote the projection of V onto a one dimensional subspace $F \cdot v$. We will consider the projective R-module $P = R \oplus eR$ and its endomorphism ring S. Properties of the rings R and S constructed in this way are formulated as the following theorem.

Theorem 5. *For every integer* $n > 1$*, the regular rings* R *and* S *constructed as above satisfy:* $M_n(R) \simeq M_n(S)$*, but* $M_k(R) \not\simeq M_k(S)$ *for* $k < n$*.*

5.10. Bibliographical Notes. Questions considered in Sect. 5.1 are discussed in detail in Curtis and Reiner [1962]. A particularly complete exposition of the theory of modular representations of finite groups, together with wide ranging applications to the theory of finite groups, can be found in Huppert and Blackburn [1982].

Passman [1977] is a rich source of material on group algebras of infinite groups, and can be regarded as a handbook on this topic. Cohn [1985] includes Cohn's solution of Mal'tsev's problem on emmbeddability of rings in division rings. Articles by Bokut' [1969], Klein [1967] and Bowtell [1967] provide a solution of the problem of Mal'tsev concerning the existence of a ring which cannot be embedded in a division ring, but whose multiplicative group embeds in a group. Gerasimov [1982] gives a matrix construction of the universal localizing homomorphism.

General theory of identities and rational identities is presented in a sufficiently complete way in the book by Rowen [1980]. In it one can also find comprehensive bibliography of the subject. A.R. Kemer's theorem was published in his work [1984]. This paper also contains a program for achieving the solution of Specht's problem for associative algebras over a field of characteristic zero. As was recently announced by A.R. Kemer, he managed to complete his work by solving that problem.

Pontryagin [1954] presents foundations of topological algebra and a description of connected locally compact fields. Criteria of normalizability of topological rings are studied in Shafarevich [1943]. Theorems concerning the possibility of endowing countable rings with topology are proved in Arnautov [1962, 1970]. The material of Sect. 5.6 can be found in V.P. Platonov's lecture at the International Congress of Mathematicians in Helsinki (Platonov [1978]).

The Brauer group of a non-commutative ring is defined in Auslander and Goldman [1960]. For a survey of results on this subject, see Orzech and Small [1975]. The most thorough guide to separable algebras and Azumaya algebras is the monograph by DeMeyer and Ingraham [1971].

Galois theory of semiprime rings was developed by Kharchenko [1977]. An exposition of this theory for prime rings can be found in Montgomery and Passman [1984]. Its applications to free algebras were obtained in Kharchenko [1978]. The monograph by Montgomery [1980] studies fixed rings under finite groups of automorphisms and presents the Bergman-Isaacs theorem. The surveys Kharchenko [1980], Fisher and Osterburg [1980] and Renault [1979] discuss the recent achievements in the investigation of automorphisms and derivations of non-commutative rings.

Regular rings were introduced by J. von Neumann in his paper [1936]. A complete exposition of von Neumann's results on the connections between complemented modular lattices and regular rings is found in von Neumann [1960] (this book was published by Halperin relatively recently, but it is based on von Neumann's manuscripts from the 1930s). These works also established first properties of regular rings. The coordinatization theorem for complemented modular lattices is described from the perspective of different methods in Skornyakov [1961]. Regular rings and related classes of rings are studied in numerous books on associative ring theory, e.g. Jacobson [1964], Lambek [1966], Faith [1973, 1976] and Stenström [1975]. Modern theory of regular rings and modules over them is contained in the voluminous monograph by Goodearl [1979]. Kaplansky's theorem on finite-dimensionality of regular Banach algebras can be found in his article [1948].

References*

Albert, A.A. [1961] Structure of algebras. Am. Math. Soc. Colloq. Publ. 24, AMS: Providence. Zbl. 23,199

Amitsur, S.A. [1966] Rational identities and applications to algebra and geometry. J. Algebra *3* (3), 304–359. Zbl. 203,40

Amitsur, S.A. [1972] On central division algebras. Isr. J. Math. *12* (4), 408–420. Zbl. 248.16006

Anan'in, A.Z. [1976] Locally finitely approximable and locally presentable varieties of algebras. Algebra Logika *16* (1), 3–23. Zbl. 381.16010. English translation: Algebra Logic *16*, 1–16 (1978)

* For the convenience of the reader, references to reviews in Zentralblatt für Mathematik (Zbl.), compiled using the MATH database, and Jahrbuch über die Fortschritte der Mathematik (Jrb.) have, as far as possible, been included in this bibliography.

Anderson, F.W., Fuller, K.R. [1974] Rings and categories of modules. Springer-Verlag: New York, Heidelberg, Berlin. Zbl. 301.16001

Andrunakievich, V.A., Ryabukhin, Yu.M. [1979] Radicals of algebras and the structure theory. Nauka: Moscow (Russian). Zbl. 507.16009

Arnautov, V.I. [1962] On topologizations of countable rings. Sib. Mat. Zh. *9* (6), 1251–1262. Zbl. 167,310. English translation: Sib. Math. J. *9*, 939–946 (1969)

Arnautov, V.I. [1970] Non-discrete topologizability of countable rings. Dokl. Akad. Nauk SSSR *191* (4), 747–750. English translation: Sov. Math. Dokl. *11*, 423–426 (1970). Zbl. 209,339

Arnol'd, V.I., Varchenko, A.N., Husein-Zade, S.M. [1982] Singularities of differential mappings. Classification of critical points, caustics and wave fronts. Nauka: Moscow. Zbl. 513.58001. English translation: Birkhäuser: Boston, Basel, Stuttgart (1985)

Artin, E., Nesbitt, C., Thrall, R. [1944] Rings with minimal condition. University of Michigan Press: Ann Arbor. Zbl. 60,77

Auslander, M., Goldman, O. [1960] The Brauer group of a commutative ring. Trans. Am. Math. Soc. *97* (3), 367–409. Zbl. 100,263

Bass, H. [1968] Algebraic K-theory. Benjamin: New York. Zbl. 174,303

Beck, I. [1972] Projective and free modules. Math. Z. *129* (3), 231–234. Zbl. 236.16020

Beidar, K.I. (= Bejdar, K.J.) [1977] Rings with generalized identities I, II. Vestn. Mosk. Univ., Ser. I *14* (2), 19–26 and (3), 30–37. Zbl. 363.16012, Zbl. 363.16013. English translation: Mosc. Univ. Math. Bull. *32* (2), 15–20 and (3), 27–33(1977)

Beidar, K.I. [1978a] Rings with generalized identities III. Vestn. Mosk. Univ., Ser. I *15* (4), 66–73. Zbl. 402.16004. English translation: Mosc. Univ. Math. Bull. *33* (4), 53–58 (1978)

Beidar, K.I. [1978b] Rings of fractions of semiprime rings. Vestn. Mosk. Univ., Ser. I *15* (5), 36–43. Zbl. 403.16003. English translation: Mosc. Univ. Math. Bull. *33* (5), 29–34 (1978)

Beidar, K.I., Latyshev, V.N., Markov, V.T., Mikhalev, A.V., Skornyakov, L.A., Tuganbaev, A.A. [1984] Associative rings. Itogi Nauki Tekh., Ser. Algebra, Topologiya, Geom. *22*, 3–115. Zbl. 564.16002. English translation: J. Sov. Math. *38*, 1855–1929 (1987)

Beidar, K.I., Mikhalev, A.V. [1985] Orthogonal completeness and algebraic systems. Usp. Mat. Nauk *40* (6), 79–115. English translation: Russ. Math. Surv. *40* (6), 51–95 (1986). Zbl. 603.06003

Beidar, K.I., Mikhalev, A.V., Slavova, K. [1980] Generalized identities and semiprime rings with involution. Usp. Mat. Nauk *35* (1), 222. Zbl. 428.16021. English translation: Russ. Math. Surv. *35* (1), 209–210 (1980

Bergman, G.M. [1976] Rational relations and rational identities in division rings I, II. J. Algebra *43* (1), 252–266, 267–297. Zbl. 307.16012 and Zbl. 307.16013

Birkhoff, G. [1967] Lattice theory (3rd ed.) Am. Math. Soc. Collog. Publ. 25. AMS: Providence. Zbl. 153,25

Bokut', L.A. [1969] On a problem of Mal'tsev. Sib. Mat. Zh. *10* (5), 965–1005. English translation: Sib. Math. J. *10*, 706–739 (1970). Zbl. 213, 319

Bokut', L.A. [1977] Associative rings, Vol.1. IGU: Novosibirsk (Russian). Zbl. 459.16001

Bokut', L.A. [1981] Associative rings, Vol.2. IGU: Novosibirsk (Russian). Zbl. 598.16001

Borho, W., Kraft, H. [1976] Über die Gelfand-Kirillov Dimension. Math. Ann. *220* (1), 1–24. Zbl. 306.17005

Bourbaki, N. [1961] Algèbre commutative. Eléments de mathématiques. Hermann: Paris. Zbl. 119,36

Bourbaki, N. [1964] Algèbre. Livre II. Hermann: Paris. Zbl. 102,272

Bowtell, A.J. [1967] On a question of Mal'tsev. J. Algebra *7* (1), 126–139. Zbl. 171,1

Braun, A. [1984] The nilpotency of the radical in a finitely generated PI ring. J. Algebra *89* (2), 375–396. Zbl. 538.16013

Cartan, H., Eilenberg, S. [1956] Homological algebra. Princeton Univ. Press: Princeton. Zbl. 75,243

Chatters, A.W., Hajarnavis, C.R. [1980] Rings with chain conditions. Research Notes in Mathematics 44, Pitman: London. Zbl. 446.16001

Cohn, P.M. [1985] Free rings and their relations (2nd ed.). Academic Press: London, New York. Zbl. 659.16001

Cozzens, J.H., Faith, C. [1975] Simple Noetherian rings. Cambridge Tracts in Mathematics 69, Cambridge Univ. Press: Cambridge. Zbl. 314.16001

Curtis, C.W., Reiner, I. [1962] Representation theory of finite groups and associative algebras. J. Wiley: New York, London. Zbl. 131,256

DeMeyer, F., Ingraham, E. [1971] Separable algebras over commutative rings. Lect. Notes Math., Vol. 181, Springer-Verlag: New York, Heidelberg, Berlin. Zbl. 215,366

Dixmier, J. [1977] Enveloping algebras. North-Holland Mathematics Library Vol. XIV, North-Holland: Amsterdam. Zbl. 339.17007

Dnestr Notebook [1982] Unsolved problems of the theory of rings and modules (3rd ed.). Mathematical Institute: Novosibirsk (Russian). Zbl. 493.16001

Drozd, Yu.A., Kirichenko, V.V. [1980] Finite-dimensional algebras. Higher School: Kiev (Russian). Zbl. 469.16001

Elizarov, V.P. [1973] Strong pretorsions and strong filters, modules and rings of quotients. Sib. Mat. Zh. *14* (3), 549–559. Zbl. 264.16002. English translation: Sib. Math. J. *14*, 380–387 (1974)

Faith, C. [1973] Algebra I: Rings, modules and categories. Springer-Verlag: New York, Heidelberg, Berlin. Zbl. 266.16001

Faith, C. [1976] Algebra II: Ring theory. Springer-Verlag: New York, Heidelberg, Berlin. Zbl. 335.16002

Fisher, J.W., Osterburg, J. [1980] Finite group actions on non-commutative rings: a survey since 1970. Ring theory and algebra III, Proc. 3rd Okla. Conf. 1979: Lect. Notes Pure Appl. Math. 55, 357–393. Zbl. 462.16024

Gelfand, I.M., Kirillov, A.A. [1966] Sur les corps liés aux algèbres enveloppantes des algèbres de Lie. Publ. Math., Inst. Hautes Etud. Sci. *31*, 5–19. Zbl. 144,21

Gerasimov, V.N. [1982] Localization in associative rings. Sib. Mat. Zh. *23* (3), 36–54. Zbl. 507.16005. English translation: Sib. Math. J. *23*, 788–804 (1983)

Gerasimov, V.N., Sakhaev, I.I. [1984] A counterexample to two conjectures on projective and flat modules. Sib. Mat. Zh. *25* (6), 31–35. English translation: Sib. Math. J. *25*, 855–859. Zbl. 588.16017

Goodearl, K.R. [1979] Von Neumann regular rings. Pitman: London. Zbl. 411.16007

Gordon, R., Robson, J.C. [1973] Krull dimension. Mem. Am. Math. Soc. *133*. Zbl. 269.16017

Herstein, I.N. [1968] Noncommutative rings. Carus Math. Monographs 15, J. Wiley: New York. Zbl. 177,58

Huppert, B., Blackburn, N. [1982] Finite groups II. Springer-Verlag: New York, Berlin, Heidelberg. Zbl. 477.20001

Jacobson, N. [1964] Structure of rings (revised edition). Am. Math. Soc. Colloq. Publ. 37, AMS: Providence. Zbl. 73,20

Jacobson, N. [1975] *PI*-algebras: An introduction. Lect. Notes Math., Vol. 441, Springer-Verlag: New York, Heidelberg, Berlin. Zbl. 326.16013

Jacobson, N. [1985] Basic algebra I (2nd ed.). Freeman: New York. Zbl. 284.16001

Jategaonkar, A.V. [1970] Left principal ideal rings. Lect. Notes Math., Vol. 123, Springer-Verlag: New York, Heidelberg, Berlin. Zbl. 192,379

Kaplansky, I. [1948] Regular Banach algebras. J. Indian Math. Soc., New Ser. *12*, 57–62. Zbl. 32,284

Kasch, F. [1977] Moduln und Ringe. B.G. Teubner: Stuttgart. Zbl. 343.16001. English translation: Lond. Math. Soc. Monogr. 17, Academic Press: New York, 1982. Zbl. 523.16001

Kemer, A.R. [1984] Varieties and Z_2-graded algebras. Izv. Akad. Nauk SSSR, Ser. Fiz.-Mat. Nauk *48* (5), 1042–1059. English translation: Math. USSR, Izv. *25*, 359–374 (1985). Zbl. 586.16010

Kemer, A.R. [1987] Finite basis property of identities of associative algebras. Algebra Logika *26* (5), 597–641. Zbl. 664.16017. English translation: Algebra Logic *26* (5), 362–397 (1987)

Kemer, A.R. [1988] Representatibility of reduced free algebras. Algebra Logika *27* (3), 274–274. Zbl. 678.16012. English translation: Algebra Logic *27* (3), 167–184 (1988)

Kertesz, A. [1968] Vorlesungen über artinsche Ringe. Akadémiai Kiado: Budapest. Zbl. 162,49

Kharchenko, V.K. [1977] Galois theory of semiprime rings. Algebra Logika *16* (3), 313–363. Zbl. 397.16037. English translation: Algebra Logic *16*, 208–246 (1978)

Kharchenko, V.K. [1978] On fixed subrings of free algebras. Algebra Logika *17*, 478–487. English translation: Algebra Logic *17*, 316–321 (1979). Zbl. 433,16004

Kharchenko, V.K. [1979] Differential identities of semiprime rings. Algebra Logika *18* (1), 86–119. English translation: Algebra Logic *18*, 58–80. Zbl. 464.16027

Kharchenko, V.K. [1980] Actions of Lie groups and Lie algebras on non-commutative rings. Usp. Mat. Nauk *35* (2), 67–90. Zbl. 435.16018. English translation: Russ. Math. Surv. *35* (2), 77–104

Kirillov, A.A. [1978] Elements of representation theory (2nd ed.). Nauka: Moscow. Zbl. 264.22011. English translation: Springer-Verlag: Berlin, Heidelberg, New York (1976)

Klein, A.A. [1967] Rings nonembeddable in fields with multiplicative semigroups embeddable in groups. J. Algebra *7* (1), 100–125. Zbl. 171,1

Kostrikin, A.I. [1977] Introduction to algebra. Nauka: Moscow. Zbl. 464.00007. English translation: Springer-Verlag: New York, Heidelberg, Berlin (1982)

Kurosh, A.G. [1962] Lectures on abstract algebra. Fizmatgiz: Moscow (Russian). Zbl. 105,249

Lambek, J. [1966] Lectures on rings and modules. Blaisdell: Waltham. Zbl. 143.264

Lang, S. [1965] Algebra. Addison-Wesley: Reading. Zbl. 193,347

L'vov, I.V. [1969] Maximality conditions in algebras with identity relations. Algebra Logika *8* (4), 449–459. Zbl. 206,47. English translation: Algebra Logic *8*, 258–263 (1971)

MacLane, S. [1963] Homology. Springer-Verlag: New York, Heidelberg, Berlin. Zbl. 133,265

MacLane, S. [1971] Categories for the working mathematician. Springer-Verlag: New York, Heidelberg, Berlin. Zbl. 232.18001

Malgrange, B. [1966] Ideals of differentiable functions. Oxford Univ. Press: London. Zbl. 177,179

Mal'tsev, A.I. [1943] On representations of infinite-dimensional algebras. Mat. Sb., Nov. Ser. *13* (2-3), 263–285 (Russian). Zbl. 60,78

Mal'tsev, A.I. [1976] Collected works. Vol.1: Classical algebra. Nauka: Moscow (Russian). Zbl. 422.01024

Mal'tsev Yu.N. [1976] On varieties of associative algebras. Algebra Logika *15* (5), 579–584. Zbl. 357.16016. English translation: Algebra Logic *15*, 361-364 (1977)

Martindale, W.S. [1969] Prime rings satisfying a generalized polynomial identity. J. Algebra *12* (4), 576–584. Zbl. 175,31

Merkurjev, A.S., Suslin, A.A. [1982] K-cohomologies of Severi-Brauer varieties and the norm residue homomorphism. Izv. Akad. Nauk SSSR, Ser. Fiz.-Mat. Nauk *46* (5), 1011–1061. English translation: Math. USSR, Izv. *21*, 307–340 (1983). Zbl. 525.18008

Milnor, J.W. [1971] Introduction to algebraic K-theory. Ann. Math. Stud. No. 72, Princeton Univ. Press: Princeton. Zbl. 237.18005

Mishina, A.P., Skornyakov, L.A. [1969] Abelian groups and modules. Nauka: Moscow (Russian). Zbl. 224.13015

Molin, F.E. [1985] Number systems. Nauka: Novosibirsk (Russian). Zbl. 607.01025

Montgomery, S. [1980] Fixed rings of finite automorphism groups of associative rings. Lect. Notes Math., Vol. 818, Springer-Verlag: New York, Heidelberg, Berlin. Zbl. 449.16001

Montgomery, S., Passman, D.S. [1984] Galois theory of prime rings. J. Pure Appl. Algebra *31* (1-3), 139–184. Zbl. 541.16031

Naimark, M.A. [1968] Normed rings (2nd ed.). Nauka: Moscow. Zbl. 175,437. English translation: Wolters Noordhoff: Groningen (1970)

Nazarova, L.A., Roiter, A.V. [1973] Categorical matrix questions and the Brauer-Thrall problem. Preprint IM–73–9. Naukova Dumka: Kiev. German translation: Mitt. Math. Semin. Giessen *115* (1975). Zbl. 315.16021

Orzech, M., Small, C. [1975] The Brauer group of commutative rings. Lect. Notes Pure Appl. Math. *11*, M. Dekker: New York. Zbl. 302.13001

Osofsky, B.L. [1973] Homological dimension of modules. CBMS-NSF Reg. Conf. Ser. Appl. Math. *12*, AMS: Providence. Zbl. 254.13015

Passman, D.S. [1977] The algebraic structure of group rings. J. Wiley: New York. Zbl. 368.16003

Pierce, R.S. [1982] Associative algebras. Springer-Verlag: New York, Heidelberg, Berlin. Zbl. 497.16001

Platonov, V.P. [1978] Algebraic groups and reduced K-theory. Proc. Int. Congr. Math., Helsinki 1978, Vol. *1*, 311–317 (1980). Zbl. 425.16018

Pontryagin, L.S [1954] Continuous groups (2nd ed.). Nauka: Moscow (Russian). Zbl. 58.260

Procesi, C. [1973] Rings with polynomial identities. M. Dekker: New York. Zbl. 262.16018

Renault, G. [1979] Action de groupes et anneaux réguliers injectifs. Lect. Notes Math., Vol. 734, 236–248, Springer-Verlag: New York, Heidelberg, Berlin. Zbl. 409.16012

Rentschler, R., Gabriel, P. [1967] Sur la dimension des anneaux et ensembles ordonnés. C.R. Acad. Sci., Paris, Sér. *A 265*, 712–715. Zbl. 155,362

Rowen, L.H. [1980] Polynomial identities in ring theory. Academic Press: New York, London. Zbl. 461.16001

Serre, J-P. [1964] Algèbre locale. Multiplicités. Lect. Notes Math., Vol. 11, Springer-Verlag: New York, Heidelberg, Berlin. Zbl. 142,286

Shafarevich, I.R. [1943] On normalizability of topological fields. Dokl. Akad. Nauk SSSR *40*, 133–135 (Russian)

Shirshov, A.I. [1984] Rings and algebras. Nauka: Moscow (Russian). Zbl. 565.01015

Skornyakov, L.A. [1961] Complemented Dedekind lattices and regular rings. Fizmatgiz: Moscow. Zbl. 115,28. English translation: Oliver & Boyd: London (1964)

Skornyakov, L.A. [1983] Elements of abstract algebra. Nauka: Moscow (Russian). Zbl. 528.00001

Suslin, A.A. [1984a] Algebraic K-theory and the norm residue homomorphism. Itogi Nauki Tekh., Ser. Sovrem. Probl. Mat. *25*, 115–207. Zbl. 558.12013. English translation: J. Sov. Math. *30*, 2556–2611 (1985)

Suslin, A.A. [1984b] Algebraic K-theory (a survey). Tr. Mat. Inst. Steklova *168*, 155–170. Zbl. 586.13012. English translation: Proc. Steklov Inst. Math. *168*, 161–177 (1986)

Stenström, B. [1975] Rings of quotients. Springer-Verlag: New York, Heidelberg, Berlin. Zbl. 296.16001

van der Waerden, B.L. [1967] Algebra II. Springer-Verlag: New York, Heidelberg, Berlin. Zbl. 192,330

van der Waerden, B.L. [1971] Algebra I. Springer-Verlag: New York, Heidelberg, Berlin. Zbl. 221.12001

van der Waerden, B.L. [1930] Moderne Algebra. Bd.1. Springer-Verlag: New York, Heidelberg, Berlin. Jrb. 56,138

van der Waerden, B.L. [1931] Moderne Algebra. Bd.2. Springer-Verlag: New York, Heidelberg, Berlin. Zbl. 2,8

von Neumann, J. [1936] On regular rings. Proc. Natl. Acad. Sci. USA *22*, 707–713. Zbl. 15,388

von Neumann, J. [1960] Continuous geometry. Princeton Univ. Press: Princeton. Zbl. 171,280

Zalesskij, A.E., Neroslavskij, O.M. [1975a] On simple Noetherian rings. Izv. Akad. Nauk SSSR, Ser. Fiz.-Mat. Nauk *1975* (5), 38–42 (Russian). Zbl. 325.16013

Zalesskij, A.E., Neroslavskij, O.M. [1977b] There exist simple Noetherian rings with zero divisors but without idempotents. Commun. Algebra *5* (3), 231–244 (Russian). Zbl. 352.16011

II. Identities

Yu.A. Bakhturin, A.Yu. Ol'shanskij

Translated from the Russian
by E. Behr

Contents

Introduction

In keeping with a philosophical definition, identity is an equality of an object, 'oneness' of it with itself; it is the preservation, in all aspects of the object's existence, of the same permanent characteristics. Formal translation of this definition into mathematical terms leads to a disappointing conclusion: an identity is an expression of the form $a = a$. Nevertheless, already the formulation and explanation of the Aristotelean concept of an identity seems to make use of a working notion of an abstract identity, i.e. an identity within certain limits – different in every case, and depending on the nature of the object. In this sense we can say that the equality $ab = ba$ of two formally different expressions ab and ba is an identity in the domain of ordinary numbers, while

$$\Delta(uv) = (\Delta u)v + u(\Delta v) \tag{1}$$

is an identity in the class of differentiable functions of one variable x, if Δf is the first derivative of f with respect to x; etc. Extending or changing of the domain of objects may lead to a violation of the identity not only from the formal point of view: for example, $ab = ba$ is not an identity in the set of matrices, while (1) ceases to be an identity when Δf denotes the second derivative of f in x.

It is quite possible that our little excursion into etymology failed to impress the reader with the importance of identities. With this in mind, we will now give some examples arising in algebra.

In a finite group of order n, the identity $x^n = 1$ is satisfied. In the ring of residues modulo a prime integer p, we have the identity $x^p = x$. The ring of operators of a two-dimensional representation of a group satisfies Hall's identity:

$$(xy - yx)^2 z = z(xy - yx)^2 .$$

In the Lie algebra of vector fields on a line the identity $(\mathrm{Alt}(x_1, x_2, x_3, x_4))(x_5) = 0$ holds, where $\mathrm{Alt}(x_1, x_2, x_3, x_4)$ is the result of applying the skew-symmetrization operation to the product $\operatorname{ad} x_1 \cdot \operatorname{ad} x_2 \cdot \operatorname{ad} x_3 \cdot \operatorname{ad} x_4$ (for a fixed vector field X, the operator $\operatorname{ad} X$ maps a vector field Y into the commutator $[X, Y]$). It is important to point out that all identities mentioned above are non-trivial, i.e. the first one does not hold in the class of all groups, the second fails in the class of all rings, etc.

Many important classes of groups, rings and algebras are defined by means of identities, i.e. they are "algebraic varieties" (see Sect. 2.2 in Chapter 1). Among them we have the class of abelian groups, the classes of nilpotent and solvable groups with bounded degree of nilpotency (resp. solvability). In these examples the word 'group' can be replaced with 'Lie algebra'. Another variety consists of representations of a group G which are *stable of degree* n; a representation (G, ρ, V) belongs to that category if the vector space V

contains a chain

$$0 = V_0 \subseteq V_1 \subseteq \ldots \subseteq V_n = V$$

such that the group G acts identically on its factor spaces. This class is determined by the identity

$$(x_1 - 1)(x_2 - 1) \ldots (x_n - 1) = 0$$

(where arbitrary operators of the form $\rho(g_1), \ldots, \rho(g_n)$, $g_i \in G$, are substituted for $x_1, \ldots, x_n$). One more item: if the group algebra of a group satisfies an identity of degree n, then the dimensions of absolutely irreducible representations of that group cannot exceed $n/2$. Another 'point for consideration' in the discussion of importance of identities is provided by the following fact: if identities of two finite simple groups coincide, then these groups are isomorphic. We will remark that this observation does not depend on the classification of finite simple groups.

We summarize the above in words of a noted mathematician A.I. Mal'tsev: "Even though identities represent the simplest closed formulæ in the language of logic, the language of identities is still rich enough to be capable of describing many subtle properties of systems and their classes" (A.I. Mal'tsev [1970]). The concept of an identity is therefore considered to be one of the deepest notions of mathematics, while identities themselves are among the most 'stable' properties of (mathematical) objects.

The theory of identities currently constitutes a greatly diverse fragment of algebra. The number of publications on this subject has long ago reached thousands. Degree of development of the theory of identities is different in different areas of algebra. There is no doubt, however, that contemporary mathematics has developed a sufficiently clear 'ideology' of this theory. It includes a small collection of standard general definitions and results; a roster of easily formulated and, as a rule, hard to solve problems; a number of methods, well-known to specialists – and hence rarely put in print; finally, a significant amount of 'folklore' material. The primary goal of the text that we are submitting to the reader's attention is to present the theory of identities from the side of ideas involved in the subject. It was not our purpose to survey all, or even the majority, of the results found in the literature. In many cases, for example, the exotic nature of results on identities is merely caused by the unusual character of algebras, whose identities are being considered. Furthermore, many characteristic phenomena of the theory appear already in the study of identities of classical algebraic systems: groups and (associative) rings; an important intermediate case is the class of Lie algebras (admittedly, these are the areas of algebra with which the authors are most familiar). Finally, in several disciplines of algebra, results about identities constitute the nucleus of the theory; they will be described in more detail in other volumes of this series – those devoted to associative and non-associative algebras, as well as algebras of non-classical signature. Even so, we too will turn to these classes whenever sufficiently important situations cannot be easily illustrated and interpreted in the case of classical algebras.

Passing on to historical notes, we will remark that with all the intrinsic attractiveness of the abstract approach to identities, such approach is not the one which lies at the origins of the theory. As usually happens, it all began with the solution of certain concrete and difficult problems. It seems that the first one of such questions was the famous *Burnside problem* about periodic groups, formulated in 1902. The question concerned finiteness of groups with a finite set of generators, which satisfy the identity $x^n \equiv 1$, where n is a fixed natural number. This problem and its numerous variants gave rise to many studies, but in its most general form it remains open to this day. For example, we do not know whether every group G with the identity $x^5 \equiv 1$ and and two generators a, b is finite. As recently as in 1968, in a paper by P.S. Novikov and S.I. Adyan, a negative solution of the Burnside problem for sufficiently large odd exponents n ($n \geq 4381$) was given. In 1958 A.I. Kostrikin proved that orders of finite groups with a fixed identity $x^p \equiv 1$ (p - a prime) and a fixed number m of generators are bounded, solving in this way the so-called *weak Burnside problem* for prime exponents. For example, when $p = 5$ and $m = 2$, the exact bound turned out to be 5^{34}. Investigation of the Burnside problem for groups stimulated analogous research in other classes of algebraic objects. Very often, statements were formulated 'by analogy' (e.g. the Kurosh problem on local finite-dimensionality of algebraic algebras, and several others). However, it turned out that the weak Burnside problem for prime exponent, for example, is equivalent to the problem of nilpotency of Engel Lie algebras, i.e. of the possibility of deducing the identity of the form $[x_1, x_2, \ldots, x_n] \equiv 0$ from the identity of the type $[x, y^{p-1}] \equiv 0$ in the class of Lie algebras with fixed number of generators over a field of characteristic $p > 0$. An important part in the development of the theory of identities was played by the problem of existence of finite basis, first posed by B.H. Neumann's doctoral dissertation in 1935. He, as well as Birkhoff, introduced the idea of a variety of groups, i.e. a class of groups satisfying a given system of identities. Drawing inspiration from classification problems which were very relevant at the early stage of the theory of infinite groups, Neumann asked whether it is true that every system of group identities is a consequence of a finite subsystem? This problem is connected with another one: is the collection of all varieties of groups a countable set? A negative solution of the finite basis problem wasn't obtained until 1969 (see Sect. 1.2 in Chapter 3). The set of group varieties turned out to have cardinality of the continuum.

A variant of the finite basis problem for associative rings is known as the Specht problem, in honor of the German algebraist Specht, who drew to it the attention of specialists in ring theory. A.R. Kemer recently announced his affirmative solution of this problem for associative algebras over a field of characteristic zero (see footnote on page 14). The question of existence of finite bases of identities of Lie algebras, similar in character to the Specht problem, remains open in the case of characteristic 0.

The reader will encounter other historical remarks and formulation of open problems while reading the main body of the chapter.

The authors gratefully acknowledge their debt to A.A. Kirillov, A.I. Kostrikin and I.R. Shafarevich for their constructive remarks which were aimed at improving the manuscript of this chapter.

Chapter 1
Elementary Concepts and Methods of the Theory of Identities

§ 1. Definition of an Identity. Examples

1.1. Definition. Let A be any algebra, i.e. a set with a collection of operations defined on it, and let $X = \{x_1, x_2, \ldots, x_n, \ldots\}$ be a countable alphabet. Repeated application of operations to letters of X results in expressions of the form $f(x_1, x_2, \ldots, x_n)$ such that by replacing the letters $x_1, x_2, \ldots, x_n \in X$ with elements $a_1, a_2, \ldots, a_n \in A$ one obtains a well-defined element $f(a_1, a_2, \ldots, a_n) \in A$. If $f(x_1, x_2, \ldots, x_n)$ and $g(x_1, x_2, \ldots, x_n)$ are two such expressions, then the formula

$$f(x_1, x_2, \ldots, x_n) \equiv g(x_1, x_2, \ldots, x_n) \tag{1}$$

is called an *identity*. We will say that the identity (1) is satisfied in A, if the equality

$$f(a_1, a_2, \ldots, a_n) = g(a_1, a_2, \ldots, a_n)$$

is valid for any $a_1, a_2, \ldots, a_n \in A$.

1.2. Example. We will consider the set $A = \mathbb{R}^3$ of three-dimensional vectors with real coordinates. It is an algebra with the operations of vector addition $u+v$, with additive inverses $-u$ and a fixed zero vector 0, scalar multiplication by any $\lambda \in \mathbb{R}$ and the vector cross product $[u, v]$ of $u, v \in \mathbb{R}^3$. Identities of this algebra, which do not involve the cross product operation, are not of any special interest: they all follow from the usual axioms of the vector space.

$$\left.\begin{array}{ll} I.1 & x_1 + (x_2 + x_3) \equiv (x_1 + x_2) + x_3\,, \\ I.2 & x_1 + x_2 \equiv x_2 + x_1\,, \\ I.3 & x + (-x) \equiv 0\,, \\ I.4 & x + 0 \equiv x\,; \end{array}\right. \qquad (I)$$

$$\begin{array}{ll} II.1 & (\lambda_1 + \lambda_2)x \equiv \lambda_1 x + \lambda_2 x\,, \\ II.2 & \lambda(x_1 + x_2) \equiv \lambda x_1 + \lambda x_2\,, \\ II.3 & (\lambda_1\lambda_2)x \equiv \lambda_1(\lambda_2 x)\,, \\ II.4 & 1 \cdot x \equiv x\,. \end{array} \qquad (II)$$

When considering identities which are related to the operation of the cross product, one should first of all isolate the 'trivial' identities, known from courses of analytic geometry ($\lambda \in \mathbb{R}$, $x, x_1, x_2, x_3 \in \mathbb{R}^3$):

$$\begin{array}{ll} III.1 & [\lambda x_1, x_2] \equiv \lambda[x_1, x_2]\,, \\ III.2 & [x_1, \lambda x_2 \equiv \lambda[x_1, x_2]\,, \\ III.3 & [x_1 + x_2, x_3] \equiv [x_1, x_2] + [x_1, x_3]\,, \\ III.4 & [x_1, x_2 + x_3] \equiv [x_1, x_2] + [x_1, x_3]\,; \end{array} \qquad (III)$$

$$\begin{array}{ll} IV.1 & [x, x] = 0\,, \\ IV.2 & [[x_1, x_2], x_3] + [[x_2, x_3], x_1] + [[x_3, x_1], x_2] \equiv 0\,. \end{array} \qquad (IV)$$

The last identity – the so-called *Jacobi identity* – is a little less trivial; its verification, however, does not present any difficulties if one notices that IV.1 easily implies anticommutativity $[x, y] = -[y, x]$, and that the Jacobian, i.e. the left-hand side of the Jacobi identity, is a trilinear skew-symmetric function of its arguments. Now by choosing an orthogonal basis e_1, e_2, e_3 of $\mathbb{R}^3$, and substituting its elements in place of x_1, x_2, x_3 in IV.2, we will see that every term of the Jacobian becomes zero. Examining the set of 'trivial' identities I, II, III and IV above we see that they represent nothing else but the usual system of axioms of the class of Lie algebras over the field of real numbers.

Naturally, when studying a concrete Lie algebra, it is also interesting to know 'non-trivial' identities, i.e. ones that hold in the given algebra but not, at the same time, in all Lie algebras. An example of such an identity for $\mathbb{R}^3$ can be obtained as follows.

We will consider the expression of the form

$$f(x_0, x_1, x_2, x_3, x_4) = \sum_{\sigma \in \mathrm{Sym}_4} \varepsilon_\sigma [[[[x_0, x_{\sigma(1)}], x_{\sigma(2)}], x_{\sigma(3)}], x_{\sigma(4)}]\,, \qquad (2)$$

where $\varepsilon_\sigma = \pm 1$ depending on the parity of the permutation σ from the symmetric group Sym_4 on four symbols (i.e. the *sign* of σ). This function is linear and skew-symmetric in the variables x_1, x_2, x_3, x_4. It follows that this expression, as any other skew-symmetric n-linear form on $\mathbb{R}^3$ with $n > 3$, evaluates to 0 – i.e. $f(x_0, \ldots, x_4) \equiv 0$ is an identity on $\mathbb{R}^3$. It remains to exhibit a Lie algebra in which this identity does not hold. We will consider the algebra $M_3(\mathbb{R})$ of 3-by-3 matrices with entries in $\mathbb{R}$, with the usual operations of matrix addition and scalar multiplication. We will replace the usual matrix multiplication with the commutator operation:

$$[A, B] = AB - BA\,. \qquad (3)$$

A straightforward verification shows that the resulting algebra satisfies all identities (I) through (IV), i.e. that one obtains a Lie algebra in this way (the so-called *general linear Lie algebra* $\mathrm{gl}(3, \mathbb{R})$). We will now compute the element $f(e_{1,1}, e_{1,2}, e_{2,2}, e_{2,3}, e_{3,3}) \in \mathrm{gl}(3, \mathbb{R})$, where $e_{i,j}$ is the matrix unit with all entries 0, except for a 1 in the (i, j)-th place. Then $e_{i,j}e_{k,l} = \delta_{j,k}e_{i,l}$, so that $[e_{i,j}, e_{k,l}] = \delta_{j,k}e_{i,l} - \delta_{l,i}e_{k,j}$ (here $\delta_{i,j}$ is the Kronecker delta). Applying this rule we see that values of those term of the sum (2) which correspond to $\sigma \neq 1$ are all zero, while the term with $\sigma = 1$ yields $e_{1,3}$. Hence $f(e_{1,1}, e_{1,2}, e_{2,2}, e_{2,3}, e_{3,3}) = e_{1,3} \neq 0$, i.e. the identity $f(x_0, x_1, x_2, x_3, x_4) \equiv 0$ is not satisfied in $\mathrm{gl}(3, \mathbb{R})$.

The reasoning described above shows that, in fact, the algebra $\mathbb{R}^3$ satisfies every identity of the form $f(x_1, \ldots, x_n; y_1, \ldots, y_m) \equiv 0$, whose left-hand side is multilinear and skew-symmetric in the variables $x_1, \ldots, x_n$, where $n \geq 4$. In order to somehow exhibit a complete system of identities satisfied by this – or, for that matter, any other – algebra, it is necessary to formalize the notion of an identity being a consequence of others. Namely, we will say that $f \equiv 0$ follows from a system of identities $g_1 \equiv 0, \ldots, g_m \equiv 0, \ldots$, if in every algebra in which $g_1 \equiv 0, \ldots, g_m \equiv 0, \ldots$ are satisfied, $f \equiv 0$ necessarily holds as well. *Equivalence of two systems of identities* is defined in an obvious way. Any subset of a system of identities, which implies all identities of that system will be called its *basis*. It should be made clear that the term 'basis' does not involve any sort of independence at all; there is, however, a natural tendency to seek the simplest possible basis of a system of identities.

Returning to the Lie algebra $\mathbb{R}^3$, we will mention that all homogeneous identities of this algebra whose degree is at least seven follow from the identity $f(x_0, x_1, x_2, x_3, x_4) \equiv 0$, where f is the polynomial (2). In order to obtain a basis for the system of all identities of this algebra it is necessary (Yu.P. Razmyslov) to include one more identity of degree 5. If b is an element of any Lie algebra A, we will use $\mathrm{ad}\, b$ to denote the operator of multiplication of elements of A by b on the right: $a(\mathrm{ad}\, b) = [a, b]$. Then the second identity in the basis mentioned above has the form

$$[x, y](\mathrm{ad}\, z)^3 \equiv [x(\mathrm{ad}\, z)^3, y] + [x, y(\mathrm{ad}\, z)^3] .$$

In other words, the cube of the operator $\mathrm{ad}\, a$, where a is any vector in $\mathbb{R}^3$, is a derivation of this algebra. (We will parenthetically remark that in a Lie algebra the operator $\mathrm{ad}\, a$ is always a derivation; it should be clear that this property does not generally extend to powers of the given operator.)

It may appear odd that a basis of identities in the Lie algebra $\mathbb{R}^3$ consists of identities of degree 5: indeed, the identities (I) through (IV), satisfied in $\mathbb{R}^3$, have degrees 1, 2 and 3. This apparent inconsistency is explained by the fact that, when talking about identities of an algebra from a given class, say that of Lie algebras, a basis of identities is – by some abuse of language – a system which defines the complete system of identities when it is appended to the axioms for the entire class.

1.3. Identities of Linear Representations. Proof of the result described above leads us to other interesting aspects of the definition of an identity. It will be convenient to regard associative and Lie algebras as special cases of a more general notion of a *linear algebra* over a field. By this we mean a vector space with a bilinear multiplication operation. If this operation is denoted by a bracket (as it is customary in the case of Lie algebras) then the class of linear algebras is defined by the system of axioms I, II and III.

Going back to the Lie algebra $L = \mathbb{R}^3$ once again, we will note the well-known isomorphism $L \otimes_{\mathbb{R}} \mathbb{C} \simeq \mathrm{sl}\,(2, \mathbb{C})$, where the *special linear algebra* $\mathrm{sl}\,(2, \mathbb{C})$ consists of complex two-by-two matrices with trace 0 (i.e. matrices $A = \begin{pmatrix} a & b \\ c & d \end{pmatrix}$ with $a, b, c, d \in \mathbb{C}$ and $\mathrm{tr}\, A = a + d = 0$) under the commutator operation (3). In other words, $\mathrm{sl}\,(2, \mathbb{C})$ is the complexification of the Lie algebra $\mathbb{R}^3$, which implies that identities with real coefficients satisfied in both of these algebras coincide.

We will now consider the algebra $\mathrm{sl}\,(2, \mathbb{C})$ as a subalgebra of the matrix algebra $M_2(\mathbb{C})$ under the commutator (3). An identity $f\,(x_1, x_2, \dots, x_n) \equiv 0$, where $f\,(x_1, x_2, \dots, x_n)$ is a complex polynomial in indeterminates $x_1, \dots, x_n$ in which multiplication represents the usual associative (but not commutative) product, is called an *identity of the pair* $(M_2(\mathbb{C}), \mathrm{sl}\,(2, \mathbb{C}))$ if $f(A_1, \dots, A_n) = 0$ for any $A_1, \dots, A_n \in \mathrm{sl}\,(2, \mathbb{C})$. One also uses the phrases *'weak identity'* and *identity of a representation* (of $\mathrm{sl}\,(2, \mathbb{C})$ in $\mathbb{C}^2$). The pair $(M_2(k), \mathrm{sl}\,(2, k))$, where k is any field, satisfies the weak identity

$$x^2 y - y x^2 \equiv 0\,. \tag{4}$$

Indeed, the Cayley-Hamilton theorem clearly implies that x^2 is a scalar matrix, whenever x is a matrix from $M_2(k)$ with zero trace. Since scalar matrices commute with all others, we see that the weak identity (4) is satisfied for the pair under consideration. One of the central points in proving a theorem on identities of $\mathrm{sl}\,(2, \mathbb{C})$ is the result saying that (4) represents a basis of weak identities of the pair $(M_2(k), \mathrm{sl}\,(2, k))$ for any field k of characteristic 0. This fact turns out to be important in finding a basis of identities of the associative algebra $M_2(k)$ as well (Razmyslov [1973]).

We will note that the associative algebra $M_n(k)$ of n-by-n matrices over a field k cannot satisfy any identity of degree less than $2n$. It is enough to verify this claim for multilinear identities (for details, see Sect. 3 which describes the process of linearization of identities, resembling the procedure of polarization of a quadratic form, which produces from it a bilinear one). In this situation every non-trivial identity of associative algebras can be written in the form

$$\sum_{\sigma \in \mathrm{Sym}_m} \alpha_\sigma x_{\sigma(1)} \cdot \ldots \cdot x_{\sigma(m)} \equiv 0\,, \quad \alpha_1 \neq 0\,. \tag{5}$$

If $m < 2n$, then a substitution, similar to the one used in Sect. 1.2 as an example of identity (2) failing for $L = \mathbb{R}^3$, demonstrates that (5) does not hold in $M_n(k)$. On the other hand, this algebra always satisfies some identity of degree $2n$. More precisely, by a theorem of Amitsur and Levitzki (see Jacobson [1956]) the *standard identity* of degree $2n$

$$S_{2n}(x_1, \ldots, x_{2n}) = \sum_{\sigma \in \mathrm{Sym}_{2n}} \varepsilon_\sigma x_{\sigma(1)} \cdot \ldots \cdot x_{\sigma(2n)} \equiv 0 \tag{6}$$

always holds in $M_n(k)$.

We will also remark that the algebra $M_2(k)$ satisfies the so-called Hall's identity

$$[[x, y]^2, z] \equiv 0\,. \tag{7}$$

This clearly follows from (4), since for any $A, B \in M_n(k)$ we have $\operatorname{tr} AB = \operatorname{tr} BA$, i.e. $[A, B] = AB - BA \in \mathrm{sl}\,(n, k)$. Our final result on identities of the algebra $M_2(k)$ over a field k of characteristic 0 is as follows: the identities (6) for $n = 2$ together with (7) form a basis of identities for the associative algebra $M_2(k)$.

We now note one unexpected fact (Vaughan-Lee; see Bakhturin [1985]): identities of the Lie algebra $M_2(k)$ (with the commutator multiplication), where k is an infinite field of characteristic 2, do not have a finite basis.

Identities of representations arise also in the study of the pair $(M_2(k), \mathrm{SL}\,(2, k))$, where $\mathrm{SL}\,(2, k)$ is the group of 2-by-2 matrices with determinant 1 over a field k. First, we note that when characteristic of k is zero, then $\mathrm{SL}\,(2, k)$ does not satisfy any non-trivial identity of the class of groups. The reason for this is that $\mathrm{SL}\,(2, k)$ contains a free subgroup F of countable rank: it can be exhibited, for example, as the commutant G' of the subgroup generated by matrices $\begin{pmatrix} 1 & 0 \\ 2 & 1 \end{pmatrix}$ and $\begin{pmatrix} 1 & 2 \\ 0 & 1 \end{pmatrix}$. Since every finitely generated group is isomorphic to a factor group of F, and an identity which fails to be satisfied in a group will fail on one of its finitely generated subgroups, it follows that $\mathrm{SL}\,(2, k)$ cannot satisfy an identity which does not hold for all groups.

The group $\mathrm{SL}\,(2, k)$ can nevertheless be considered as a subset of the algebra $M_2(k)$ of endomorphisms of a two-dimensional vector space k^2. A weak identity of the pair $(M_2(k), \mathrm{SL}\,(2, k))$ (a representation identity) is a formula of the type

$$f(x_1, \ldots, x_n, x_1^{-1}, \ldots, x_n^{-1}) \equiv 0\,, \tag{8}$$

where f is an associative polynomial. An identity (8) holds for the pair $(M_2(k), \mathrm{SL}\,(2, k))$ if $f(A_1, \ldots, A_n, A_1^{-1}, \ldots, A_n^{-1}) = 0$ for every choice of $A_1, A_2, \ldots, A_n \in \mathrm{SL}\,(2, k)$. Clearly, (7) is one of such identities. There are, however, essentially weaker identities as well, i.e. ones that do not reduce to ordinary identities. A basis of identities of the pair being discussed is (Kushkulei [1978]):

$$[(x_1 + x_1^{-1}), x_2] \equiv 0\,.$$

1.4. Concluding Remarks. Closing this introductory section we will remark that in the case of matrices larger than 2-by-2, the problem of describing a basis for identities of related Lie algebras, groups and representations is still open. One exception is the recent announcement by A.V. Yakovlev about the existence of finite bases of identities of an associative matrix algebra of any given size. Another question which remains unsolved to this day is the fundamental one contained in the general Specht's problem: to prove that identities of any associative algebra over a field of characteristic zero has a finite basis (modulo the system of identities defining the class of associative algebras). Quite recently Kemer, in a series of papers, announced a positive solution of this problem. An analogous question about identities of groups was answered in the negative in the works of A.Yu. Ol'shanskij, S.I. Adyan and M.R. Vaughan-Lee. The simplest system of group identities which cannot be reduced to a finite one has the following form (Yu.G. Kleiman [1973], R.M. Bryant [1973]):

$$x_1^8 \equiv 1\,,\ (x_1^2 x_2^2)^4 \equiv 1\,, \dots,\ (x_1^2 \cdot \dots \cdot x_n^2)^4 \equiv 1,\ \dots\,.$$

It is clear that every identity of this system is a consequence of all preceding ones. However, there is no finite basis of identities for this system. In the example of a Lie algebra with no finite basis of identities which was presented earlier (Sect. 1.3), the system of identities itself has a very unwieldy form. The lack of finite bases of identities of semigroups (including finite ones) is also a commonplace phenomenon.

§ 2. Formulation of Problems. Working Notions

2.1. General Problems. The theory of identities of algebras has two interrelated aspects: identities and algebras. This results in two 'global' tasks:

1) to describe algebras with given identities, and

2) to describe identities of given algebras.

In the first case the point is to study the structure of an algebra, which satisfies an identity or a system of identities of some particular form (interesting, as a rule, not only from the viewpoint of the theory of identities). Quite often the structure of an algebra is greatly affected by the mere fact of it satisfying a non-trivial identity – as in the case of associative rings. There are important examples in which an algebra from some natural class is completely determined by its identities.

The second problem has to do with determining the identities of concrete algebras or classes of algebras. One has to point out in the beginning that the problem formulated in this way is exceedingly difficult. For example, in many situations it isn't even clear whether a system of identities has a finite basis (the finite basis problem). Furthermore, questions about identities are generally separated from questions about algebras in which these identities are satisfied – even though, obviously, for any system of identities one can find a class of algebras which is defined by that system. It is at this point where algebraic varieties enter the stage. The language of varieties essentially erases the distinction between algebras and their identities or, more precisely, allows to pass freely from one of these notions to the other. This is the reason why work in the theory of identities is usually conducted in the framework of varieties.

In spite of what was said above, we have divided the main part of our survey into two segments. Chapter 2 will deal with the structure of algebras with identities, while Chapter 3 will be devoted to the study of identities themselves. The remainder of Chapter 1 discusses elementary methods of the theory of identities.

2.2. Working Notions. We will use the term 'algebra' in a broad sense, describing a set with operations. Groups, rings, algebras in the usual sense – are all special cases of this notion.

Definition. A class of algebras with identical sets of operations (e.g. groups, rings, linear algebras – i.e. standard algebras over a fixed field etc.) is a *variety* if it consists of all algebras satisfying a given system of identities.

For example, the class of all groups is a variety of algebras with the operations of multiplication, inverse and a neutral element, defined by the system of axioms

$$1)\ xe \equiv ex \equiv x\,; \quad 2)\ xx^{-1} \equiv x^{-1}x \equiv e\,; \quad 3)\ (xy)z \equiv x(yz)\,. \tag{9}$$

The class of all abelian groups is also a variety, defined by the axioms (9) together with the commutativity identity

$$xy \equiv yx\,. \tag{10}$$

In fact, when discussing varieties of groups, the identities (9) are assumed to be automatically satisfied. It is therefore common to say that the variety of abelian groups is determined by the single relation (10). Analogous conventions are adopted in considering other large classes of algebras as well (e.g. rings, linear algebras over fields etc.).

One of the examples of varieties of rings are varieties of nil rings of bounded nilpotency index. Each of them is determined by simple relations of the form $x^n \equiv 0$, where n is some integer ≥ 1.

We will now suppose that V is a certain system of group identities, which is expressed in the form $V = \{v_\alpha(x_1,\dots,x_n) \equiv 1 \,|\, \alpha \in I\}$ (it is easily seen that every system of identities of groups is equivalent to a system of this kind). By $V(G)$ we will denote the subgroup of G generated by all elements $v_\alpha(g_1,\dots,g_n)$, where $\alpha \in I$ and $g_1,\dots,g_n \in G$. It is clear that elements of $V(G)$ lie in the kernel of every homomorphism of G into any group from the variety defined by the system V. It follows that $V(G)$ is the smallest normal subgroup of G among all those normal subgroups H for which the factor group G/H satisfies all identities of V. $V(G)$ is called the *verbal subgroup corresponding to the system of identities* V (or *to the set of words* $\{v_\alpha(g_1, \ldots, g_n) \,|\, \alpha \in I\}$). Examples of verbal subgroups are common; the commutant G' of a group G is a verbal subgroup corresponding to the identity $xyx^{-1}y^{-1} \equiv 1$ or, in other words, to the variety of abelian groups.

Verbal ideals of rings are defined in a similar way. In this case every identity can be written as $f(x_1, \ldots, x_n) \equiv 0$, where $f(x_1, \ldots, x_n)$ is a non-commutative and – in the most general case – non-associative polynomial in indeterminates $x_1, \ldots, x_n$. Given a system V of identities in this form, we define the *verbal ideal* $V(R)$ of a ring R as the ideal generated by the set of all evaluations of polynomials from V on R.

Similar definitions are more complicated in cases when identities cannot be reduced to the form $v(x_1, \ldots, x_n) \equiv 1$, as in groups, or $f(x_1, \ldots, x_n) \equiv 0$, as in the case of rings. Such situation arises in the setting of semigroups – i.e. algebras with a single associative binary operation. If S is a semigroup and $\mathfrak{V}$ is a variety of semigroups, then the set of all homomorphic images of S which belong to $\mathfrak{V}$ contains a 'largest' semigroup S_0. The semigroup S_0 is obtained by introducing on S the weakest equivalence relation $\sim$ for which

$$u(a_1, \ldots, a_n) \sim v(a_1, \ldots, a_n),$$

where $a_1, \ldots, a_n$ are arbitrary elements of S and $u(x_1, \ldots, x_n) \equiv v(x_1, \ldots, x_n)$ is any identity from the system V which defines the variety $\mathfrak{V}$. This equivalence turns out to be a *congruence*, i.e. the set of equivalence classes becomes a semigroup S_0 under the induced operation of multiplication of their representatives. In this situation, S_0 maps naturally onto every homomorphic image of S lying in $\mathfrak{V}$. The congruence $\sim$ is uniquely determined by the set of identities V (i.e. by the variety $\mathfrak{V}$) and is referred to as the *verbal congruence* corresponding to V. It is denoted by ρ_V, so that: $x \,\rho_V\, y \Leftrightarrow x \sim y$ and $S_0 = S/\!\sim\; = S/\rho_V$.

Verbal congruence in any class of algebras with arbitrary operations is defined in an analogous way.

Turning again to the case of groups, we will remark that if $x, y \in G$ and V is a set of identities in the reduced form, then $x \,\rho_V\, y \Leftrightarrow xV(G) = yV(G)$. It follows that the verbal congruence corresponds to the decomposition of the group G into cosets of its normal subgroup $V(G)$.

Definition. Let X be a non-empty set, $\mathfrak{V}$ – a variety of algebras. An algebra $F(X, \mathfrak{V})$ from that variety is *a free algebra in* $\mathfrak{V}$ *with the free generating set* X if every mapping $\varphi : X \to A$, where $A \in \mathfrak{V}$, extends uniquely to a homomorphism $\bar{\varphi} : F(X, \mathfrak{V}) \to A$ (here, as usual, by a homomorphism we mean a mapping compatible with all operations in the signature of algebras from $\mathfrak{V}$; see below).

It is clear that if a free algebra of a variety exists, it must be unique up to isomorphism acting as identity on the set X.

The free algebra of a variety is an extraordinarily important object, since every algebra of that variety can be viewed as a factor algebra of some free algebra with a 'sufficiently large' X. Cardinality of X is called the *rank* of the algebra $F(X, \mathfrak{V})$. In majority of cases, the rank of a free algebra in a variety is an invariant of that algebra. In Sect. 4.1, however, we will encounter a situation in which this invariance fails.

Existence of the free algebra in a variety is proved in two stages. We will assume that algebras under consideration are equipped with a collection of operations (or *signature*) Ω. Ω_n will denote the subset of Ω consisting of all n-ary operations ($n \geq 0$). An operation from Ω_0 merely represents the fixing of a certain element of the algebra on which this operation is defined. When $n > 0$, an operation $\omega \in \Omega_n$ (i.e. an n-ary operation) is simply a mapping $\omega : \underbrace{A \times \ldots \times A}_{n \text{ times}} \to A$, where A is the algebra on which ω acts. For example, in case of (multiplicative) groups the signature $\Omega = \Omega_{\text{gr}}$ has the form $\Omega_{\text{gr}} = \Omega_0 \cup \Omega_1 \cup \Omega_2$, with $\Omega_0 = \{\omega_0\}$, $\Omega_1 = \{\omega_1\}$ and $\Omega_2 = \{\omega_2\}$, where ω_0 corresponds to the selection of the neutral element e of a group, $\omega_1(a) = a^{-1}$, and $\omega_2(a, b) = ab$. One often writes simply $\Omega_{\text{gr}} = \{e, {}^{-1}, \cdot\}$. Applying these definitions to the task at hand, the first stage consists of the construction of a free algebra $F_\Omega(X)$ of signature Ω with a free generating set X. Its elements, called *words*, are defined by the following requirements:

(a) every element $x \in X$ is a word;

(b) every symbol of a 0-ary operation (i.e. operation from Ω_0) is a word;

(c) if $w_1, \ldots, w_n$ are words and ω is an n-ary operation for $n > 0$, then $\omega(w_1, \ldots, w_n)$ is also a word.

Each operation $\omega \in \Omega$ acts on the set $F_\Omega(X)$ according to property (c). It has to be noted that, say, the free algebra $F_{\Omega_{\text{gr}}}(X)$ with the group-theoretic signature $\Omega_{\text{gr}} = \{e, {}^{-1}, \cdot\}$ without the context of group axioms is not a free group – it isn't even a group. A free group $G(X)$ is a free algebra of the variety with signature Ω_{gr}, defined by the system of identities (9). This algebra is obtained as the factor algebra of $F_{\Omega_{\text{gr}}}$ modulo the verbal congruence ρ_V, defined by the system (9) of identities 1), 2) and 3) (strictly speaking, each of the identities 2) and 3) should be broken up into two separate identities).

In the example cited above we already described the second part of the construction of the free algebra with a generating set X in a variety $\mathfrak{V}$ of algebras with signature Ω. Namely, this algebra is the factor algebra of $F_\Omega(X)$ modulo the verbal congruence corresponding to the defining system of identities of the variety $\mathfrak{V}$. As concerns the verbal congruence itself, it is simply the intersection of all those congruences ρ for which $u_\alpha \, \rho \, v_\alpha$ for all $\alpha \in I$ – cf. the definition given in the examples dealing with groups and semigroups. The second fundamental property of the algebra $F(X, \mathfrak{V})$ which follows from this construction is this: every relation $u(x_1, \ldots, x_n) =$

$v(x_1, \ldots, x_n)$ between the free generators of $F(X, \mathfrak{V})$ is an identity of the variety $\mathfrak{V}$.

It is worth pointing out that the set of all algebras of given signature Ω is also a variety. It corresponds to an empty system of identities. The algebra $F_\Omega(X)$ is a free algebra of this variety, freely generated by the set X. Indeed, given a mapping $\varphi : X \to A$ from X to an algebra A of signature Ω, the homomorphism $\bar{\varphi} : F_\Omega(X) \to A$ merely represents the evaluation of a word $\omega \in F_\Omega(X)$ on elements of A corresponding to the letters from X under the mapping φ.

Free algebras in varieties are frequently encountered outside the context of theory of identities. Three such algebras – the free group, free associative ring and the free Lie algebra – will be discussed in Sect. 4.2, which is devoted especially to them. Here we will present only the extremely simple constructions of a free semigroup and a free associative algebra with a free generating set X. It is quite often that a free algebra is built not 'from above', as in the general construction, but 'from below', using known models. If X is a non-empty set then the free semigroup $S(X)$ is simply the set of all words in the alphabet X with the operation of juxtaposition of words. If $\varphi : X \to A$, A – a semigroup, is any mapping then $\bar{\varphi}(x_1, \ldots, x_n) = \varphi(x_1) \ldots \varphi(x_n)$ is a homomorphism from $S(X)$ to A which extends φ. The free associative algebra $A(X)$ is the vector space over the base field, whose basis is $S(X)$, and with natural operations.

2.3. Birkhoff's Theorem. One of the fundamental results of the theory of identities is Birkhoff's theorem (Birkhoff [1935]).

Theorem. *A non-empty class of algebras $\mathfrak{N}$ of fixed signature is a variety if and only if it is closed under passing to subalgebras, homomorphic images, and Cartesian products of its elements.*

Because of the importance of this theorem, we will include its proof here. The essence of reasoning is fully preserved even if one assumes, for example, that $\mathfrak{N}$ has the group signature. Structure of the proof is very helpful in understanding the nature of algebras from a variety determined by identities of some algebra (or of a collection of algebras). Cartesian product is defined in Sect. 4.2 in Part I of this volume.

It is evident that the closure of a given class under the operations specified above is a vital part of this theorem; for brevity, we will call such a class 'closed'. We will first assume that $\mathfrak{N}$ is a closed class. Consider the set V of all identities of the form $v \equiv 1$ (in countably many variables), satisfied in all groups from $\mathfrak{N}$. Let a group G satisfy all identities of $\mathfrak{N}$. We need to show that $G \in \mathfrak{N}$. Granting that, it will be clear that $\mathfrak{N}$ is a variety defined by the system of identities V.

We will consider the set X whose cardinality is equal to that of G. Let $F(X)$ be the free group with the free set of generators X, and let $\mathfrak{C}$ be any subclass of $\mathfrak{N}$ which yields the same set of identities V as $\mathfrak{N}$ does.

Let us now consider the set $\mathfrak{M}$ of all homomorphic images (up to isomorphism) of the group $F(X)$ in groups from $\mathfrak{C}$, and the set Φ of all homomorphisms from $F(X)$ onto groups from $\mathfrak{M}$. Let $G_\varphi = \varphi(F(X))$ for $\varphi \in \Phi$. Let B be the subgroup of the Cartesian product $C = \prod_{\varphi\in\Phi} G_\varphi$ generated by all elements f_x (where $x \in X$) such that $f_x(\varphi) = \varphi(x)$ for all $\varphi \in \Phi$. It will suffice to show that $B \simeq F(X, \mathfrak{V})$, where $\mathfrak{V}$ is the variety defined by the system of identities V. Indeed, in that case a bijection $\psi : X \to G$ will extend to a homomorphism $\bar{\psi} : F(X, \mathfrak{V}) \to G$. This will show that G is a homomorphic image of the subgroup B of the Cartesian products C of groups from $\mathfrak{R}$, i.e. $G \in \mathfrak{R}$, which will complete the proof.

In order to exhibit the required isomorphism we will take any mapping $\chi : X \to D \in \mathfrak{V}$. Then there exists a homomorphism $\bar{\chi} : F(X) \to D$ such that $\bar{\chi}|_X = \chi$. We will now compare the kernel of $\bar{\chi}$ with the kernel of the homomorphism $\bar{\varepsilon} : F(X) \to C$, which extends the mapping $\varepsilon : x \mapsto f_x$ for every $x \in X$. If $\varepsilon(v(x_1, \ldots, x_n)) = 1$, then $v(f_{x_1}, \ldots, f_{x_n}) = 1$ and for every $\varphi \in \Phi$, by definition of the function f_x, we have $v(\varphi(x_1), \ldots, \varphi(x_n)) = 1$ – i.e. $\varphi(v(x_1, \ldots, x_n)) = 1$. Since $\phi \in \Phi$ was arbitrary, we see that $v \equiv 1$ is an identity of groups from $\mathfrak{C}$, and hence holds for groups in $\mathfrak{V}$. Because D is an element of $\mathfrak{V}$, we have $v(\chi(x_1), \ldots, \chi(x_n)) = 1$, or $\bar{\chi}(v(x_1, \ldots, x_n)) = 1$. It follows that $\ker \bar{\chi} \supset \ker \bar{\varepsilon}$. This allows to define a mapping $\tilde{\chi} : B \to D$ by putting $\tilde{\chi}(f_x) = \chi(x)$ for all $x \in X$. B is therefore a free group in the variety $\mathfrak{V}$, and the theorem has been proved.

Using the construction presented above we can, to some extent, describe those algebras which satisfy the same identities as a certain fixed algebra A. The collection of all such algebras is a variety $\mathfrak{V}$, called the *variety generated by the algebra* A (written as $\mathfrak{V} = \operatorname{var} A$). An important consequence of the theorem which was just proved is the fact that if A is a finite algebra, then every finitely generated algebra from $\operatorname{var} A$ is finite. Indeed, let us analyze the above proof in the case $\mathfrak{R} = \mathfrak{V}$, $\mathfrak{C} = \{A\}$. Then every homomorphism $\varphi \in \Phi$ is determined by the image of a subset X in the finite algebra A. It is easy to see that $|F(X, \mathfrak{V})| \le |A|^{|A|^{|X|}}$, where $|M|$ is the cardinality of a set M. However, every finitely generated algebra B from $\mathfrak{V}$ is a homomorphic image of some such algebra $F(X, \mathfrak{V})$, which proves the corollary.

For example, in the variety of groups generated by the symmetric group Sym_3 of order $6 = 3!$, the order of a group generated by two elements is bounded from above by the constant $6^{6^2} = 6^{36}$. This estimate turns out to be very coarse, since quite simple calculations show that there is a far smaller bound, $2^2 \cdot 3^5 = 972$, and that the free group of this variety generated by two elements is isomorphic to the subgroup of Sym_{15} generated by the permutations $(1\ 2)(4\ 5)(7\ 8\ 9)(10\ 11\ 12)$ and $(2\ 3)(7\ 8)(4\ 5\ 6)(13\ 14\ 15)$.

Free algebra of a variety is an algebra whose structure reflects only the most general properties of algebras from that variety. Thus it is not surprising that in terms of the usual structure properties, the free algebra in a variety generated by a given algebra often sharply differs from the generating algebra itself. This difference manifests itself in a particularly clear way in the case of linear algebras over an infinite field. For example, if $A = \mathbb{R}^3$ is the Lie algebra of vectors (see Sect. 1.2), then A is simple and finite-dimensional; in particular, $A^2 = [A, A] = A$. On the other hand, if $\mathfrak{V} = \operatorname{var} A$, then the $\mathfrak{V}$-free Lie algebra $L = F(\{x_1, x_2\}, \mathfrak{V})$ has infinite dimension; explicit description of the commutator $[\,,\,]$ in L is difficult; L is graded by the subspaces L_n of non-associative homogeneous polynomials of fixed degree n relative to $\{x_1, x_2\}$, and $L^n \subset L_n \oplus L_{n+1} \oplus \ldots$, i.e. $L \supset L^2 \supset L^3 \supset \ldots$; moreover,

$$\bigcap_{n=1}^{\infty} L^n = \{0\} \tag{11}$$

(generalized nilpotency). We will note that dimensions of the subspaces L_n can be calculated: $\dim L_1 = 2$; if $n = 2m+1 > 1$, then $\dim L_n = m(m+1)$, while if $n = 2m$ then $\dim L_n = \frac{1}{2}m(m+1)$ (see Drensky [1984]).

Property (11) clearly implies that $\operatorname{var} A$ is also generated by its nilpotent algebras: $\operatorname{var} A = \operatorname{var}(\{L/L^n \mid n = 1, 2, \ldots\})$. In connection with this we will introduce the notion of *approximation.* Let G be an algebra, and let $\mathcal{D}$ be a class of algebras of the same signature as G. We say that G is *approximated by algebras from* $\mathcal{D}$ if, for every distinct elements $x, y \in G$, there exists an epimorphism $\varphi : G \to D \in \mathcal{D}$ such that $\varphi(x) \neq \varphi(y)$.

Birkhoff's theorem has another important corollary: if $\mathfrak{V} = \operatorname{var}(\mathcal{D})$, then $\mathfrak{V}$-free algebras are approximated by the set of algebras from $\mathcal{D}$ and their subalgebras. It is clear that the converse holds as well: approximability of a $\mathfrak{V}$-free algebra of countable rank by algebras from a class $\mathcal{D}$ implies that the variety $\mathfrak{V}$ is generated by its algebras belonging to $\mathcal{D}$. For example, the variety of all groups is generated by finite p-groups, by nilpotent torsion-free groups etc.

2.4. Additional Remarks. It is clear that describing systems of identities, modulo their equivalence, for a given signature Ω, is the same as describing varieties of algebras with that signature. If we are only interested in systems of identities containing a certain fixed subsystem (e.g. for the group signature – the system of group axioms), then we are studying varieties contained in a certain fixed variety (in our example, in the variety of all groups). There is one more aspect of this situation. Let $F_\Omega(X)$ be a free algebra of signature Ω on a countable set of free generators $X = \{x_1, x_2, \ldots\}$. Then varieties of signature Ω and verbal congruences of the algebra $F_\Omega(X)$ are in an inclusion-reversing one-to-one correspondence. If $\mathfrak{V}$ is a fixed variety, then there is also a similar correspondence between verbal congruences in the $\mathfrak{V}$-free algebra $F(X, \mathfrak{V})$ and subvarieties of the variety $\mathfrak{V}$.

In particular, there is an inclusion-reversing 1-1 correspondence between varieties of groups and the verbal subgroups of the free group of countable rank. In view of the fact that every variety of abelian groups is determined by identities in one variable, we have a 1-1 relationship between varieties of abelian groups and verbal subgroups of the group of integers. The last theme is this: there is a bijection between the set of positive integers and the collection of varieties of abelian groups, in which inclusion between varieties corresponds to divisibility of integers. This illustrates how the apparently formal relationship between varieties and congruences, normal subgroups and ideals presents great opportunities for investigation of identities. The point here is, first and foremost, that this connection allows to reduce questions about classes of algebras, or about systems of distinguished identities, to questions about the internal structure of a single, naturally arising algebra. Within this framework, the process of obtaining all consequences of an identity $u \equiv v$ becomes a formal procedure of passing to the verbal closure, i.e. of constructing the minimal verbal congruence ρ in the algebra $F(X, \Omega)$, for which $u \rho v$. For example, in the case of linear algebras, an identity $u(x_1, \ldots, x_n) \equiv 0$ is a consequence of an identity $v(x_1, \ldots, x_n) \equiv 0$ if and only if the element $u(x_1, \ldots, x_n)$ lies in the verbal ideal I generated by the element $v(x_1, \ldots, x_n)$. At the same time, the ideal T generated (as an ideal) by all elements of the form $v(f_1, \ldots, f_n)$ (where $f_i \in F(X, \mathfrak{V})$) has the property that $F(X, \mathfrak{V})/T$ belongs to the variety generated by the identity $v \equiv 0$. Since T is the smallest such ideal, we see that $I = T$. This clearly implies that a *homogeneous identity* of degree n in a linear algebra cannot yield, as its consequence, a homogeneous identity of degree $m < n$ (see definition in Sect. 3.1).

We will conclude this section by describing two important parameters of a system of identities – namely, its basis rank and axiomatic rank. In this fragment we will use the language of invariants of varieties. And so let $\mathfrak{V}$ be a variety. We will assume that $\mathfrak{V}$ can be defined by means of a system of identities in a bounded number of variables. Let r be the smallest possible number of variables appearing in such a system. We say that $r = r_a(\mathfrak{V})$ is the *axiomatic rank* of the variety $\mathfrak{V}$. If $\mathfrak{V}$ cannot be described by a system of identities in finitely many variables, then we say that the axiomatic rank of $\mathfrak{V}$ is infinite, written as $r_a(\mathfrak{V}) = \infty$. Definition of this parameter requires one clarification. As a rule, we restrict ourselves to considerations within bounds of a certain variety $\mathfrak{M}$ of a given signature (for example, the variety of all groups). The axiomatic rank of a variety $\mathfrak{V}$ is then understood to be the smallest number of variables in defining identities of $\mathfrak{V}$ modulo those that determine $\mathfrak{M}$. It is obvious that a variety of infinite axiomatic rank does not have a finite basis of identities; the converse, however, is not always true. For example, for any infinite system of group identities in variables $x_1, \ldots, x_n, \ldots$, by replacing every variable x_n with the commutator $[y, x^n] = y^{-1}x^{-n}yx^n$ we arrive at a system of identities whose axiomatic rank is 2. The property of 'having no finite basis' is preserved in that process. One of the examples of a variety with infinite axiomatic rank is the variety of Lie algebras generated by the Lie algebra $\mathrm{sl}\,(2, k)$ over an infinite field k of characteristic 2.

The other important parameter is the basis rank of a variety $\mathfrak{V}$. If $\mathfrak{V}$ can be generated by an algebra with a finite number of generators, and if r is the minimal such number, then we say that $r = r_b(\mathfrak{V})$ is the *basis rank* of the variety $\mathfrak{V}$. If $\mathfrak{V}$ cannot be defined by identities of a finitely generated algebra, then we say that the basis rank of $\mathfrak{V}$ is infinite, and write $r_b(\mathfrak{V}) = \infty$ (see Sect. 3.1 in Chapter 2 for more details).

In connection with the question how a variety is generated by its algebras, an important rôle is played by the concept of a *critical algebra.* If A is an algebra, B – its subalgebra and $\rho \subset B \times B$ – a congruence in B, then the factor algebra B/ρ is called a *subfactor of the algebra* A. The subfactor $A/=$ is called *improper*, while all other subfactors are called *proper.* In particular, if K is a normal subgroup of a subgroup $H \subseteq G$, the group H/K is a subfactor of G.

Definition. An algebra A is called *critical* if it does not belong to the variety generated by its proper subfactors.

Let σ be the intersection of all non-trivial congruences ρ_α, $\alpha \in I$, of the algebra A. If σ is the trivial congruence, i.e. $\bigcap_{\alpha \in I} \rho_\alpha = \Delta$, then there exists an injection $A \hookrightarrow \prod_{a \in I} A/\rho_\alpha$ (defined by assigning to an element $a \in A$ the function f_a such that $f_a(\alpha) = [a]_{\rho_\alpha}$). In this case, $A \in \operatorname{var}(\{A/\rho_\alpha \,|\, \alpha \in I\})$, i.e. A is not a critical algebra. It follows that a critical algebra A must possess the minimal non-trivial congruence. Such algebra is called *subdirectly irreducible* or *monolithic.* For example, the cyclic group G of order n is monolithic if and only if $n = p^s$, where p is a prime integer and $s = 1, 2, \ldots$. Such a group G also turns out to be critical: the reason for that is that in each of its proper subfactors satisfies the identity $x^{p^{s-1}}$, which does not hold in G. However, it would be a mistake to conclude that every monolithic algebra is necessarily critical. Critical algebras play a particularly important part in the investigation of *locally finite varieties*, i.e. those in which finitely generated algebras are finite. The reason for this can be expressed in the following observation: every locally finite variety is generated by its critical algebras; the property of being locally finite implies that the variety is generated by a collection of its finite algebras, for example the free algebras of ranks $1, 2, 3, \ldots$ (every variety is generated by such a set); next, using induction on cardinality of a finite algebra, one can pass to its critical subfactors. Among locally finite varieties we have the varieties generated by a single finite algebra. When algebras in question are linear algebras over a field, then local finiteness can be replaced with the property of being locally finite-dimensional – or, in other contexts, different conditions similar in effect. The problem of describing critical algebras in natural categories appears to be very difficult. As an example, we will show that finite simple groups, associative rings, Lie algebras etc. are all critical algebras (Bakhturin [1985], H. Neumann [1967]).

§ 3. Identities in Linear Algebras

3.1. Homogeneous and Multilinear Identities. If k is a field and X – a non-empty set, then $k\langle X \rangle$ will denote the set of all non-associative polynomials in X with coefficients from k. The algebra $k\langle X \rangle$ is a linear algebra over the field k, with a basis consisting of all possible monomials in X, and with a distributive operation of multiplication under which product of two monomials is defined as their juxtaposition, with parentheses placed as needed. For example, $x_1 \cdot x_2 = x_1x_2$, $x_1 \cdot x_2x_3 = x_1(x_2x_3)$, $(x_1x_2)x_1 \cdot (x_2x_1)(x_3x_4) = ((x_1x_2)x_1)((x_2x_1)(x_3x_4))$, and so on. The algebra $k\langle X \rangle$ is a free algebra in the variety of algebras of signature $\Omega = \{+, -, 0, \lambda(\lambda \in k), \cdot\}$ determined by the system of axioms (I–III) (see beginning of Sect. 1), in which $\lambda \in \mathbb{R}$ is replaced with $\lambda \in k$ and $\cdot$ takes place of $[\,,\,]$. If A is a linear algebra over k, i.e. an algebra with signature Ω satisfying the system of axioms described above, then every mapping $\varphi : X \to A$ extends to a homomorphism $\bar{\varphi} : k\langle X \rangle \to A$, by letting

$$\bar{\varphi}(f(x_1, \ldots, x_n)) = f(\varphi(x_1), \ldots, \varphi(x_n)) .$$

The right-hand side is well-defined, since the axioms (I–III) allow to compute in A the value of any non-associative polynomial. As was already mentioned before, every identity of a linear algebra can be written in the form $f(x_1, \ldots, x_n) \equiv 0$, where $f(x_1, \ldots, x_n)$ is a non-zero non-associative polynomial.

A polynomial $f(x_1, \ldots, x_n)$ will be called *normal*, if it is a linear combination of monomials which all involve every variable from a certain set, fixed for the given polynomial. Using substitution of zeroes for some of the variables, it is easy to see that every system of identities is equivalent to a *normal* one, i.e. to a system of identities of the form $f(x_1, \ldots, x_n) \equiv 0$, where f is a normal polynomial.

Another important type of identities are the homogeneous ones.

Definition. An identity $f(x_1, \ldots, x_n) \equiv 0$ is *homogeneous* if $f(x_1, \ldots, x_n)$ is a homogeneous polynomial, i.e. a linear combination of monomials whose degrees are all equal.

In a number of important cases it is possible to reduce the study of identities to the investigation of homogeneous ones.

A third type, multi-homogeneous identities, is a special case of the two mentioned above. An m-tuple $\alpha = (n_1, \ldots, n_m)$ is the multidegree of a monomial $w(x_1, \ldots, x_m)$ if, for $i = 1, \ldots, m$, the integer n_i is the number of factors x_i in w.

Definition. A polynomial $f(x_1, \ldots, x_m)$ is *multi-homogeneous* if it is a linear combination of monomials with identical multidegrees.

Example 1. The polynomial $f(x_1, x_2, x_3) = x_1x_2 - x_1x_3$ is homogeneous, but not normal.

Example 2. $f(x_1, x_2, x_3) = (x_1x_2)x_3 - x_3(x_1x_2) + 2x_3(x_1^2x_2^2)$ is normal, but not homogeneous.

Example 3. $f(x_1, x_2, x_3) = (x_1x_3)x_2^2 + 3(x_2x_3)(x_2x_1)$ is multi-homogeneous with multidegree $(1, 2, 1)$.

Definition. A polynomial $f(x_1, \ldots, x_m)$ is *multilinear* if it is multihomogeneous of multidegree $\underbrace{(1, \ldots, 1)}_{m}$.

Accordingly, one may talk about multilinear identities. We have the following theorem (already mentioned in Sect. 1.3):

Theorem. *Every non-trivial identity has a non-trivial multilinear consequence.*

A central technique used in the proof of this fact is the well-known procedure of *linearization.* We first select a variable x which appears in the left member of the given identity $f(x_1, \ldots, x_n) \equiv 0$. We will now write the left-hand side as $f(x_1, \ldots, x_n) = f_0 + f_1 + \ldots + f_t$, where f_0 does not depend on x, f_1 is linear in x, etc, and f_t is homogeneous with degree t in x. If $f_0 \neq 0$ then substituting $x = 0$ we find that $f_0 \equiv 0$ is a non-trivial consequence of the original identity. Since f_0 involves fewer variables than f does, we may apply induction on n. Next, if $f_0 = 0$ and $t = 1$, then f is already linear in x and we may proceed to consider other variables. If, on the other hand, $t > 1$, then fixing all variables except x and writing f as a polynomial depending only on x, we will consider the polynomial $g(y, z) = f(y + z) - f(y) - f(z)$, where y and z are some new indeterminates. It is easily seen that the total degree of the polynomial g in the variables y and z equals the degree of f in x, but neither its degree in y nor in z exceeds $t - 1$ (g is constructed in such a way that its monomials of degree t in y and in z cancel out). Further, let $a_1xa_2x \ldots a_txa_{t+1}$ and $b_1xb_2x \ldots b_txb_{t+1}$ be two distinct monomials of $f(x_1, \ldots, x_n)$ with degree t in x (naturally, with some placement of parentheses). Then these monomials, and only these, yield two distinct monomials $a_1za_2y \ldots a_tya_{t+1}$ and $b_1zb_2y \ldots b_tyb_{t+1}$, which implies that $g(y, z) \equiv 0$ is a non-trivial consequence of the original identity. It is clear that repeated application of this process reduces the number of those variables which appear in the polynomial with maximal degree, while at the same time preserving its total degree; this necessarily leads to a multilinear polynomial in a finite number of steps. The theorem has been proved.

We will remark that the consequence of the original identity which was constructed above is non-trivial in the class of all linear algebras. When deriving a multilinear identity from a non-trivial identity within some narrower class of algebras, additional precautions need to be taken in order to obtain a non-trivial result, but the process is essentially the same. For example, linearizing the identity $x^3 \equiv 0$ in the class of associative algebras, we follow the sequence

$$x^3 \xrightarrow{\text{rel. to } x} zyy + yzy + yyz + yzz + zyz + zzy$$

$$\xrightarrow{\text{rel. to } y} zuv + zvu + uzv + vzu + uvz + vuz\,.$$

Re-labeling the variables we see that the identity $\sum_{\sigma\in\mathrm{Sym}_3} x_{\sigma(1)}x_{\sigma(2)}x_{\sigma(3)} \equiv 0$ is a consequence of $x^3 \equiv 0$. Similarly, $x^n \equiv 0$ yields the identity $\sum_{\sigma\in\mathrm{Sym}_n} x_{\sigma(1)}\dots x_{\sigma(n)} \equiv 0$. Conversely, substituting equal values for all indeterminates of this identity, we obtain the identity $n!x^n \equiv 0$. It now becomes clear that in case of a base field whose characteristic does not divide $n!$, the identity $x^n \equiv 0$ and its multilinear consequence $\sum_{\sigma\in\mathrm{Sym}_n} x_{\sigma(1)}\dots x_{\sigma(n)} \equiv 0$ are equivalent. Reasoning along these lines it is possible to obtain an important theorem: every system of identities over a field of characteristic 0 is equivalent to a multilinear system.

For an arbitrary infinite field a weaker version of this theorem holds: every system of identities is equivalent to a multi-homogeneous one. Indeed, let us represent $f(x_1,\dots,x_n)$ as before: $f = f_0 + f_1 + \dots + f_t$, where f_i is homogeneous of degree i with respect to a variable x; exploiting the assumption that the base field is infinite, we will make the substitutions $x \mapsto \lambda_0 x, x \mapsto \lambda_1 x, \dots, x \mapsto \lambda_t x$ with $\lambda_0, \lambda_1, \dots, \lambda_t$ being distinct elements of the base field. We will thus obtain the set of consequences

$$\begin{cases} f_0 + \lambda_0 f_1 + \lambda_0^2 f_2 + \dots + \lambda_0^t f_t \equiv 0\,, \\ f_0 + \lambda_1 f_1 + \lambda_1^2 f_2 + \dots + \lambda_1^t f_t \equiv 0\,, \\ \dots \\ f_0 + \lambda_t f_1 + \lambda_t^2 f_2 + \dots + \lambda_t^t f_t \equiv 0\,. \end{cases}$$

If $f_0, f_1, \dots, f_t$ are regarded as unknowns, the system obtained above has a unique solution $f_0 \equiv 0, f_1 \equiv 0, \dots, f_t \equiv 0$, because its Vandermonde determinant is non-zero. This means that $f \equiv 0$ may be replaced with the system of identities $\{f_0 \equiv 0, f_1 \equiv 0, \dots, f_t \equiv 0\}$, which is homogeneneous with respect to x. The proof is completed by induction on the number of 'non-homogeneous' indeterminates.

3.2. Gradations of Free Algebras. We will now describe several consequences of the theorem, proved above, dealing with the structure of a free algebra in a variety over an infinite field. We will first note that if F_n is the space spanned by all monomials of degree n from $k\langle X \rangle$, and F_α – the span of all monomials of multidegree α, then the following direct sum decompositions are obvious:

$$k\langle X \rangle = \bigoplus_{n=1}^{\infty} F_n\,, \quad k\langle X \rangle = \bigoplus_{\alpha} F_\alpha\,.$$

We also have $F_n F_m \subseteq F_{n+m}$ and $F_\alpha F_\beta \subseteq F_{\alpha+\beta}$. Let V be the verbal ideal corresponding to a variety $\mathfrak{V}$. In the case of an infinite base field, the results cited in Sect. 3.1 allow us to conclude that $V = \bigoplus_\alpha V_\alpha$ and $V = \bigoplus_{n=1}^{\infty} V_n$, where $V_\alpha = V \cap F_\alpha$ and $V_n = V \cap F_n$. The factor algebra $T = k\langle X \rangle / V = F(X, \mathfrak{V})$ can therefore be written as $T = \bigoplus_{n=1}^{\infty} T_n$ or $T = \bigoplus_\alpha T_\alpha$, with $T_n T_m \subseteq T_{n+m}$ and $T_\alpha T_\beta \subseteq T_{\alpha+\beta}$; moreover, $T_n \simeq F_n/V_n$ and $T_\alpha \simeq F_\alpha/V_\alpha$ as vector spaces (in this situation one says that T is an algebra with gradation, or a *graded algebra*). An important feature of the gradation $T = \bigoplus_\alpha T_\alpha$ is the fact that all subspaces T_α are finite-dimensional over the base field. Free algebras in varieties defined by systems of multilinear identities are equipped with similar gradations. In particular, the free associative algebra, the free Lie algebra and many others are graded by the degree and multidegree relative to any set of free generators.

The above-mentioned property of free algebras in multilinear varieties (i.e. those for which $V = \bigoplus_\alpha V_\alpha$), and of free algebras in varieties over infinite fields in particular, allows to draw a number of conclusions about the structure of such algebras as well as about the identities of such varieties. Let $A = F(X, \mathfrak{V})$ be a free algebra in such a variety $\mathfrak{V}$. Then:

1) the chain of powers $A \supseteq A^2 \supseteq A^3 \supseteq \ldots$ intersects in the zero algebra;

2) the algebra A is approximated by finite-dimensional nilpotent algebras, i.e. for every non-zero $x \in A$ there exists an ideal N_x such that $x \notin N_x$ and A/N_x is a finite-dimensional nilpotent algebra;

3) if the set X is finite, then A is a *Hopfian algebra*, i.e. it cannot be mapped onto itself by a homomorphism with non-trivial kernel (this should not be confused with a *Hopf algebra* which, by the way, does not appear anywhere in this volume);

4) let Y be a set of elements of A, linearly independent modulo the square of A. Then Y is a free generating set (in the context of the variety $\mathfrak{V}$) of the subalgebra generated by it;

5) every multilinear variety can be defined by an *independent set of identities* (in which no identity is a consequence of the others).

3.3. Action of the Group $\mathrm{GL}(n,k)$**.** In the study of identities of linear algebras, what seems to be the deepest approach is based on understanding of the action of classical groups $\mathrm{GL}(n,k)$ and Sym_n on the homogeneous summands of free algebras. This gives a chance to apply the well-developed representation theory of these groups (in what follows, it will be assumed that k is a field of characteristic zero). We begin by defining actions of the groups in question. Let $X = \{x_1, \ldots, x_n\}$ be a finite set, and let $T = F(X, \mathfrak{V})$ be the free algebra in the variety $\mathfrak{V}$ over k, with the set of free generators X. If $\mathcal{A}$ is a linear operator on the vector space $\langle X \rangle$ with basis X, and T_n – the n-th homogeneous component of the algebra T, then we define an action of the operator $\mathcal{A}$ on an element $f(x_1, \ldots, x_n) \in T_n$ by setting

$$\mathcal{A} \circ f(x_1, \ldots, x_n) = f(\mathcal{A}(x_1), \ldots, \mathcal{A}(x_n)) .$$

It is clear that $\mathcal{A}f \equiv 0$ is a consequence of the identity $f \equiv 0$. If $\mathcal{A} \in \mathrm{GL}(n,k)$, then the identities $\mathcal{A}f \equiv 0$ and $f \equiv 0$ are equivalent. We will suppose that a homogeneous identity $f \equiv 0$ of degree m in variables $x_1, \dots, x_n$ is a consequence of a homogeneous polynomial $g \equiv 0$ of degree m. Then f belongs to the $\mathrm{GL}(n,k)$-submodule of T_m generated by the polynomial g. Indeed, f lies in the span of evaluations of the polynomial g on linear combinations of the elements $x_1, \dots, x_n$, i.e. in the span of elements of the form $g(B(x_1), \dots, \mathcal{B}(x_n))$ – where $\mathcal{B}$ is an endomorphism (a linear operator) of the vector space $\langle X \rangle$. Every such endomorphism is a composition of automorphisms, or elements of $\mathrm{GL}(n,k)$, with endomorphisms which leave all variables x_i fixed, except x_1 – which is mapped to 0. It is therefore enough to verify that $g(0, x_2, \dots, x_n)$ is in the $\mathrm{GL}(n,k)$-submodule generated by $g(x_1, \dots, x_n)$. In order to do so, we will write $g = g_0 + \dots + g_m$, where g_i is a linear combination of monomials of degree i in x_1 $(i = 0, 1, \dots, m)$. Choosing $m+1$ distinct elements $\lambda_0, \dots, \lambda_m$ and applying the automorphisms $\mathcal{A}_i : x_1 \mapsto \lambda_i x_1, x_2 \mapsto x_2, \dots, x_n \mapsto x_n$, we see that the $\mathrm{GL}(n,k)$-submodule being considered contains all elements of the form $g_0 + \lambda_i g_1 + \dots + \lambda_i^m g_m$, where $i = 0, 1, \dots, m$. As before, this implies that $g_0, \dots, g_m$ themselves are in that submodule. Since $g_0 = g(0, x_2, \dots, x_n)$, our assertion has been proved. It follows that pairwise non-equivalent systems of identities of degree m in variables $x_1, \dots, x_n$ are in a one-to-one correspondence with $\mathrm{GL}(n,k)$-submodules of T_m. This provides a complete description of homogeneous systems of identities of any degree.

Example 1. We will determine subvarieties of the variety $\mathfrak{V}$ of associative algebras over a field k of characteristic 0, satisfying the identity $x_1x_2x_3 \equiv 0$. This is equivalent to the investigation of systems of identities of degree 2 in two variables x_1 and x_2. Let $T = (\{x_1, x_2\}, \mathfrak{V})$. We have $T = T_1 \oplus T_2$, where $T_1 = \langle x_1, x_2 \rangle$ and $T_2 \simeq T_1 \otimes T_1$. This last isomorphism is a G-module homomorphism, with $G = \mathrm{GL}(2,k)$. Well-known decomposition formulas for the tensor square of the canonical two-dimensional representation of $G = \mathrm{GL}(2,k)$ imply that T_2 is a direct sum of a one-dimensional irreducible module with a three-dimensional one: namely, the symmetric S^2T_1 and the exterior $\Lambda^2 T_1$ squares of the space T_1. The generator $x_1 \otimes x_2 + x_2 \otimes x_1$ of the symmetric square S^2T_1 maps to $x_1x_2 + x_2x_1$ in T_2, while the exterior square yields the polynomial $x_1x_2 - x_2x_1$. Since identities of degree 0 or 1 define only trivial subvarieties, we arrive at the conclusion that $\mathfrak{V}$ has five subvarieties: 1) $\mathfrak{V}$ itself; 2) the null variety (determined by the identity $x_1 \equiv 0$); 3) subvariety of commutative algebras (corresponding to $x_1x_2 - x_2x_1 \equiv 0$); 4) subvariety of anticommutative algebras (corresponding to $x_1x_2 + x_2x_1 \equiv 0$); and 5) the subvariety of algebras with trivial multiplication (defined by $x_1x_2 \equiv 0$).

In the special case of associative algebras, as in the above example, the m-th homogeneous component A_m of the free associative algebra $A(x_1, \dots, x_n)$ is a tensor product $A_m \simeq \underbrace{A_1 \otimes \dots \otimes A_1}_{m}$. This isomorphism is an isomorphism of $\mathrm{GL}(n,k)$-modules. The classical theory of tensor representations of the general linear group says that a description of GL-submodules can be obtained by means of representation theory of the symmetric group Sym_m, acting on $\underbrace{A_1 \otimes \ldots \otimes A_1}_{m}$ on the right via permutations of factors. For such description it is necessary to introduce so-called *Young diagrams.*

Let $m = m_1 + \ldots + m_t$ be a partition of the integer m such that $m_1 \geq m_2 \geq \ldots \geq m_t > 0$. We will associate with it a diagram with t rows of blocks, whose row number i consists of m_i blocks. For example, the partition $3 = 2 + 1$ gives rise to the diagram

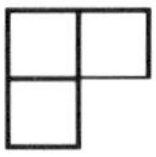

Filling such a diagram with integers $1, 2, \ldots, m$ in an arbitrary way we obtain a *Young tableau d*. With each tableau we associate two subgroups of the group Sym_m: C_d and R_d. Elements of C_d permute the entries in each column of d, while elements of R_d permute the entries in each row. Let

$$e_d = \sum_{\tau \in R_d, \sigma \in C_d} \varepsilon_\sigma \tau \sigma$$

be an element of the group algebra of Sym_m, where ε_σ is the parity of the permutation σ constructed from the tableau d. Then every irreducible $\mathrm{GL}(n, k)$-submodule of A_m has, up to isomorphism, the form $A_m e_d$ for a suitable tableau d; the whole module A_m is a direct sum of such submodules; $A_m e_d$ is the zero submodule if and only if the number t of rows of d is greater than m; and finally, $A_m e_d \simeq A_m e_{d'}$ if and only if d and d' are obtained from the same diagram. Dimension of the $\mathrm{GL}(n, k)$-module $V_{n,d} = A_m e_d$ is computed according to the formula

$$\dim V_{n,d} = \prod_{n \geq i > j \geq 1} \frac{m_j - m_i + i - j}{i - j}.$$

The approach described here applies not only to classification of identities of low degrees in associative algebras, but also to the case of other classes of linear algebras. It is merely necessary to keep in mind that the homogeneous component of degree m of a free algebra with generators $x_1, \ldots, x_n$ can be regarded as a factor module of the $\mathrm{GL}(n, k)$-module $\underbrace{A_m \oplus \ldots \oplus A_m}_{s}$, where s is the number of all possible placements of parentheses which cannot be transformed one into another by means of any identity holding in the variety of linear algebras under consideration. For example, in the case of Lie algebras we have $s = 1$, since using the Jacobi identity together with anticommutativity every monomial can be expressed as a linear combination of "left-normalized" monomials (i.e. those which have the recursively defined form x_i, or $[m, x_i]$ for some left-normalized monomial m). One of the areas in which these techniques are applied is the study of the Hilbert-Poincaré series (see Chapter 3, Sect. 5.1), which is an important numerical characteristic of a variety.

3.4. Action of the Group Sym_n**.** If one is only interested in identities themselves, then it is often convenient to restrict oneself to the consideration of multilinear identities and action of the symmetric group on them. Having fixed a set of variables $x_1, \ldots, x_n$ and the set $P_n(\mathfrak{V})$ of all multilinear polynomials in those variables in the free algebra generated by $x_1, \ldots, x_n$ in a variety $\mathfrak{V}$, we will define an action of the group Sym_n on $P_n(\mathfrak{V})$ in the following way: let $\sigma \in \mathrm{Sym}_n$ and $f(x_1, \ldots, x_n) \in P_n(\mathfrak{V})$; then

$$\sigma(f(x_1, \ldots, x_n)) = f(x_{\sigma(1)}, \ldots, x_{\sigma(n)}) \, .$$

If d is a Young tableau and $R = k[\mathrm{Sym}_n]$ – the group algebra of the symmetric group, then every irreducible submodule of $P_n(\mathfrak{V})$ is isomorphic to a module of the form Re_d, where e_d is the element defined in the previous section. Dimension of the module Re_d can be determined by means of the so-called 'hook rule':

$$\dim Re_d = \frac{n!}{\prod_{i,j} h_{i,j}} \, ,$$

in which $h_{i,j}$ is the *length* of the hook with corner at the (i,j)-th block of the diagram d (dimension does not depend on the placement of entries in the diagram). For example, if

$d = \begin{array}{|c|c|c|}\hline 1&2&3\\ \hline 4&5 \\ \cline{1-2}\end{array}$, then $h_{1,1}$ is the length of the hook $\begin{array}{|c|c|c|}\hline 1&2&3\\ \hline 4 \\ \cline{1-1}\end{array}$, i.e. $h_{1,1} = 4$;

$h_{2,1}$ is the length of $\begin{array}{|c|c|c|}\hline 4&2&3\\ \hline\end{array}$, i.e. 2; $h_{1,2}$ is the length of $\begin{array}{|c|c|}\hline 2&3\\ \hline 5 \\ \cline{1-1}\end{array}$, i.e. 3;

$h_{1,3}$ refers to $\begin{array}{|c|}\hline 3\\ \hline\end{array}$ and equals 1; finally, $h_{2,2}$ is the length of $\begin{array}{|c|}\hline 3\\ \hline\end{array}$, i.e. 1. It follows that $\dim Re_d = 5! \cdot (4 \cdot 2 \cdot 3 \cdot 1 \cdot 1)^{-1} = 5$. Moreover, $P_n(\mathfrak{V})$ is a direct sum of submodules isomorphic to the modules Re_d, and the set of non-equivalent systems of multilinear identities of degree n is in a one-to-one correspondence with the set of Sym_n-submodules. In the case when $P_n(\mathfrak{V})$ is generated, as a Sym_n-module, by some collection of monomials $v_1, \ldots, v_p$, for every system of identities we can write out "canonical" identities whose set is equivalent to the given system. In order to do this it is enough to consider the polynomials $e_d v_i$, where d varies over the set of Young tableaux and $i = 1, 2, \ldots, p$. Then either $e_d v_i = 0$, or $Re_d v_i$ is an irreducible submodule of $P_n(\mathfrak{V})$. Since every non-zero submodule of $P_n(\mathfrak{V})$ is a direct sum of irreducible ones, every system of multilinear identities is equivalent to a system of the form $\sum \lambda_{d,i} e_d v_i \equiv 0$. The structure of identities is particularly clear when $P_n(\mathfrak{V})$ does not have irreducible submodules occurring with multiplicities greater than 1. One can then fix a finite set of canonical identities $e_d v_i \equiv 0$, and every system of identities will be equivalent to a certain subset of that set. For example, this situation arises for identities of Lie algebras of degrees 1, 2, 3, 4 and 5. In the presence of multiple irreducible modules, the set of systems of identities has a significantly more complex nature. Subsystems corresponding to a given Young diagram of multiplicity m are in a $1-1$ correspondence with subspaces of an m-dimensional vector space over the base field k.

In the study of associative, Lie and many other classes of algebras the Sym_n-module $P_n(\mathfrak{V})$ turns out to be cyclic. For example, if $\mathfrak{V}$ is the variety of all associative algebras, then $P_n(\mathfrak{V})$ is simply isomorphic to the regular Sym_n-module $R = k[\mathrm{Sym}_n]$; in particular, the multiplicity with which Re_d occurs in $P_n(\mathfrak{V})$ is equal to its dimension. When $\mathfrak{V}$ is the variety of all Lie algebras, $P_n(\mathfrak{V})$ is again a cyclic, but no longer free, Sym_n-module. There is a formula which allows to determine the multiplicities of the canonical submodules Re_d. When $n \neq 4, 6$, the modules Re_d have non-zero multiplicity for all diagrams d – with the exception of those consisting of a single row or a single column.

As an example of application of the technique which was just presented, we will describe subvarieties in the variety $\mathfrak{V}$ of Lie algebras over a field k of characteristic zero, generated by the Lie algebra L of all matrices $\begin{pmatrix} a & b \\ 0 & 0 \end{pmatrix}$, $a, b \in k$. The commutator of two such matrices has the form $\begin{pmatrix} 0 & c \\ 0 & 0 \end{pmatrix}$, and a commutator of two commutators equals zero. This means that L satisfies the identity $(x_1x_2)(x_3x_4) \equiv 0$. Applying the Jacobi identity and anticommutativity (see formulas IV.1 and IV.2 in Sect. 1.2), we obtain

$$((x_1x_2)x_3)x_4 \equiv ((x_1x_2)x_4)x_3 \,. \tag{12}$$

Identity (12), together with anticommutativity and the Jacobi identity, allow to reduce every polynomial from $P_n(\mathfrak{V})$ to a linear combination of left-normalized monomials of the form

$$x_ix_1x_{i_1}\dots x_{i_{n-2}}\,, \text{ where } i \neq 1 \text{ and } i_1 \leq \dots \leq i_{n-2}\,.$$

This means that $\dim P_n(\mathfrak{V}) \leq n-1$. We will consider the following Young tableau d:

2	3	...	n
1			

The element $e_d(x_1, \dots, x_n)$ has the form

$$\sum_{\sigma \in \mathrm{Sym}\{2,\dots,n\}} \left(x_1x_{\sigma(2)}\dots x_{\sigma(n)} - x_{\sigma(2)}x_1x_{\sigma(3)}\dots x_{\sigma(n)}\right) =$$

$$= 2 \sum_{\sigma \in \mathrm{Sym}\{2,\dots,n\}} x_1x_{\sigma(2)}\dots x_{\sigma(n)} \,.$$

As a consequence of the identity $e_d(x_1, \ldots, x_n) \equiv 0$ with $x_1 = x, x_2 = \ldots = x_n = y$, we obtain $xy^n \equiv 0$. This identity does not hold in L – for example when $x = \begin{pmatrix} 0 & 1 \\ 0 & 0 \end{pmatrix}$ and $y = \begin{pmatrix} 1 & 0 \\ 0 & 0 \end{pmatrix}$. It follows that $e_d(x_1, \ldots, x_n)$ is a non-zero element of $P_n(\mathfrak{V})$, and hence $Re_d(x_1, \ldots, x_n)$ is a non-zero irreducible Sym_n-submodule of $P_n(\mathfrak{V})$. According to the hook law, its dimension equals $n-1$. This means that $P_n(\mathfrak{V})$ itself is an irreducible Sym_n-module; imposing any additional identity of degree n would therefore make all polynomials of degree n become identically zero, and so the same is true for all polynomials of degrees greater than n. We thus see that the collection of subvarieties of $\mathfrak{V}$ is a chain, and that every proper subvariety is defined by an identity

$$x_1 x_2 \ldots x_n \equiv 0$$

(which implies index of nilpotency not greater than n).

Other important areas of application of Young diagrams – which include distributive lattices of varieties, Capelli identities, bases of identities in concrete algebras, etc. – will be encountered in the following chapters of this text.

§ 4. Free Algebras

We begin this section with an example of a non-classical variety, in order to demonstrate that the internal structure of a free algebra of a variety may be highly non-trivial, and that its investigation may lead to the solution of problems far removed from the theory of identities.

4.1. Automorphisms of Free Cantor Algebras. Let $n \geq 2$ be an integer, and let Ω_n be a signature consisting of unary operations $\alpha_1, \ldots, \alpha_n$ and an n-ary operation λ. $\mathfrak{V}_n$ will denote the variety of algebras with signature Ω_n, defined by the system of identities

$$x\alpha_1 \ldots x\alpha_n \lambda \equiv x\,,$$

$$x_1 \ldots x_i \ldots x_n \lambda \alpha_i \equiv x_i\,, \quad i = 1, \ldots, n\,.$$

Example of an algebra from $\mathfrak{V}_2$ is provided by Cantor's renumbering process. Numbering pairs of natural numbers as follows

$$\begin{array}{lllll} 1 & 2 & 4 & 7 & \ldots \\ 3 & 5 & 8 & \ldots & \\ 6 & 9 & \ldots & & \\ 10 & \ldots & & & \\ \ldots & & & & \end{array} \qquad ,$$

Fig. 1

we will define a binary operation λ on the set of natural numbers $\mathbf{N}$ by assigning to each pair (i, j) the integer appearing at the intersection of the i-th row and j-th column of this table. Further, we will define $x\alpha_1$ to be the number of the row which contains x and, analogously, $x\alpha_2$ will be the number of the column in which x appears.

Let now X be a non-empty set, and $F_n(X)$ – the free algebra of the variety $\mathfrak{V}_n$ with X as its free set of generators. Every element of the algebra $F_n(X)$ has a unique presentation as a so-called standard form, i.e. an element $x\alpha_{i_1} \dots \alpha_{i_k}$ (where $x \in X$, $k \geq 0$ and $1 \leq i_j \leq n$ for $j = 1, 2, \dots, k$), or – inductively – $w_1 \dots w_n\lambda$, where w_i are standard forms such that there is no standard form u satisfying $w_i = u\alpha_i$ for all $i = 1, \dots, n$.

Definition. $F_n(X)$ is called a *free n-Cantor algebra.*

Structure and automorphisms of algebras of this type were studied by Jónsson, Tarski, Świerzkowski and D.M. Smirnov (see articles in: *Solvable and infinite simple groups. Collection of articles.* Series: Matematika. Novoe v zarub. nauke. Vol. 21. Mir: Moscow). One of the characteristic properties of a free Cantor algebra is the fact that, unlike in the classical free algebra case, the rank of such an algebra, i.e. cardinality of a free generating set, is not an invariant. Indeed, for every $x \in X$ the algebra $F_n(X)$ is freely generated by the set $(X \setminus \{x\}) \cup \{x\alpha_i \mid 1 \leq i \leq n\}$ (simple extension); $F_n(X)$ is also freely generated by the set $(X \setminus \{x_1, \dots, x_n\}) \cup \{s_1 \dots x_n\lambda\}$ (simple contraction). Repeatedly applying extensions and contractions we conclude that $F_n(X) \simeq F_n(Y)$ if $|X| \equiv |Y| \bmod (n-1)$. Among other properties of the algebra $F_n(X)$ we have the so-called Schreier property: every subalgebra of $F_n(X)$ is a free algebra in $\mathfrak{V}_n$. Moreover, a free generating set of any subalgebra of $F_n(X)$ extends to a free set of generators of the whole algebra.

In the study of free Cantor algebras the main focus of interest are their automorphisms. We will denote by $G_{n,r}$ the group of automorphisms of the algebra $F_n(X)$, where $|X| = r$. An important characteristic of elements of $G_{n,r}$ is their *depth.* If $\{\theta_1, \dots, \theta_s\} \subseteq G_{n,r}$, then there exists a unique minimal extension Y of the free generating set X such that for $i = 1, \dots, s$, $Y\theta_i$ lies in the subalgebra of $F_n(X)$ generated by the set X together with operations $\alpha_1, \dots, \alpha_n$. If Y is a d-fold extension of X, i.e. $|Y| = r + (n-1)d$, then we say that the set $\{\theta_1, \dots, \theta_s\}$ has depth d. Complicated combinatorial arguments show that the group $G_{n,r}$ is generated by elements of depth not greater than 3, and that it suffices to consider only those relations between them which have the form $\theta_1 \cdot \ldots \cdot \theta_s = 1$, where the sets $\{\theta_1, \dots, \theta_s\}$ have depth not greater than 6 and bounded cardinality s. This directly implies that $G_{n,r}$ can be defined by a finite set of relations (these should not be confused with identity relations, none of which but trivial ones being satisfied in $G_{n,r}$). Furthermore, we will remark that $G_{n,r}$ is an infinite group, since it contains the finite symmetric groups of arbitrarily large orders.

We conclude that $G_{n,r}$, as well as any of its subgroups of finite index, is an infinite finitely presented group. By means of the concept of extension one can define parity of elements of the group $G_{n,r}$, whenever n is an odd integer. We then set $G^+_{n,r} = G_{n,r}$ if n is even, and define $G^+_{n,r}$ to be the set of all even automorphisms, if n is odd. It is not difficult to show that $G^+_{n,r}$ is generated by elements of finite order, and that every non-trivial normal subgroup of $G^+_{n,r}$ contains all elements of finite order, i.e. for any $n \geq 2$ and $r \geq 1$ the group $G^+_{n,r}$ is a finitely presented simple group. Investigation of conjugacy classes of elements in the groups $G_{n,r}$ yields the following result: if groups $G^+_{m,r}$ and $G^+_{n,s}$ are isomorphic, then $m = n$ and $\gcd(n-1, r) = \gcd(n-1, s)$.

In this way the study of automorphisms of free Cantor algebras leads to the proof of existence of infinitely many (pairwise non-isomorphic) finitely presented infinite simple groups (Higman).

In many areas of mathematics an important rôle is played by the free algebra in the variety of all semigroups – the semigroup of all words in some alphabet. Other related classical objects include: free groups, free associative algebras (algebras of non-commutative polynomials), and free Lie algebras. Magnus theory makes it possible to consider these objects from a unified point of view.

4.2. Magnus Theory. Parallelism between the theories of groups, associative rings and Lie algebras was noticed long ago. Sometimes (even frequently) it was a consequence of reasoning 'by analogy', and sometimes it followed from constructions of one kind or another. In the theory of identities of the three classes of algebras mentioned above, the important point is that their free algebras represent, as it were, three sides of a single object. This significant fact is the basis for defining a Magnus algebra.

Let k be a commutative ring with 1 (e.g. a field, or the ring of integers), X – a non-empty set. We will consider the algebra $A(X)$ of non-commutative associative polynomials in X with coefficients from k. Every element u of $A(X)$, as was noted in Sect. 3.3, is a sum of homogeneous polynomials in X:

$$u = u_0 + u_1 + \ldots + u_s\,, \text{ where } \deg u_i = i \text{ for } i = 0, 1, \ldots, s\,.$$

We will now introduce the set $\widehat{A(X)}$ of infinite sums $v = \sum_{n=0}^{\infty} v_n$, where v_n is a homogeneous polynomial of degree n. We define multiplication on this set by the formula

$$\sum_{n=0}^{\infty} u_n \sum_{n=0}^{\infty} v_n = \sum_{n=0}^{\infty} w_n \text{ where } w_n = u_0 v_n + u_1 v_{n-1} + \ldots + u_n v_0\,.$$

It is clear that we obtain an associative algebra, in which $A(X)$ is contained as a subalgebra. The algebra $\widehat{A(X)}$ is called a *Magnus algebra*. We denote by M the set of series from $\widehat{A(X)}$ with no constant term. Then the set $G(X) = 1 + M$ is a group; indeed, the series $1 + u$ $(u \in M)$ has a well-defined inverse $u^{-1} = 1 - u + u^2 - u^3 + \ldots$. The group $G(X)$ is called the

Magnus group. If k is a field of characteristic zero, then it is easy to show that for every $u \in M$ we have two series $\exp(u) = 1 + u + \frac{u^2}{2!} + \frac{u^3}{3!} + \dots$ and $\ln(1+u) = u - \frac{u^2}{2!} + \frac{u^3}{3!} - \dots$, such that the mappings $\exp : M \to G(X)$ and $\ln : G(X) \to M$ are mutually inverse bijections.

Theorem

1) Relative to the ordinary addition and multiplication operations, the set X generates in $\widehat{A(X)}$ a free associative algebra.

2) Relative to addition and the commutator $[a,b] = ab - ba$, the set X generates in $\widehat{A(X)}$ a free Lie algebra; $A(X)$ coincides with its universal enveloping algebra[1].

3) Let Y be the set of elements of the form $1 + x$ with $x \in X$. Then relative to the multiplication operation, Y generates a free subgroup of $G(X)$.

4) If k is a field of characteristic zero, then relative to the Baker-Campbell-Hausdorff operation $u \circ v = \ln(\exp(u)\exp(v))$ the set X freely generates in $\widehat{A(X)}$ the free group $F(X)$.

We will use $\widehat{L(X)}$ to denote the set of series $v = \sum_{n=0}^{\infty} v_n$, whose homogeneous components lie in $L(X)$. It is particularly important that $\widehat{L(X)}$ turns out to be a group under the operation in (4). A precise statement of this fact is known as the Baker-Campbell-Hausdorff formula involving a series of commutators of increasing degrees, whose first terms have the form

$$u \circ v = u + v + \frac{1}{2}[u,v] + \frac{1}{12}[u,v,v] + \frac{1}{12}[v,u,u] + \dots \,. \tag{13}$$

The significance of formula (13) lies in the fact that in every Lie algebra for which the right side is well-defined, it yields a new operation under which the Lie algebra becomes a group. This essentially classical result permits to establish a correspondence between varieties of Lie algebras over the field of rationals, and varieties of groups. Indeed, if $\mathfrak{V}$ is such a variety of Lie algebras and X – a countable set of indeterminates then, as was remarked in Sect. 3.2, the free algebra $F(X, \mathfrak{V})$ is graded by degree relative to the set X. We will complete the algebra F to the algebra of series $\hat{F}$, and introduce on it a commutator operation in a way analogous to the one employed in the construction of $\widehat{A(X)}$. Then the series (13) is defined for any $u, v \in \hat{F}$, and we obtain a group $\hat{F}^0$. Its subgroup generated by the set X turns out to be free in some variety of groups $\mathfrak{V}^0$. This process gives rise to a mapping $\mathfrak{V} \mapsto \mathfrak{V}^0$ from the set of varieties of Lie algebras over $\mathbb{Q}$ to the set of varieties of groups. It is neither surjective (the free groups in varieties obtained in this fashion are torsion-free), nor injective (the variety of all groups has in its preimage every variety generated by a finite-dimensional non-solvable algebra; due to D.I. Eidelkind, see Bakhturin, Slin'ko and Shestakov [1981]). It does, however,

---

[1] See Sect. 1.17 in Part I of this volume.

have a left inverse transformation. Namely, given a group G with a lower central series[2]

$$G = \gamma_1(G) \supset \gamma_2(G) \supset \ldots \supset \gamma_s(G) \supset \ldots . \tag{14}$$

For every s we will take the smallest subgroup $I_s(G)$ containing $\gamma_s(G)$, and such that $G/I_s(G)$ is torsion-free ($I_s(G)$ is then called the *isolator* of the subgroup $\gamma_s(G)$). We obtain a series

$$G = I_1(G) \supset I_2(G) \supset \ldots \supset I_s(G) \supset \ldots , \tag{15}$$

whose terms (like terms of the series (14)) satisfy the relation $[I_s(G), I_r(G)] \subset I_{s+r}(G)$. The group

$$L(G) = I_1(G)/I_2(G) \oplus \ldots \oplus I_s(G)/I_{s+1}(G) \oplus \ldots$$

is an abelian torsion-free group, and it is equipped with a multiplication operation which makes it into a Lie ring: for homogeneous elements $\xi = xI_{s+1}(G)$ and $\eta = yI_{r+1}(G)$ (with $x \in I_s(G)$ and $y \in I_r(G)$) we set

$$\xi\eta = [x,y]I_{s+r+1}(G) \in I_{s+r}(G)/I_{s+r+1}(G) ,$$

where $[x, y]$ is the usual commutator in the group G. In order to verify that the introduced operations are well-defined, and that they satisfy anticommutativity as well as the Jacobi identity, we use the following commutator identities, valid in any group:

$$\begin{gathered} [x,y] = [y,x]^{-1} , \quad [xy,z] = [x,z]^y[y,z] , \\ [[x,y^{-1}],z]^y[[y,z^{-1}],x]^z[[z,x^{-1}],y]^x = 1 , \end{gathered} \tag{16}$$

where $u^v = v^{-1}uv$.

Forming a tensor product with $\mathbb{Q}$ over $\mathbb{Z}$ we obtain a Lie algebra $\mathcal{L}(G) = L(G) \otimes_{\mathbb{Z}} \mathbb{Q}$ over the field $\mathbb{Q}$. A variety of groups will be called a *variety of Lie type* if for every $\mathfrak{V}$-free group G the intersection of terms of the chain (15) equals $\{1\}$, i.e. if $\mathfrak{V}$ is generated by its nilpotent torsion-free groups. In particular, for every variety $\mathfrak{V}$ of Lie algebras over $\mathbb{Q}$, $\mathfrak{V}^0$ is such a variety. If to every variety of groups $\mathfrak{V}$ we assign the variety $\mathcal{L}(\mathfrak{V})$ of Lie algebras over $\mathbb{Q}$, generated by all algebras $\mathcal{L}(G)$ with $G \in \mathfrak{V}$, then it turns out that $\mathcal{L}(\mathfrak{V})^0 = \mathfrak{V}$ whenever $\mathfrak{V}$ is a variety of Lie type and, when $\mathfrak{V}$ is a nilpotent

---

[2] Terms of the *lower central series* of any group G are defined inductively: $\gamma_1(G) = G$, $\gamma_i(G) = [\gamma_{i-1}(G), G]$ for $i > 1$, where $[K, L]$ is the subgroup generated by commutators $[x,y] = x^{-1}y^{-1}xy$ ($x \in K$, $y \in L$).

variety of Lie algebras, we also have $\mathcal{L}(\mathfrak{V}^0) = \mathfrak{V}$ (due to K.K. Andreyev; see Bakhturin [1985, §8.4]).

This correspondence can be used to solve a number of problems on the structure of certain important types of groups: free, free solvable with given index of solvability, and others. In general, we say that G is a *Magnus group* if its lower central series has trivial intersection and its factors $\gamma_s(G)/\gamma_{s+1}(G)$ are torsion-free. An important, already classical, result of Magnus and Witt states that the free group $F(X)$ is a Magnus group. In this case, factors of the lower central series are free abelian groups. If $|X| = d$, then rank $l_d(n)$ of the abelian group $\gamma_n(F(X))/\gamma_{n+1}(F(X))$ can be calculated according to Witt's formula

$$l_d(n) = \frac{1}{n}\sum_{m|n}\mu(m)d^{n/m}, \tag{17}$$

in which $\mu(m)$ is the Möbius function – i.e. $\mu(1) = 1$, while $\mu(p_1^{k_1}\dots p_s^{k_s}) = (-1)^s$ if $k_1 = \dots = k_s = 1$ and 0 otherwise.

In light of the correspondence discussed above it is not surprising that the same formula (17) also describes the dimension of the n-th homogeneous component of a free Lie algebra of rank d. By presenting the logical sequence involved, we will show that this formula is in fact first proved for Lie algebras, using part (2) of the theorem quoted above. Definition of commutators of basis elements in a free group F and a free algebra L is also fully analogous. In the first case basic commutators of weight s freely generate (as an abelian group) $\gamma_s(F)$ modulo the subgroup $\gamma_{s+1}(F)$; in the second case they form a basis of the homogeneous component L_n. We will now give M. Hall's definition of *basic commutators* of a group F with free generators $\{x_i\}_{i\in I}$.

Definition. 1) $c_i = x_i$ are *basic commutators* of weight $w(x_i) = 1$.

2) Suppose that basic commutators of weight less than n have been defined. Then basic commutators of weight n are $c_k = [c_i, c_j]$, where (a) c_i and c_j are basic commutators and $w(c_i) + w(c_j) = n$, and (b) $c_i > c_j$, and if $c_i = [c_s, c_t]$ then $c_j \geq c_t$.

3) Basic commutators of weight n are greater than commutators of lower weights, while the order among themselves is arbitrary.

At this point we will recall an important property of free groups (and free Lie algebras): every subgroup (subalgebra) of a free group (free Lie algebra) is itself free. In the group case this is the elementary Nielsen-Schreier theorem, and in the case of Lie algebras – the Shirshov-Witt theorem. In the context of associative algebras the analogous result does not hold.

It will be useful to keep in mind *Schreier's formula* $[F : H] = \frac{r(H)-1}{r(F)-1}$; here $[F : H]$ is the index of a subgroup H in the free group F, and $r(H)$, $r(F)$ are respective ranks of H and F. In particular, in a group with n generators a subgroup of index m can be generated by at most $m(n-1)+1$ elements.

The property of being a Magnus group has been established by A.L. Shmel'kin for the group which is free in the variety $\mathfrak{U}^l$ of solvable groups with index of solvability bounded by a fixed integer l (the free solvable group of index l). For convenience and brevity, we will call a variety of groups a *Magnus variety* if its free groups are of Magnus type. We will recall (see Sect. 2.1 in Chapter 3) that the product $\mathfrak{U}\mathfrak{V}$ of two varieties of groups is the class of extensions of normal subgroups from $\mathfrak{U}$ by factor groups from $\mathfrak{V}$.

A.L. Shmel'kin and his students established that if $\mathfrak{U}$ and $\mathfrak{V}$ are Magnus varieties, then their product $\mathfrak{U}\mathfrak{V}$, intersection $\mathfrak{U} \wedge \mathfrak{V}$ and union (or join) $\mathfrak{U} \vee \mathfrak{V}$ are also Magnus varieties. One of the natural questions in the theory of Magnus (or Lie) varieties is the following one, due to A.L. Shmel'kin: suppose that a Magnus variety $\mathfrak{V}$ of groups is distinct from the variety of all groups; is it true that $\mathfrak{V}$ is solvable, i.e. consists of solvable groups?

4.3. On Combinatorial Approach to the Study of Free Groups in Varieties. In order to give a sufficiently complete description of the free algebra in a given concrete variety $\mathfrak{V}$ defined by a system of identities V, it is usually necessary to find some canonical way of denoting its elements. This allows to fairly easily settle questions about consequences of the system V, as well as other problems. However, it is by far not always possible to represent elements of the free algebra in as simple a manner as in the case of the variety of all groups, all semigroups, all linear algebras, or all associative algebras. Already in the context of Lie algebras the Hall basis (see Sect. 4.2) requires a non-trivial definition; even so, finding a gradation over a field of characteristic zero (see Sect. 3.2) often helps to describe the structure of the $\mathfrak{V}$-free linear algebra by constructing bases for its homogeneous components. In the case of group varieties, a similar possibility occurs significantly less often. Computations turn out to be manageable, for example, for free nilpotent groups, in which there exists a canonical representation of elements in terms of basic commutators, and also for free solvable groups, whose elements are presentable in wreath products of free abelian groups. The same can be said about free groups in products of such "good" varieties (see Sect. 2.1 in Chapter 3). In other situations, obtaining answers to concrete questions often makes it necessary to consider specially chosen groups, in which it is possible to evaluate words of the free group.

Defining a group with m generators, free in a variety $\mathfrak{V}$, by means of identities $f_i(x_1, \ldots, x_{n_i}) \equiv 1$ is equivalent to describing it by means of defining relations $f_i(v_1, \ldots, v_{n_i}) = 1$ for all words $v_1, v_2, \ldots$ in the alphabet $\{x_1^{\pm 1}, \ldots, x_m^{\pm 1}\}$. It therefore seems natural to attempt to carry out a combinatorial analysis of consequences of relations introduced in this manner.

Direct study of such consequences, frequently successful in similar problems of the theory of semigroups, is usually impeded by serious obstacles resulting from a loss of characteristic features of words of f_i in their consequences. For example, it is known that the identity $x^6 \equiv 1$ implies solvability of degree 3 (M. Hall [1959]). In practice, however, the task of explicitly expressing the commutator $[[[x_1, x_2], [x_3, x_4]], [[x_5, x_6], [x_7, x_8]]]$ in terms of a product of sixth powers of some words in the (absolutely) free group can be accomplished only with help of a good computer.

A successful analysis of consequences of the identity $x^n \equiv 1$ led P.S. Novikov and S.I. Adyan to the solution of the bounded Burnside problem (Novikov and Adyan [1968]). They showed infiniteness of the free group $B(m,n)$ of rank $m \geq 2$ in the variety $\mathfrak{B}_n$ of groups with period n, where n is an odd integer ≥ 665. Existence of algorithms for determining equality and conjugacy of words from $B(m,n)$ was proved as well. It was also established that every subgroup, and a centralizer of every non-trivial element of $B(m,n)$, are cyclic. Variety $\mathfrak{B}_n$ (for large odd n) turned out to be the first example of a *non-regular variety* of groups, distinct from the variety of all groups $\mathcal{O}$, i.e. a variety whose free group of any rank contains a subgroup isomorphic to a free group of greater rank. Relying on this research, V.L. Shirvanyan even established embeddability of the group $B(\infty, n)$ of countable rank in the group $B(2,n)$.

The method developed in solving the Burnside problem also allowed S.I. Adyan to find answers to other questions about topics such as periodic groups and torsion-free groups (see e.g. Sect. 1.3, Chapter 3). Application of this method is connected with a significant amount of technical work, needed in inductive proofs of a large number of lemmas. The reader will find a thorough exposition of this technique and its applications in Adyan [1975].

In many cases however, it is possible to replace an identity with a manageable system of defining relations. For example, in Ol'shanskij [1982], which contains a much shorter proof of the Novikov-Adyan theorem on groups $B(m,n)$ (at the cost of weakening the bound on n), the system of defining relations $v_i^n = 1$ in the group $B(m,n)$ has a simple inductive description: word v_i is chosen as a shortest word of infinite order in the group

$$\langle\, x_1, \ldots, x_m \,|\, v_1^n = 1, \ldots, v_{i-1}^n = 1 \,\rangle\ .$$

A new way of investigating groups with many defining relations (in particular, groups with identities), employing a universal topological interpretation of the process of inferring consequences of the defining relations, was suggested by van Kampen in 1933. Without getting involved in precise definitions and technique at this point (see Sect. 4 in Chapter 2), we will present a simple example demonstrating that every group G with identity $x^3 \equiv 1$ also satisfies $[[y,x],x] \equiv 1$ or, equivalently, $[x, y^{-1}xy] \equiv 1$ – i.e. that every element of G is contained in some abelian normal subgroup of G. Indeed, the words y^3, $(xy^{-1})^3$ and $(x^{-1}y^{-1})^3$, which are read off of the diagram in Figure 2 by traversing a triangular or hexagonal loop, all become identically 1 in G (when travelling against the direction of an arrow, we take the inverse of the corresponding letter).

Traversing the contour of the entire diagram beginning at vertex O, we obtain as a consequence the word $x^{-1}y^{-1}x^{-1}yxy^{-1}xy$, which corresponds to $[x, y^{-1}xy] = 1$.

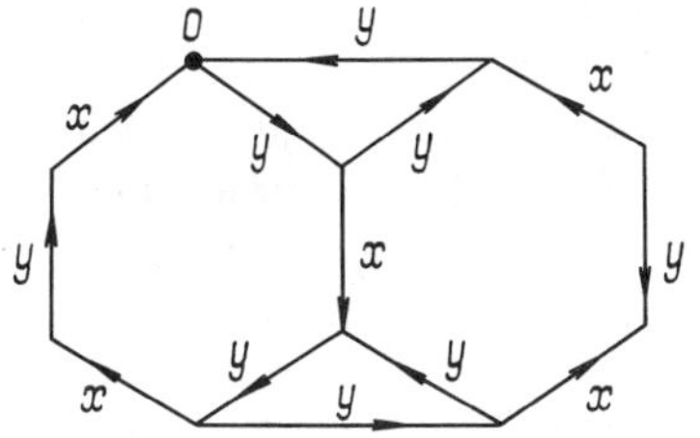

Fig. 2

4.4. Supplementary Remarks. Free groups and free Lie algebras over a field have the important property that each subgroup (resp. subalgebra) is free itself. This statement is not valid for free associative algebras (even of rank 1: the subalgebra generated by elements x^2 and x^3 is not free). A variety of algebras in which subalgebras of free algebras are also free, is called a *Schreier variety* (in honor of Schreier – a co-author, with Nielsen, of the theorem on freeness of subgroups of free groups; in the Lie algebra case the theorem is due to A.I. Shirshov and E. Witt). The problem of describing Schreier varieties appears to be very popular. All Schreier varieties (this property is quite exotic) have been identified in the case of groups, Lie algebras, as well as several other classes (Bakhturin [1985], Lyndon and Schupp [1977]). In the beginning of this section we emphasized the importance of studying groups of automorphisms of free algebras. In the context of linear algebras, Lie algebras of their derivations – i.e. linear transformations δ such that $\delta(xy) = \delta(x)y + x\delta(y)$ – are also of great interest. For example, the Lie algebra of derivations of the free associative and commutative algebra generated by $x_1, \ldots, x_n$ over a field k of characteristic zero is the infinite-dimensional *simple Cartan Lie algebra* W_n (for each $n = 1, 2, \ldots$). For a base field of characteristic p, we obtain the simple Lie algebras W_n by considering the algebra of derivations of the free associative-commutative algebra with a set of free generators $\{x_1, \ldots, x_n\}$ and an additional identity $X^p \equiv 0$. The algebras W_n in this situation are finite-dimensional and simple; they, together with their certain subalgebras, comprise the 'non-classical' fragment of the recently completed classification of finite-dimensional simple p-Lie algebras over an algebraically closed field of characteristic $p > 0$ (Block and Wilson [1984]; see also Sect. 2.3 in Chapter 2).

§ 5. Some Related Topics and Applications

5.1. Structure of Finite-Dimensional Division Rings. The problem of describing the structure of finite-dimensional central division algebras, i.e. division algebras over a field which coincides with the algebra's center, remained unsolved throughout the 40's. One of the conjectures on the subject was that every such algebra D with center k can be obtained by starting with a Galois extension of the field k with Galois group G and a co-cycle $\rho : G \times G \to K^* = K \setminus \{0\}$, and forming a vector space over K with basis $\{u_g \mid g \in G\}$ on which multiplication is defined by

$$\Big(\sum_g \alpha_g u_g\Big)\Big(\sum_h \beta_h u_h\Big) = \sum_{g,h} \alpha_g g(\beta_h)\rho(g,h)u_{gh} .$$

Such construction is called a *crossed product*. A counterexample to this conjecture was obtained by Amitsur in 1972 (see Jacobson [1975]). The principal object used by Amitsur was the so-called universal division algebra $UD(k,n)$ of degree n, arising in the core of the theory of PI-algebras. In order to construct it, we need to consider the algebra of polynomials $\Xi = k[\xi_{i,j}^{(l)}]$ in a countable set of indeterminates $\{\xi_{i,j}^{(l)} \mid 1 \leq i,j \leq n \text{ and } l = 1,2,\ldots\}$. We will denote by $K\{\xi\}$ the subalgebra of the algebra $M_n(\Xi)$ of square $n \times n$ matrices over Ξ, generated by the matrices $\xi^{(l)} = \big(\xi_{i,j}^{(l)}\big)$ with $l = 1,2,\ldots$. The algebra $K\{\xi\}$ is the *algebra of generic* $n \times n$ *matrices* over the field k. It isn't difficult to show that $K\{\xi\}$ is isomorphic to a free algebra of countable rank in the variety generated by the associative algebra of $n \times n$ matrices over k. Another important property of $K\{\xi\}$ is that it contains no zero divisors. In this situation, constructing the total ring of fractions of $K\{\xi\}$ yields the division ring $UD(k,n)$. In 1972, Amitsur proved the following

Theorem. *If n is divisible by a square of an odd prime, or by 8, then $UD(\mathbb{Q},n)$ is not a crossed product.*

For details, see Part I.

5.2. Non-Commutative Algebraic Geometry. We will fix a positive integer n and consider the ring of generic $n \times n$ matrices $K_m\{\xi\}$ with m generators $\xi^{(1)},\ldots,\xi^{(m)}$ over a field k. This ring is called the *free affine algebra of rank* m. This is a natural generalization of the ring of polynomials in m indeterminates. With the goal of generalizing the m-dimensional affine space $\mathcal{A}^m$ in mind, we will consider the set M_n^m of m-tuples of matrices of order n. Fixing polynomials $f_i(\xi^{(1)},\ldots,\xi^{(m)}) \in K_m\{\xi\}$, we arrive at the notion of a generalized affine algebraic variety as the subset of M_n^m consisting of all m-tuples $(A_1,\ldots,A_m)$, for which $f_i(A_1,\ldots,A_m) = 0$ for all i. One of the first results of the "non-commutative algebraic geometry" was a direct generalization of Hilbert's Nullstellensatz, due to Amitsur [1974].

Theorem. *Let $K_m\{\xi\}$ be the ring of generic $n \times n$ matrices in m indeterminates ($m \geq 1$), G – a subset of $K_m\{\xi\}$, $I(G)$ – the ideal generated by G and f – a polynomial from $K_m\{\xi\}$, which vanishes on the affine variety defined by the set G. Then for some positive integer s we have $f^s \in I(G)$.*

An important result in the theory of PI-algebras states that, for every PI-algebra A over a field k, there exists an integer n for which $f(x_1, \ldots, x_m) \equiv 0$ is an identity of $M_n(k)$ if and only if there is an integer s such that $f(x_1, \ldots, x_m)^s \equiv 0$ is an identity in A.

The fundamental work being done in this direction has evolved around affine algebras – which is the name given to finitely generated PI-algebras. Every affine algebra A is a factor algebra of an algebra $R = K_m\{\xi\}$ modulo some ideal I. An ideal I is *prime* if the inclusion $J_1 J_2 \subseteq I$, where J_1, J_2 are ideals of R, implies that J_1 or J_2 is contained in I. It is clear that in this case the factor algebra $A = R/I$ (a prime affine algebra) plays the rôle of the ring of regular functions on an irreducible affine variety. According to Posner's theorem (see Sect. 3.1, Chapter 2), a prime affine algebra A possesses a classical PI-ring of fractions Q. Let Z be the center of Q. Z is then a field, and the transcendence degree $\operatorname{trdeg}_k Z$ of the field Z over k is equal to the maximal length of chains of prime ideals of the ring A (V.T. Markov). Later, Schelter showed that lengths of non-refinable chains of prime ideals are all equal. Together with Markov's result, this leads us to a good definition of the dimension of an irreducible variety.

An important part in the study of any affine ring A is played by its *Jacobson radical* $J(A)$. A fundamental result in this area was obtained by Razmyslov, who proved nilpotency of the radical $J(A)$ (the fact that $J(A)$ is a nil ideal follows from Amitsur's theorem quoted above). This result, for a field of characteristic zero, was later strengthened by Kemer, who relied on Razmyslov's method to show that the Jacobson radical of any finitely generated PI-algebra A over k is nilpotent. The strongest theorem, due to Braun, asserts nilpotency of the Jacobson radical in the case of arbitrary finitely generated PI-algebras over Noetherian commutative rings (Razmyslov [1974], Kemer [1980b], Braun [1982]).

The study of graded PI-algebras leads to generalizations of projective algebraic varieties. Several works devoted to this subject follow a scheme-theoretic approach.

5.3. Identities and Representations. Certain questions in the theory of representations of groups and algebras involve identities. We will suppose, for example, that L is a Lie algebra over a field k, for which all irreducible representations have a finite, bounded dimension. As is well-known, a similar statement holds in this case for the universal enveloping algebra $\mathcal{U}(L)$. The intersection of kernels of irreducible representations of an associative algebra coincides with its Jacobson radical which, in the case of an enveloping algebra $\mathcal{U}(L)$, equals zero. The algebra $\mathcal{U}(L)$ is therefore approximated by its images in irreducible representations. These images, in turn, are subalgebras in algebras of endomorphisms of vector spaces with dimension bounded by some integer n. As we know, this means that every homomorphic image – and so the algebra $\mathcal{U}(L)$ itself – satisfies an identity of degree at most $2n$ (see Sect. 1.3 in this chapter). According to Sect. 1.4 of Chapter 2, the algebra

L then contains an abelian subalgebra H of finite codimension (in the case of a field of characteristic zero, we even have $L = H$), and there exists a positive integer m such that every inner derivation $\operatorname{ad} x$, $x \in L$, is a root of some non-zero polynomial of degree not greater than m. In order for this necessary condition to become sufficient as well, one has to impose a bound on the dimension d of the abelian subalgebra H: namely, d should be strictly smaller than cardinality of the base field. Making use of properties of special Lie algebras (Sect. 1.3, Chapter 2) we obtain the following result.

Theorem (Bakhturin [1985]). *Over an algebraically closed field of characteristic $p > 0$, all irreducible representations of a finitely generated Lie algebra L have a finite and bounded dimension if and only if L is finite-dimensional.*

Similar reasoning can be applied to the case of a countable base field.

The approach described above cannot be directly carried over to the group situation because, generally speaking, the group algebra $k[G]$ is not Jacobson semisimple. However, in the context of many interesting problems $k[G]$ turns out to be a PI-algebra, which permits to apply a theorem of Passman (see Sect. 1.4, Chapter 2). For example, it is well-known that the group algebra of a finite group G over a field of characteristic zero, as well as when the characteristic is a prime $p > 0$ not dividing the order of G, is Jacobson semisimple (Maschke's theorem). In particular, every irreducible G-module over k is injective, i.e. splits as a direct summand in any G-module in which it is contained as a submodule. In passing to infinite groups, where semisimplicity breaks down, it would be interesting to describe those groups, for which at least injectivity of irreducible modules still holds.

For countable groups the answer to this question is known (Farkas and Snider [1974], Hartley [1977]).

Theorem. *Let k be a field of characteristic $p \geq 0$ and G – a countable group. Then every irreducible $k[G]$-module is injective if and only if G is a periodic group with an abelian subgroup of finite index, and, in the case $p \neq 0$, G contains no elements of order p.*

The primary application of PI-theory is found in the proof of necessity of the hypotheses of this theorem. It is interesting, however, that even relatively simple fragments of Hartley's paper are formulated in the language of identities. For example, let k be a field, G – an almost abelian group (i.e. G contains an abelian subgroup of finite index), and V – an irreducible G-module with centralizer E. Then $\dim_E V < \infty$. Indeed, using the sufficiency part of Passman's theorem we see that $k[G]$ is a PI-algebra, and hence so is its every subalgebra and factor algebra. By Schur's lemma, E is a division ring. If $\dim_E V = \infty$ then by the Jacobson density theorem, for every n, $k[G]/\operatorname{Ann} V$ contains a subalgebra B whose homomorphic image is isomorphic to the matrix algebra $M_n(E)$ – which would then have to satisfy the identities of $k[G]$. According to Kaplansky's theorem however, $M_n(E)$

does not satisfy an identity of degree $< 2n$. This contradiction proves our assertion.

One more topic, connected with the representation problem, is related to M. Artin's theorem on the description of Azumaya algebras. We focus on central separable algebras, i.e. semisimple algebras over a field k whose center coincide with k, and which remain semisimple under every extension of the base field. In Artin [1969] and Procesi [1972] it was shown that a ring R is an Azumaya algebra of rank n^2 over its center if and only if it satisfies all identities of the full algebra of $n \times n$ matrices, and none of its homomorphic images satisfy all identities of matrices of smaller order. This condition is equivalent to the statement that R satisfies all identities of $M_n(k)$, and has no representations in a vector space of dimension less than n over any extension K of the base field k. In newer, streamlined proofs of this result, as well as in many other problems involving algebras with identities, an ever increasing rôle is played by central polynomials, discovered by Razmyslov [1973b] and Formanek [1972].

5.4. Central Polynomials. In 1957 Kaplansky asked the question whether, for every $n \geq 2$, there exists an associative polynomial $f(x_1, \ldots, x_m)$ whose evaluations on elements of the full matrix algebra $M_n(k)$ (k – a field) all lie in the center of that algebra, not all of them being zero. In the case of a finite field k an answer was obtained by Latyshev and Shmel'kin [1969]. The problem was solved in the more complicated situation of an infinite field by Razmyslov and Formanek in 1972. We will explicitly describe the *central polynomial* of Razmyslov. To do so, we set $A_i(ax_ib) = bx_ia$, where a, b are monomials which do not depend on x_i. Further, we define a multilinear polynomial d_m to be

$$d_m = \sum_{\sigma \in \mathrm{Sym}_m} \varepsilon_\sigma x_{\sigma(1)} y_1 x_{\sigma(2)} y_2 \ldots x_{\sigma(m-1)} y_{m-1} x_{\sigma(m)} \, .$$

Razmyslov shows that values of the polynomial $f = A_{n^2}(d_{n^2})$ on matrices of trace zero from $M_n(k)$ are scalar matrices, not all of them being zero. Replacing each variable in $f(x_1, \ldots, x_{n^2}, y_1, \ldots, y_{n^2-1})$ with a commutator (i.e. changing x_i to $[x_i', x_i'']$ and y_j to $[y_j', y_j'']$, we obtain the desired central polynomial. It is clear that the degree of this polynomial equals $4n^2-2$. With $n = 2$ we get 14. At the same time, the polynomial $[x, y]^2$ of degree 4 is also central (see Sect. 1.3). Recently, central polynomials of degree n^2 have been constructed for all $n \geq 2$. Even this is not the lowest possibility: for example when $n = 3$, the minimal degree of a central polynomial is 8 (see also Part I).

In the investigation of central polynomials, it is natural to formulate statements in the language of pairs. Let us suppose that (U, G) is a pair consisting of an associative algebra U and its Lie subalgebra (relative to the commutator operation) G. Then a polynomial $f(x_1, \ldots, x_n)$ is called central if, for every choice of $g_1, \ldots, g_n \in G$, $f(g_1, \ldots, g_n)$ is a central element of U, but $f(x_1, \ldots, x_n) \equiv 0$ is not a weak identity of the pair (U, G) (see Sect. 1.3). We will write $R_{m,n}$ to denote the set of multilinear polynomials of degree mn, in which all indeterminates can be divided into n subsets, each of cardinality m, in such a way that the polynomial is skew-symmetric with respect to variables which belong to the same subset. For instance, the polynomial

$$\sum_{\sigma \in \mathrm{Sym}_3} \sum_{\tau \in \mathrm{Sym}_3} \varepsilon_\sigma x_{\sigma(1)} y_{\tau(1)} y_{\tau(2)} x_{\sigma(2)} y_{\tau(3)} x_{\sigma(3)}$$

belongs to $R_{3,2}$, while $[x_1, x_2][y_1, y_2][z_1, z_2] \in R_{2,3}$.

The next result is due to Yu.P. Razmyslov (Razmyslov [1983]).

Theorem. *Let G be a semisimple Lie algebra of finite dimension m over an algebraically closed field of characteristic zero. Assume that its associative enveloping algebra U is simple and has non-trivial center. Then for some positive integer n, the set $R_{m,n}$ contains a central polynomial of the pair (U, G).*

One of the important corollaries is the following

Theorem. *Let W_t be the Lie algebra of all derivations of the algebra of commutative polynomials $\mathcal{E}$ in t indeterminates over a field k of characteristic zero. W_t can be regarded as a left $\mathcal{E}$-module, and the algebra $\mathcal{E}$ can be identified with a subalgebra of $\mathrm{End}_k W_t$. Then for $m = 2t + t^2$ there exists a polynomial f in $R_{m,n}$ such that for any $w_1, \ldots, w_l \in W_t$ $(l = mn)$ we have $f(\mathrm{ad}\, w_1, \ldots, \mathrm{ad}\, w_l) \in \mathcal{E}$, and the mapping $f \circ \mathrm{ad} : \underbrace{W_t \otimes \ldots \otimes W_t}_{l} \to \mathcal{E}$ is surjective.*

From the point of view of applications, it should be noted that the last theorem provides a fully constructive way of recovering the algebra of infinitely differentiable functions on a smooth n-dimensional manifold (in the sense of differential geometry) from the Lie algebra of vector fields.

Chapter 2
Algebras with Identities

§ 1. Effect of an Identity

The presence in an algebra of an identity which is non-trivial in some fixed class of algebras, is reflected in the structure of that algebra. This assertion can be tautological or, in the other extreme, false, depending on what is understood by the structure of an algebra. On the one hand, in principle, the mere fact that the identity is present can be regarded as an element of the structure. On the other hand, however, combinatorial complexity of periodic groups (even with bounded period, and even with finitely many generators) as well as similarity between them and the free groups show that, in the group situation, an identity of the form $x^n \equiv 1$ with large n does not greatly influence the structure in the usual sense.

1.1. *PI*-Algebras

Definition. An associative ring (or algebra) with a non-trivial identity is called a *PI-ring* (*PI-algebra*), from the words 'polynomial identity'.

Perhaps the most noticeable impact of an identity on the structure of an algebra can be seen in the case of associative rings. One of the reasons for this is provided by Kaplansky's theorem (see Jacobson [1956]): the algebra of $n \times n$ matrices does not satisfy any identity of degree less than $2n$. This statement is proved by directly substituting matrix units in a multilinear identity of degree $d < 2n$ (cf. Sect. 1.2 in Chapter 1). Since every associative ring A factored modulo its Jacobson radical $J(A)$ is a subdirect product of primitive rings, while the latter are isomorphic to rings of matrices over division rings, it is not difficult to obtain the following result: every associative ring with identity and containing no nil ideals is isomorphic to a subring of a ring of matrices of finite order over a suitable commutative ring. It is useful to note that a partial converse of this statement also holds – the ring of matrices $M_m(k)$ over a commutative ring k satisfies the standard identity of degree $2m$ (see Sect. 1.3, Chapter 1, and Jacobson [1975]):

$$S_{2m}(x_1, \ldots, x_{2m}) = \sum_{\sigma \in \mathrm{Sym}_{2m}} \varepsilon_\sigma x_{\sigma(1)} \cdot \ldots \cdot x_{\sigma(2m)} \equiv 0 \, .$$

When the hypothesis about lack of nil ideals is omitted, the above theorem is no longer true. For example, the exterior algebra $G = \Lambda(V)$ on an infinite-dimensional vector space (*Grassman algebra*) over any field satisfies the commutator identity $[[x, y], z] \equiv 0$, but no standard identity holds in it – and hence the embedding described above cannot exist.

Among the classical results on *PI*-algebras we have a theorem of Levitzki (see Jacobson [1956]): every nil *PI*-algebra is locally nilpotent. In other words, if A is finitely generated and if for every $a \in A$ there exists an integer m for which $a^m = 0$, then there exists an integer n such that the identity $x_1 x_2 \ldots x_n \equiv 0$ holds in A. This theorem is related to results on the Jacobson radical of an associative *PI*-algebra, as well as statements about algebraic algebras. The long development of this theory culminated in the theorem on nilpotency of the Jacobson radical of a finitely generated *PI*-algebra over a field (see Sect. 5.2, Chapter 1).

When the restriction on the number of generators is dropped, the theorem is no longer valid; namely, the Jacobson radical of the commutative power series ring $k[[t]]$ over a field k coincides with the ideal of series with no constant term. At the same time, $k[[t]]$ contains no zero divisors.

We will remark that in the theory of finitely generated PI-algebras an extremely important part is played by the notion of height, and by the height theorem due to A.I. Shirshov (see Zhevlakov et al. [1978]). The theorem can be formulated as follows: there exists a function $h(x, y)$ such that in every PI-algebra with identity of degree m, each monomial w in r variables $a_1, \ldots, a_r$ is a linear combination of monomials of the form $w' = v_1^{m_1} v_2^{m_2} \ldots v_l^{m_l}$, where $l \leq h(m, r)$, all the v_i's are monomials in $a_1, \ldots, a_r$ of degree less than m, and the number of occurrences of every a_j $(j = 1, \ldots, r)$ in w' and in w is identical.

Such $h(x, y)$ is called a *height function.* This theorem implies, in particular, that a finitely generated algebraic PI-algebra over a field is finite-dimensional. The term *algebraic algebra* describes an algebra A in which every element is a root of some non-zero polynomial $f(t)$ with coefficients from the base field.

An important consequence of Shirshov's height theorem is *polynomial growth* of a finitely generated PI-algebra. If an algebra A is generated by a set $\{a_1, \ldots, a_r\}$ over a field k, then taking A_1 to be the span of the generating set and setting $A_n = A_{n-1} + A_{n-1}A + AA_{n-1}$, we obtain $A = \bigcup_{n=1}^{\infty} A_n$. The function $f(n) = \dim A_n$ is called the *growth function* of the algebra A relative to the specified set of generators. We say that growth $f(n)$ is *polynomially bounded* if there exists a polynomial $p(x)$ for which $f(n) \leq p(n)$ for all n. We will note that growth of the free associative algebra on two generators is *exponential*, i.e. there is a function $q(x) = a^x$ with $a > 1$ such that $f(n) \geq q(n)$ for all n, and that the property of having polynomial or exponential growth does not depend on the choice of a finite set of generators.

The theory of associative PI-algebras also includes results of a more global character, related to the passage to the classical ring of fractions.

Definition. Let A be a ring and S – the set of regular elements of A (i.e. non-zero divisors); a ring A_S is a *right classical ring of fractions of* A if A embeds in A_S in such a way that every element of S becomes invertible, and every element of A_S has the form as^{-1} for some $a \in A$ and $s \in S$.

The example of a free associative algebra of rank 2 shows that there exist rings with no zero divisors, which do not have a classical ring of fractions. However, an old result of Amitsur says that every PI-ring with no zero divisors has a classical ring of fractions. There are also more general theorems dealing with existence and nature of rings of fractions (see Sect. 3.2). It should be remarked that generalizations follow not only the path of replacing the ring of fractions with a more universal construction, but also that of considering more general kinds of identities. One such generalized identity of a ring A can be defined as a formula of the type

$$\sum_{i=1}^{n} w_{i,1}X_{i,1}w_{i,2}X_{i,2}\dots w_{i,s}X_{i,s}w_{i,s+1} \equiv 0\,,$$

where $X_{i,j}$ are non-commuting indeterminates and the $w_{i,j}$'s are elements of A. It is natural to call such an identity non-trivial, if its left side does not always vanish when the variables are replaced by elements from some ring containing a homomorphic image of A (in which case the $w_{i,j}$'s are also replaced with their homomorphic images). We will mention in passing that there are other, more or less natural, generalizations of identities. In the study of identities in fields, for instance, it is convenient to consider *rational identities*, i.e. expressions obtained from indeterminates by means of operations of addition, multiplication and forming reciprocals.

1.2. Linear Groups with Identity. Evidently, the question about the structure of an algebra with identity is meaningful only in those classes of algebras for which some satisfactory general structure theory exists at all. The classes of all groups and all Lie algebras barely meet this criterion. In contrast, the class of associative rings with the theory of radicals and semisimplicity can be counted among such well-structured cases.

It is therefore not surprising that structure theory evolves in those classes of groups and Lie algebras with identities, which are somehow connected with associative algebras. Among them we have linear groups and special Lie algebras over a field k. A *linear group* G is a subgroup of $\mathrm{GL}\,(n,k)$ for some natural n. Taking the closure of G in the Zariski topology and passing to an algebraically closed extension of k, we obtain an algebraic linear group $\tilde{G}$. Let $\tilde{G}_0$ be the connected component of $\tilde{G}$, and R – its solvable radical. The factor group $\tilde{G}_0/R$ is a connected semisimple algebraic group. If $\tilde{G}_0/R$ is non-trivial then, according to Chevalley's theorem, it contains a subgroup isomorphic to $\mathrm{SL}\,(2,K)$ or $\mathrm{PSL}\,(2,K)$. As was mentioned in Sect. 1.3 of Chapter 1, each of those groups contains a free non-abelian subgroup, i.e. it cannot satisfy a non-trivial identity. It follows that $\tilde{G}_0 = R$, and hence $\tilde{G}_0$ is solvable. Since $\tilde{G}/\tilde{G}_0$ is a finite group, we see that $\tilde{G}$ is almost solvable. Clearly, the same reasoning is valid in relation to the group G as well: every linear group with identity contains a solvable (normal) subgroup of finite index. This statement can be sharpened with help of the Lie-Kolchin theorem, according to which a connected solvable linear group over an algebraically closed field consists of matrices which are simultaneously triangularizable. It is then easy to see that the commutant of such a group is nilpotent. And so we have

Theorem (Platonov [1967]). *Let $G \subseteq \mathrm{GL}\,(n,k)$, where k is any field of characteristic zero. If G is a group with identity, then there exists a normal series $G \rhd H \rhd N \rhd \{1\}$, in which $|G/H| < \infty$, H/N is abelian, and N is nilpotent.*

1.3. Lie Subalgebras of *PI*-Algebras. A *special Lie algebra L* over a field k is defined as a Lie subalgebra of an associative PI-algebra A. As was already remarked in Sect. 5.2 of Chapter 1, for every associative PI-algebra A there exists an integer n such that $f(x_1, \ldots, x_m) \equiv 0$ is an identity of the matrix algebra $M_n(k)$ if and only if $f(x_1, \ldots, x_m)^d \equiv 0$, for some positive d, is an identity in A. Because the algebra $M_n(k)$ is finite-dimensional, it satisfies an identity of the form

$$s_m(x_0, x_1, \ldots, x_m) = \sum_{\sigma \in \mathrm{Sym}_m} \varepsilon_\sigma [x_0, x_{\sigma(1)}, \ldots, x_{\sigma(m)}] \equiv 0 \, .$$

According to the above remark, A then satisfies $s_m(x_0, x_1, \ldots, x_m)^d \equiv 0$. We will note that if $u^d = 0$, then for every v we have $[v, \underbrace{u, \ldots, u}_{2d-1}] = 0$ (with suitable placement of brackets). It follows that A, and hence L, satisfies the identity

$$[x, s_m(x_0^{(1)}, \ldots, x_m^{(1)}), \ldots, s_m(x_0^{(2d-1)}, \ldots, x_m^{(2d-1)})] \equiv 0 \, .$$

The fcat that this identity is non-trivial is verified with use of matrix units (as in Sect. 1.2, Chapter 1). This implies that special Lie algebras are Lie algebras with identities (the converse is not true: a solvable Lie algebra of degree of solvability 3 may already fail to be special).

Special Lie algebras with finitely many generators over a field of characteristic zero have a particularly 'nice' structure. A Lie algebra L will be called *algebraic*, if for every $a \in L$ one can find a non-zero polynomial $f_a(t)$ such that $f_a(\operatorname{ad} a) = 0$. There are several results which parallel the classical theorems on finite-dimensional Lie algebras.

Theorem. *Let L be a finitely generated special Lie algebra over a field of characteristic zero. Then:*

1) *If L is algebraic, then it is finite-dimensional. In particular, if L is an Engel algebra (i.e. $f_a(t) = t^{d(a)}$), then it is nilpotent (an analog of Engel's theorem). This result, due to V.N. Latyshev, does not depend on the characteristic of the base field.*
2) *If L is solvable, then its commutant $[L, L]$ is nilpotent (an analog of Lie's theorem).*
3) *If L is almost solvable, i.e. $L \triangleright R$ (where R is the solvable radical) with L/R finite-dimensional, then $L = G \oplus R$ for a semisimple algebra G (an analog of Levi's theorem).*

Parts (2) and (3) of the theorem can be found in Bakhturin [1985].

When the assumption of a finite set of generators is dropped, the above statements are no longer true. We will again consider an infinite-dimensional Grassman algebra G (see Sect. 1.1). Namely, the algebra L of matrices

$\begin{pmatrix} a & b \\ 0 & 0 \end{pmatrix}$, where $a, b \in G$, is solvable, but its commutant is not nilpotent (Bakhturin [1985]).

1.4. Identities and Representations of Groups and Lie Algebras. Even though imposition of an arbitrary identity on a group or a Lie algebra does not usually produce immediate results concerning its structure, it clearly should not be assumed that the rôle of identities in these theories is negligible. For example, we will see that in the study of irreducible representations of groups, the problem of describing the groups whose irreducible representations have a finite and bounded degree is 'almost' equivalent to the task of identifying those groups whose group algebras satisfy at least one non-trivial identity. In the case of Lie algebras, the situation is similar (see Sect. 5.3, Chapter 1).

Description of groups whose group algebras satisfy a non-trivial identity quite naturally depends on the base field under consideration. In the case of a field of characteristic zero, we have the following theorem of Isaacs and Passman.

Theorem (Isaacs and Passman [1964]). *Group algebra $k[G]$ of a group G over a field k of characteristic zero is a PI-algebra if and only if G contains an abelian normal subgroup H of finite index.*

Sufficiency of this condition can be explained as follows. The group algebra $S = k[G]$ can be viewed as a right module over $R = k[H]$ via right multiplication by elements from R. Because H has finite index in G, this module is finitely generated. The algebra $k[H]$ is commutative, and so it is a PI-algebra. According to the Procesi-Small theorem (Procesi and Small [1968]), the ring of endomorphisms of a finitely generated module over a PI-algebra is a PI-algebra itself. Assigning to each $s \in S$ the operator of left multiplication of elements of $k[G]$ by s, we obtain a homomorphic embedding of S into $\mathrm{End}_R S$, which shows that S is a PI-algebra as claimed. Similar reasoning is used in proofs of a number of other theorems in this vein. The proof of necessity is more complicated. An important part of it relies on the examination of subsets $\Delta_n(G)$ $(n = 1, 2, \ldots)$, consisting of elements of G with at most n distinct conjugacy classes, together with the subgroup $\Delta(G) = \bigcup_{n=1}^{\infty} \Delta_n(G)$. One of the fundamental results states that if $k[G]$ satisfies an identity of degree m, then $[G : \Delta(G)] \leq m/2$. Moreover, the commutant $[\Delta(G), \Delta(G)]$ is finite. Later we will discuss similar subsets in the setting of Lie algebras.

The case of group algebras with identity over a field of characteristic $p > 0$ has been studied by Passman. He showed that the group algebra $k[G]$ of a group G over such field is a PI-algebra if and only if G contains a chain of normal subgroups $G \triangleright H \triangleright K \triangleright \{1\}$ such that G/H is finite, H/K is abelian and $|K| = p^s$ for some integer s (Passman [1972]).

We will mention that currently there are generalizations of these results to the case of group rings over a PI-ring which is not necessarily a field, the case of semigroup algebras which are not group algebras, etc.

Recall that the *universal enveloping algebra* $\mathcal{U}(G)$ of a Lie algebra G is an associative algebra with 1, containing the Lie algebra as a subalgebra (with respect to the commutator operation $[a, b] = ab - ba$), with basis consisting of 1 and the monomials $g_{i_1} g_{i_2} \dots g_{i_s}$, where $i_1 \leq i_2 \leq \dots \leq i_s$, $\{g_i\}_{i \in I}$ is a basis of the Lie algebra G, and the index set I is linearly ordered. An important property of the algebra $\mathcal{U}(G)$ is that every G-module is a module over $\mathcal{U}(G)$ and conversely. In this sense, the universal enveloping algebra of a Lie algebra is analogous to the group algebra of a group. Existence of an infinite-dimensional, absolutely irreducible representation of any finite-dimensional Lie algebra over a field of characteristic zero and the fact that the Jacobson radical of the enveloping algebra of such a Lie algebra is always zero, combined with results presented in the previous section, yield a theorem due to Latyshev (see Bakhturin [1985]).

Theorem. *Over a field of characteristic* 0, *the algebra* $\mathcal{U}(G)$ *is a PI-algebra if and only if* G *is an abelian Lie algebra.*

In the case of a base field of characteristic $p > 0$, the situation appears to be more complex. We have the following result of Bakhturin [1985]:

Theorem. *The universal enveloping algebra* $\mathcal{U}(G)$ *of a Lie algebra* G *over a field of characteristic* $p > 0$ *is a PI-algebra if and only if* G *contains an abelian ideal* H *of finite codimension, and the adjoint representation of* G *is algebraic (i.e. for every* $x \in G$ *there exists a non-zero polynomial* $f(t)$ *for which* $f(\operatorname{ad} x) = 0$*).*

As in the case of groups, an important part is played by the subset $\Delta(G)$ of a Lie algebra G, which consists of those elements g for which $\dim G/C_G(g) < \infty$, where $C_G(g) = \{h \in G \mid [g, h] = 0\}$ is the centralizer of g in G; it is clear that, as before, $\Delta(G) = \bigcup_{n=0}^{\infty} \Delta_n(G)$, where $\Delta_n(G) = \{g \in G \mid \dim G/C_G(g) \leq n\}$. Reasoning which involves passing to the ring of fractions (see Sect. 1.1) allows to conclude that whenever $\mathcal{U}(G)$ satisfies an identity of degree d, every set of n elements from G is linearly dependent modulo $\Delta_{n^2}(G)$, where $n = \left[\frac{d^2}{4}\right]$. The study of a Lie algebra H which satisfies a condition $H = \Delta_t(H)$ for some t can be simplified by applying the following general theorem due to P.M. Neumann. In the statement of this result, we consider a given bilinear transformation $\varphi : U \times V \to W$ of vector spaces U, V, W over a field k. The *breadth* $b(x)$ of an element $x \in U$ is the integer $\dim V/\operatorname{Ann}_V x$, where $\operatorname{Ann}_V x = \{y \in V \mid \varphi(x, y) = 0\}$. Breadth of an element $y \in V$ is defined in a similar way.

Theorem. *If there exist positive integers* r, s *such that* $b(x) \leq r$ *for all* $x \in U$ *and* $b(y) \leq s$ *for all* $y \in V$, *then* $\dim_k \varphi(U, V) \leq rs$.

The theorem implies that, for the Lie algebra H mentioned above, $\dim H^2 \leq t^2$. Combinatorial arguments then imply that the adjoint representation is indeed algebraic.

This method has already been applied in the investigation of similar questions in the theory of Lie superalgebras.

§ 2. Around Burnside

2.1. Burnside-Type Problems. At the first glance, it may seem that algebras satisfying identities in one variable have the simplest structure. Even here, however, we arrive at interesting and very difficult problems which have been the focus of algebraists' attention for decades. In the group case, we have the Burnside's identity $x^n \equiv 1$. For linear algebras (with associative power operation), we encounter algebraicity conditions $f(x) \equiv 0$, where f is some polynomial with coefficients from the ring of scalars. The procedures for isolating homogeneous components allow, in the case of infinite base fields, to pass to identities of the form $x^n \equiv 0$. It should be noted that in the context of anticommutative algebras, such as Lie algebras, this identity is meaningless. It is therefore common to consider identities involving two variables x, y, and one of them – say, y – in the first power. Denoting the operator of *right multiplication* by x as usual by $\operatorname{ad} x$, we see that the above identity for Lie algebras takes the form $yf(\operatorname{ad} x) \equiv 0$, where f is as before. Separating homogeneous components (if it is possible) we obtain the identity $(\operatorname{ad} x)^n \equiv 0$ (*Engel's identity*). While considering other classes of algebras, it is more convenient to write $(R_x)^n \equiv 0$, where R_x is again the operator of right multiplication by x.

A fundamental problem related to algebras with identities of such 'Burnside type' is their local finiteness. By a locally finite algebra we mean an algebra whose finitely generated subgroups (subrings, subalgebras) are finite.

2.2. Burnside Varieties. *Exponent* (or *period*) of a group G (a variety of groups $\mathfrak{M}$) is the minimal positive integer n such that $x^n \equiv 1$ is an identity in the group G (or, respectively, in the variety $\mathfrak{M}$). If such an integer does not exist, the exponent of G (resp. of $\mathfrak{M}$) is defined to be zero. Every variety of exponent n is contained in the variety $\mathfrak{B}_n$ of all groups with identity $x^n \equiv 1$, and every variety of exponent 0 contains the variety of all abelian groups $\mathfrak{A}$.

The *bounded Burnside problem* (Burnside [1902]) can be stated as the question of finiteness of a finitely generated group of exponent n. An equivalent formulation asks: is every group from $\mathfrak{B}_n$ locally finite? Evidently, it suffices to settle the question of whether the $\mathfrak{B}_n$-free group $B(m,n)$ of rank m is finite.

Solution of the Burnside problem is trivial for $n = 2$, since an easy verification shows that $\mathfrak{B}_2 \subset \mathfrak{A}$ and $|B(m,2)| = 2^m$. In 1902 Burnside proved finiteness of the group $B(m,3)$, whose order was found to be $3^{f(m)}$, where $f(m) = m + \binom{m}{2} + \binom{m}{3}$. In 1940, I.N. Sanov showed that the variety $\mathfrak{B}_4$ is locally finite, and M. Hall obtained the same result for the variety $\mathfrak{B}_6$ in 1957 (Hall [1959]).

The integer 5 is the smallest exponent for which the bounded Burnside problem remains open. It is also unknown whether groups of exponent 2^k, $k = 3, 4, \ldots$, are locally finite. For large odd exponents, the bounded Burnside problem has been solved in the negative (see Sect. 4.3, Chap. 1).

Another important question about the structure of groups from $\mathfrak{B}_n$ is the so-called *weak Burnside problem.* It concerns the existence of an integer-valued function $f(m,n)$ such that the order of any finite group with m generators and satisfying the identity $x^n \equiv 1$ is bounded by $f(m,n)$. This question has an affirmative answer in the case when n is not divisible by a square of a prime number. This solution is based on the case of prime $n = p$, which was fully investigated in the works of A.I. Kostrikin, see Kostrikin [1986] (the task of explicitly identifying $f(m,n)$ is extremely difficult; the value $f(2,5) = 5^{34}$ was found with help of a computer). This partial solution made use of a reduction to Lie algebras with the identity $(\operatorname{ad} x)^{p-1} \equiv 0$ (see Sect. 2.3). The case of composite n is attributed to P. Hall and G. Higman, who exploited the notion of a p-length of a group as well as the classification of finite simple groups.

We will remark that a solvable periodic group is always finite. It is therefore important to study solvability of groups with identity $x^n \equiv 1$. Many efforts were made, in particular, to determine solvability of groups satisfying $x^4 \equiv 1$ (the Hall-Higman problem). A definitive solution – construction of a non-solvable group with this identity – was obtained by Razmyslov [1978]. Earlier, in the course of solving a problem posed by A.I. Kostrikin, Razmyslov showed that degrees of solvability of finite groups with identity $x^p \equiv 1$, for a prime $p \geq 5$, are not bounded. This statement can also be formulated in the language of varieties. Let $\mathfrak{R}_p$ be the variety generated by all finite groups of exponent p (i.e. with identity $x^p \equiv 1$). According to Kostrikin's result, $\mathfrak{R}_p$ is a locally finite variety. Razmyslov's theorem now asserts that $\mathfrak{R}_p$ is not a solvable variety for $p \geq 5$ (Razmyslov [1972]). As in the case of the theorem of Kostrikin, Razmyslov's results quoted above contain, as an integral part, theorems on Lie algebras (see Sect. 2.3). Open problems in this area include the weak Burnside problem for exponents p^n, where $p \geq 3$ is a prime integer and $n \geq 2$, as well as 2^m when $m \geq 3$.

2.3. Engel Lie Algebras. A Lie algebra L is called an *Engel algebra* if, for every $x \in L$, there exists a natural number n such that $(\operatorname{ad} x)^n = 0$. Let us suppose, for example, that L is the Lie algebra of nilpotent linear operators on a vector space V. We will write $\operatorname{ad} x = r_x - l_x$, where r_x and l_x are, respectively, the operators of right and left multiplication by x (obtaining operators which do not necessarily belong to L). Since r_x and l_x commute, powers of their difference can be expressed by means of Newton's binomial formula; if $x^m = 0$ then we have

$$(\operatorname{ad} x)^{2m-1} = \sum_{k=0}^{2m-1} \binom{2m-1}{k} (-1)^k (r_x)^k (l_x)^{2m-1-k} =$$

$$= \sum_{k=0}^{2m-1} \binom{2m-1}{k} (-1)^k r_{x^k} l_{x^{2m-1-k}} .$$

In this sum, for every k, $x^k = 0$ or $x^{2m-1-k} = 0$, so that the resulting operator is zero. The classical Engel's theorem states that if L is a finite-dimensional Lie algebra of nilpotent linear operators on V, then the space V contains a non-zero eigenvector common to all operators from L (see Bakhturin [1985]). In particular, if V is finite-dimensional then in some basis all operators from L have an upper triangular form. In this case, $x_1 x_2 \dots x_n = 0$ for all $x_1, \dots, x_n \in L$ with $n = \dim V$. If L is a finite-dimensional Engel algebra, then $(\operatorname{ad} x_1) \dots (\operatorname{ad} x_n) = 0$ for $n = \dim L$. Because $\operatorname{ad}[x_1, x_2] = \operatorname{ad} x_1 \operatorname{ad} x_2 - \operatorname{ad} x_2 \operatorname{ad} x_1$, it follows that in a finite-dimensional Engel Lie algebra every product $[x_1, \dots, x_{n+1}]$ (with arbitrary placement of brackets) equals zero. L is therefore a nilpotent Lie algebra. Another important source of Lie algebras satisfying the Engel condition are Lie rings associated with finite p-groups. And so, if a finite p-group G satisfies the identity $x^p \equiv 1$, then the Lie ring $L(G)$, constructed from the series (14) in Chapter 1, satisfies the identity $(\operatorname{ad} x)^{p-1} \equiv 0$. We will note that the number of generators of $L(G)$ coincides with the number of generators of G, while the order of G equals the cardinality of $L(G)$. By showing nilpotency of degree c, depending on the number of generators, for Lie rings of characteristic p with identity $(\operatorname{ad} x)^{p-1} \equiv 0$, we at the same time obtain boundedness of the order of $L(G)$, and hence of G. Such result was indeed proved by A.I. Kostrikin in the late 50's; more precisely, we have

Theorem. *A Lie algebra with identity* $(\operatorname{ad} x)^n \equiv 0$ *over a field of characteristic zero or* $p > n$ *is locally nilpotent.*

On that occasion a new technique was developed (see Sect. 2.4), which proved useful in studying other important questions – e.g. those concerning classification-type problems. Achievements in the area of Engel Lie algebras are presented in the new monograph by Kostrikin [1986]. For group-theoretic consequences see Sect. 2.2.

A natural problem connected with this result is the question of nilpotency of an arbitrary Lie ring satisfying the Engel condition. In its general formulation, the problem has a negative solution given long ago (1955) by Cohn. Namely, there exists a non-nilpotent metabelian Lie ring of prime characteristic p, which satisfies the identity $(\operatorname{ad} x)^{p+1} \equiv 0$. An example of a finitely generated Engel Lie algebra over an arbitrary field, and in which the exponents in the Engel condition on its elements do not have a common bound, is derived from the general construction due to E.S. Golod [1964] (see also Sect. 2.5).

A notable landmark encountered in the course of investigation of Engel Lie algebras is Razmyslov's example of a non-nilpotent Lie algebra with the condition $(\operatorname{ad} x)^{p-2} \equiv 0$ over any field of prime characteristic $p > 0$. Construction of this example gave rise to a new method involving so-called α-functions on 2-words, which turned out to be applicable to a number of other problems of the theory of varieties (see e.g. Bakhturin [1985] and Kostrikin [1986]).

Braun observed that the bound $p > n$ in Kostrikin's results can be improved to $p > n-2$ (see Kostrikin [1986]). The fate of further study of Engel Lie algebras turned out to be much different for base fields of zero and positive characteristic. In the latter case, the problem of local nilpotency remains unsolved (as we saw before, global nilpotency fails). On the other hand, making heavy use of a result of Kostrikin (see Sect. 2.4), E.I. Zel'manov very recently proved the following

Theorem. *Over a field of characteristic zero, the identity* $(\operatorname{ad} x)^n \equiv 0$ *implies nilpotency of the Lie algebra.*

This theorem was preceded by results of S.P. Mishchenko who showed, in particular, that an Engel variety of Lie algebras over a field of characteristic zero, whose identities have growth bounded by an exponential function (see Sect. 5.1, Chapter 3), is a nilpotent variety. Varieties of this type include, among many others, varieties generated by Lie algebras of vector fields on a smooth manifold of finite dimension, special varieties of Lie algebras, etc. Zel'manov's theorem includes, as a special case, a theorem of Higgins on nilpotency of solvable Engel algebras over a field of characteristic zero. In the case of solvable Lie algebras over a field of characteristic $p > n$, nilpotency of Lie algebras satisfying $(\operatorname{ad} x)^n \equiv 0$ remains a useful tool (Bakhturin [1985]).

Even though the Engel problem has a positive solution in the case of characteristic zero, the task of estimating the degree of nilpotency of Engel Lie algebras in simpler but non-trivial situations is still outstanding. Such estimates for the condition $(\operatorname{ad} x)^4 \equiv 0$ were already obtained by Higgins [1954] and Heineken [1963]. In the case $(\operatorname{ad} x)^5 \equiv 0$, nothing is known as yet.

Every finitely generated algebraic Lie algebra over a field of characteristic zero satisfying a non-trivial identity, as was shown by E.I. Zel'manov, is finite-dimensional. By an *algebraic Lie algebra* we understand a Lie algebra in which for every element x there exists a polynomial $f(t)$ in one indeterminate which vanishes on the operator $\operatorname{ad} x$. Other results, dealing with classification problems, have also been obtained by Zel'manov ([1983], [1984b]). For example, he showed that every simple Jordan algebra is classical. Apart from this, he obtained a description of simple graded (possibly infinite-dimensional) Lie algebras over a field of characteristic $p = 4n+1$ or 0, which have the form $L = \bigoplus_{i=-n}^{n} L_i$ with $\sum_{i \neq 0} L_i \neq \{0\}$. An important tool used in Zel'manov's proofs is his generalization of a theorem, due to A.I.

Kostrikin, on 'envelopes of thin sandwiches', i.e. non-zero elements x of a Lie algebra, for which $(\operatorname{ad} x)^2 = 0$: a Lie algebra over a base ring in which 2 and 3 are invertible, and generated by a finite set of envelopes of thin sandwiches, is nilpotent. We will also mention that an algebra without envelopes of thin sandwiches is called an algebra without strong degeneracy. Every Lie algebra contains a unique maximal ideal (the *Kostrikin radical*), modulo which the algebra has no strong degeneracy.

Kostrikin's technique and ideas have been applied in many publications on modular Lie algebras, i.e. Lie algebras and Lie p-algebras over a field of characteristic $p > 0$, primarily in attempts at classification of simple finite-dimensional algebras. Such classification has been completed for p-algebras over an algebraically closed field (Block and Wilson [1984]), and it confirms a conjecture of A.I. Kostrikin and I.P. Shafarevich. The case of ordinary Lie algebras, i.e. those without p-operations, is significantly more difficult, and the classification problem remains open for this situation.

We will mention one more open question, important in the theory of group varieties: suppose that a Lie algebra G satisfies an identity which fails in $\mathrm{sl}(2, \mathbb{C})$; is it true that G is then solvable?

An affirmative answer to this question would also provide a solution of the problem of A.L. Shmel'kin on varieties of groups of Lie type (see Sect. 4.2 in Chapter 1).

2.4. The Sandwich Method. In this section we present some characteristic particulars of the method, developed by A.I. Kostrikin [1958] for the purpose of solving the weak Burnside problem in the case of a prime exponent. As we already remarked in Sections 2.2 and 2.3, the core of that reasoning consists of a proof of local nilpotency of Lie algebras with the identity $(\operatorname{ad} x)^{p-1} \equiv 0$ over a field of characteristic $p > 0$. In fact, Kostrikin proves the following, markedly stronger theorem (see Kostrikin [1986]).

Theorem. *Every Lie algebra, satisfying an identity* $(\operatorname{ad} x)^n \equiv 0$ *over a field of characteristic zero or* $p < n$, *contains a non-zero abelian ideal.*

This fact suffices to settle the problem of local nilpotency of such Lie algebras. The reason for this is that every Lie algebra L contains the largest locally nilpotent ideal $R(L)$. Kostrikin shows that under the hypotheses of the above theorem, the factor algebra $L/R(L)$ does not contain non-zero locally nilpotent ideals. It is clear that the theorem now implies that $L = R(L)$, i.e. L is locally nilpotent. Because L is finitely generated, we see that it is in fact nilpotent.

One of characteristic points in Kostrikin's work is the identification of a Lie algebra L with a Lie subalgebra of an associative algebra $E(L)$ of linear operator on L. Indeed, when proving the existence of a non-trivial abelian ideal in L by contradiction, we can suppose that the center of L is zero; in this case, the mapping $x \mapsto \operatorname{ad} x$, for $x \in L$, yields the desired identification. Using this, we will note that for any $x_0, \ldots, x_m \in L$ the element $x_1x_2\dots x_m$ is an operator acting on the algebra L, and $[x_0, x_1, \dots, x_m]$ is an element of L obtained from x_0 via the action of this operator, i.e. $[x_0, x_1, \dots, x_m] = x_0(\operatorname{ad} x_1)\dots(\operatorname{ad} x_m)$. If $[x_0, x_1, \dots, x_m] = 0$ for every $x_0 \in L$ then, by our hypothesis, $x_1x_2\dots x_m = 0$. We will now try to write down the conditions for an abelian ideal I to be generated by an element c of the algebra L. Every element of I is a linear combination of monomials of the form $[cx_1, \dots, x_k]$. This means that it is necessary and sufficient to have $[[cx_1, \dots, x_k], [cx_{k+1}, \dots, x_l]] = 0$ for every $x_1, \dots, x_l$. Applying the anticommutative and Jacobi identities we obtain triviality of commutators of the form $[x_{i_1}, cx_{i_2}, \dots, x_{i_s}, cx_{i_{s+1}}, \dots, x_{i_l}]$. This, according to the remark above, is equivalent to the validity of relations

$$cx_1\dots x_m c \equiv 0\,, \quad m = 0, 1, 2, \dots \tag{1}$$

for all $x_1, \dots, x_m \in L$. We will note that the identity $(\operatorname{ad} x)^n \equiv 0$, in our notation, takes the form $x^n \equiv 0$, and so applying linearization we may assume that the parameter m in (1) ranges between 0 and $n-1$.

A fundamental combinatorial problem arises now: finding such a non-zero element c of a Lie algebra L with the condition $(\operatorname{ad} x)^n \equiv 0$ and with the characteristic of the base field restricted as above, which would satisfy (1). This task should be more properly called 'constructing' rather than 'finding', since the desired element c in the end turns out to depend on the element used in the first stage of this procedure in a relatively canonical (albeit branched-out) fashion. For notational convenience, an element c is called a *sandwich* of thickness r if, for every $x_1, \dots, x_r \in L$, we have $cx_1\dots x_r c = 0$, but $cy_1\dots y_{r+1}c \neq 0$ for some $y_1, \dots, y_{r+1}$. Thickness of a sandwich is an odd integer. For example, if $c^2 = 0$, then for every $x, y \in L$ we have

$$0 = [x[yc^2]] = [xyc^2] - 2[xcyc] + [xc^2y] = -2[xcyc]\,,$$

i.e. $cyc = 0$. Sandwiches of thickness 1 are called thin, while those of greater thickness – thick. Thin sandwiches turn out to be nilpotent elements of index 2. Even though the whole algebra L consists of nilpotent elements for which $x^n = 0$, the task of identifying elements with index of nilpotency 2 is highly complicated. The process of constructing nilpotent elements of decreasing index, and hence sandwiches of increasing thickness, was called by Kostrikin a process of descent, since it is based on passing to elements with ever more restrictive conditions imposed on them. Descent to nilpotent elements of index 3 is realized in a uniform way: if v is an element of index m, where $4 \le m \le p-1$ (this assumption is not necessary for characteristic zero), then for some $u \in L$ we have $[uv^{m-1}] \neq 0$, while $[uv^{m-1}]^{m-1} = 0$, so that $[u, v^{m-1}]$ is an element with index of nilpotency at most $m-1$. The transition from nilpotent elements of index 3 to sandwiches is significantly more complex: if $b^3 = 0$, then for every $u \in L$ one considers elements of the form $g_m = g_m(u) = [u[bu]^m b^2]$. Induction on m shows that $g_m^2 = b^2(u^2b^2)^{m+1}$, i.e. $g_{n-1}^2 = 0$. Take

t to be minimal with the property that $g_{t-1}^2 = b^2(u^2b^2)^t = 0$. If $t = 1$, then we obtain a sandwich $[ub^2]$. Otherwise (i.e. when $t > 1$), it can be proved that there exists an element $f \in L$ such that some element $b_0 = g_m(f)$ with index of nilpotency not exceeding 3 has the additional property $b_0^2(u^2b_0^2)^s = 0$ (in which $s = [t/2]$), satisfied for all $u \in L$. Descent of the parameter t yields an element $b_1 \neq 0$ which satisfies $b_1^3 = 0$, and $b_1^2u^2b_1^2 = 0$ for all $u \in L$. If b_1 is not a sandwich, then the reasoning applied to b_0 above can also be applied to $c = [ub_1^2] \neq 0$. We will also mention that if the base field has characteristic zero, or $p > n + \left[\frac{n}{2}\right]$, then the construction of this sandwich is much simpler.

The process of descent to sandwiches of arbitrary thickness is divided into two qualitatively distinct stages. The first one consists of constructing any thick sandwich from thin sandwiches; the other represents construction of sandwiches of arbitrary thickness. The first stage is so complicated that the second one seems comparatively simple. The procedure for producing arbitrarily thick sandwiches is based on the following: if c is a sandwich of thickness $2m - 1$, $m \leq (p-3)/2$ and $a \in L$ is an element for which $c_0 = [ca^{2m+1}c]$ is non-zero, then c_0 also has thickness at least $2m - 1$. As far as thickness is concerned, the new sandwich c_0 is no better than c; however, if $m \geq 2$ then c_0 satisfies the relation

$$c_0u^{2m}c_0v^{2m}c_0 = 0\,, \tag{2}$$

for all $u, v \in L$. If c_0 is not a sandwich of length $2m + 1$, then for some $b \in L$ the element $c_1 = [c_0b^{2m+1}c_0] \neq 0$ is already a sandwich whose thickness is $2m + 1$ or greater. This assertion holds for $p > 7$ and $7 \leq 2m + 3 < p$ (the remaining cases can be disposed of by additional, simpler, reasoning).

The reader has undoubtedly noticed that at various stages the desired effect (construction of an element of smaller nilpotency index or a sandwich with greater thickness) is achieved by repeated application of a procedure, which initially does not produce the required result. This regularity manifests itself in a particularly clear manner in the process of passing from thin sandwiches to the thick one (i.e. the step from $m = 1$ to $m = 2$ in the notation of the above paragraph). On the one hand, the author exhibits several procedures for passing from one kind of sandwiches to others, involved in certain identities (in the sense of (2)). On the other hand, it is shown that when some fixed thin sandwiches appear in certain identities of that type, then a thick sandwich can be created out of them. For example, if L contains a thin sandwich c for which $cu_1^2cu_2^2c\ldots cu_m^2c = 0$ for all $u_1, u_2, \ldots, u_m \in L$, then L also contains a thick sandwich (when $p > 5$). In the most natural case of $m = 2$, such sandwich is obtained by repeated (not more than three times) application of the transition $c \mapsto [ca^3c]$ for suitable $a \in L$. Another similar situation concerns identifying a thin sandwich c, for which the equality $[cu^3c][cv^3c] = 0$ is valid for all $u, v \in L$.

To proceed further, it is necessary to consider pairs of thin sandwiches $c_1, c_2 \in L$, linked by relations of the form $[c_1c_2] = 0$ and $c_1c_2 \neq 0$, and such that the equalities $c_1 u_1 \dots u_m c_1 c_2 = c_2 u_1 \dots u_m c_1 c_2 = 0$ (with $m \leq n$) are identically satisfied for all $u_1, \dots, u_m \in L$. The maximal m for which these last relations hold is called the thickness of the pair (c_1, c_2). The most complex task, technically speaking, is the construction of a pair with thickness 1, which can be achieved under the assumption that there are no thick sandwiches. This construction is not merely more tedious, but it also requires introduction of certain new notions – such as an extension of a sandwich, an extremal subalgebra, a sandwich-type algebra etc. It is easier to construct a thick pair (i.e. one with $m \geq 2$) in an algebra with identity $(\operatorname{ad} x)^n \equiv 0$ and $p > 7$. Finally, given a thick pair (c_1, c_2) in a Lie algebra L, the author considers the ideal I_c generated by the sandwich $c = [a(c_1 + c_2)^2] \neq 0$, with some $a \in L$. For all $g, h \in I_c$, the identity

$$cg^2ch^2c = 0$$

holds in L. This allows to find in L/I_c a sandwich c_0 of thickness $p - 4$. Its preimage in L cannot satisfy an identity $c_0 h_2 c_0 = 0$ for all $h \in I_c$, for otherwise for any $u, v \in L$ we would have $[c_0 u^3 c_0][c_0 v^3 c_0] = 0$, and according to a remark made above L would then contain a thick sandwich. This means that there are $h_0, b_0 \in L$ such that the element $e_0 = [c_0 b_0 h_0^2 c_0]$ is non-zero; then $e_0 h^2 e_0 = 0$ for all $h \in I_c$, and hence e_0 is a thick sandwich. This contradiction completes the proof of the existence of a non-zero abelian ideal.

2.5. Nil Rings. According to Levitzki's theorem (see Sect. 1.1), an associative ring with an identity $x^n \equiv 0$ is locally nilpotent. In the case of algebras over a field of characteristic zero (or $p > n$), an even stronger theorem of Nagata-Higman[1] is valid (see, for example, Zhevlakov et al. [1978]):

Theorem. *Under the assumptions made above, an algebra A satisfying $x^n \equiv 0$ is nilpotent with index of nilpotency not exceeding $2^n - 1$.*

Higgins' proof of this result goes as follows. Let A^* be the algebra obtained from A by formally adjoining a unity 1. Since for all $a, b \in A$ we have $(a + b)^n = 0$, by separating homogeneous components (restrictions on the characteristic are used at this point) we obtain

$$f(a, b) = \sum_{i=0}^{n} a^i b a^{n-1-i} = 0\,.$$

Moreover, $f(a, 1) = na^{n-1}$. We will consider the element

$$g(a, b, c) = \sum_{i,j=0}^{n-1} a^i c b^j a^{n-i-1} b^{n-j-1}\,, \text{ where } a, b, c \in A\,.$$

[1] It recently transpired that this theorem was first proved in Dubnov and Ivanov [1943].

We now have

$$g(a,b,c)=\sum_{j=0}^{n-1} f(a,cb^j)b^{n-j-1}=0\,,$$

but on the other hand

$$g(a,b,c)=\sum_{i=0}^{n-1} a^icf(b,a^{n-1-i})=na^{n-1}cb^{n-1}\,.$$

This implies $a^{n-1}cb^{n-1}=0$ (exploiting restrictions on the characteristic again). We will denote by N the ideal of A generated by elements of the form a^{n-1}, with $a\in A$. Because of the above, $NAN=\{0\}$. The factor algebra $\bar{A}=A/N$ satisfies the identity $x^{n-1}\equiv 0$. We can proceed by induction, and assume that $\bar{A}^l=\{0\}$ for $l=2^{n-1}-1$. This means that $A^l\subseteq N$, so that $A^{2l+1}\subseteq NAN=\{0\}$. Since $2l+1=2^n-1$, the theorem has been proved.

We will mention that it is impossible to weaken the hypotheses about the characteristic of the base field. Over a field k of characteristic $p>0$ there exist non-nilpotent algebras, in which the p-th power of every element is zero. As an example, we can take the subalgebra of the algebra of polynomials $k[x_1,x_2,\ldots,x_n,\ldots]$ with no constant term, and form its factor algebra R modulo the ideal I generated by the elements $x_1^p,\ldots,x_n^p,\ldots$. Because of the properties of binomial coefficients, $(u+v)^p=u^p+v^p$ holds in this ring, which demonstrates the validity of $x^p\equiv 0$. At the same time, the product $x_1x_2\ldots x_n$ is a non-zero element for every $n>0$. We also note that using the algebra R it is easy to give an example of a non-nilpotent Lie ring satisfying the identity $(\operatorname{ad} x)^{p+1}\equiv 0$. Namely, if M is a cyclic R-module with generator a and R is regarded as an abelian Lie algebra, then by setting $L=R\oplus M$ with the commutator $[r,m]=rm$ we obtained the desired Lie ring. One more important observation: according to Razmyslov's results, over a field of characteristic zero, $x^n\equiv 0$ even implies $x_1\ldots x_{n^2}\equiv 0$.

The topics described here are related to the Kurosh problem: is it true that every algebraic algebra A is locally finite? A negative answer was provided by E.S. Golod [1964], who constructed, for every integer $d\geq 2$, an example of an infinite-dimensional algebra generated by d elements, in which every algebra generated by $d-1$ elements is nilpotent – and hence finite-dimensional. If it is also assumed that A is a PI-algebra, then the Kurosh problem has a positive solution, following from Shirshov's height theorem (see Sect. 1.1).

We will quote one more result, dealing with associative algebras and Lie algebras at the same time. Let A be an associative algebra over a field of characteristic zero, and let L be its Lie algebra (with respect to the commutator operation). We will suppose that A is generated by L as an associative algebra. In this situation, a result of S.P. Mishchenko [1984] holds: assume that the pair (A,L) satisfies a weak identity $x^n\equiv 0$; then A is a nilpotent algebra. If $A=L$, then we in particular obtain the Nagata-Higman theorem in characteristic zero.

The Kurosh problem has been formulated and considered in a number of other classes of linear algebras. For instance, in the case of alternative algebras (i.e. those in which any two elements generate an associative subalgebra), A.I. Shirshov proved that algebraic PI-algebras are locally finite-dimensional. He also obtained an analogous result for *special Jordan algebras* (i.e. subalgebras of associative algebras over a field of characteristic not equal 2, with the operation $x \circ y = \frac{1}{2}(xy + yx)$). These results imply nilpotence of corresponding finitely generated algebras with identity $x^n \equiv 0$. E.I. Zel'manov [1983] generalized Shirshov's theorem to the case of arbitrary Jordan algebras. Dropping the assumption of a finite number of generators causes this result to fail (examples due to Zhevlakov, Shestakov and Dorofeev in Zhevlakov et al. [1978]). Studies in this direction are being carried out also in other classes of non-associative algebras.

§ 3. Identities and Constructions

3.1. Extensions of Algebras with Identity. In Birkhoff's theorem (Sect. 2.3, Chapter 1) one can notice the preservation of identities under the simplest operations on algebras: passing to a subalgebra and factor algebra, as well as forming a Cartesian product of algebras (with a fixed identity). It is natural to ask which constructions, applied to algebras with identities, result again in an algebra with identities? Among them we have extensions of groups and rings, passing to a ring of matrices over a ring, forming the adjoint associative algebra of a linear Ω-algebra, passing to the adjoint group of an associative ring or the commutator algebra of some algebra, forming tensor product of algebras, and many others.

In the case of groups, the *product* $\mathfrak{U}\mathfrak{V}$ *of varieties* $\mathfrak{U}$ and $\mathfrak{V}$ is the class of those groups which contain a normal subgroup from $\mathfrak{U}$, modulo which the factor group belongs to $\mathfrak{V}$. The fact that $\mathfrak{U}\mathfrak{V}$ is a variety can be most easily verified by means of Birkhoff's theorem. For example, if $G \in \mathfrak{U}\mathfrak{V}$, i.e. G contains a normal subgroup $K \in \mathfrak{U}$ for which $G/K \in \mathfrak{V}$, and if H is a subgroup of G, then $H \cap K \in \mathfrak{U}$ and $H \cap K$ is a normal subgroup of H; by one of the isomorphism theorems $H/H \cap K \simeq HK/K \subseteq G/K$, so that the factor group lies in $\mathfrak{V}$. The verification for a factor algebra and a Cartesian product is equally simple.

In many important classes of algebras the *kernel congruence* $a \sim b \Leftrightarrow \varphi(a) = \varphi(b)$ induced by a homomorphism φ coincides with a decomposition relative to some subalgebra of a special kind. This is the case for groups, rings, linear Ω-algebras (i.e. linear algebras with a system Ω of additional multilinear operations). In all of these situations it is possible to define a product of two varieties just as it was done above for groups.

The study of products of varieties constitutes an important chapter of the theory of group varieties, varieties of Lie algebras, and – to a lesser extent –

varieties of associative algebras. Certain topics in this area can be found in Chapter 3, Sect. 2.

3.2. Identities of Related Rings. Perhaps the most thoroughly studied problem in this area is to find an identity in an associative algebra constructed from other associative algebras. One of the main reasons for that is the fact that once an associative algebra satisfies any non-trivial identity, it automatically has to satisfy 'very many' others. This assertion is formalized in an important theorem of Regev [1971].

Theorem. *Let A be an associative algebra over a field k. We will denote by P_n the set of all multilinear polynomials of degree n in $x_1, \ldots, x_n$, and by $P_n(A)$ – its subset consisting of those $f \in P_n$ for which $f \equiv 0$ is an identity in A. If A satisfies an identity of degree d then $\dim P_n/P_n(A) \leq (d-1)^{2n}$ for every $n \geq 1$.*

For comparison, we will note the obvious equality $\dim P_n = n!$. A particularly simple combinatorial proof of the theorem stated above was given by Latyshev. One of important consequences of this result is a solution of a problem from Jacobson's book "Structure of Rings".

Theorem. *Tensor product of PI-algebras is a PI-algebra.*

There is also the following related result.

Theorem. *The ring of square $n \times n$ matrices $M_n(R)$ over a PI-ring R is a PI-ring.*

Similarly, transition to the ring of endomorphisms of a finitely generated module over a *PI*-ring yields a *PI*-ring as well (Leron and Vapne [1970], Procesi and Small [1968]). As a rule, passing to the ring of fractions is also a 'well-behaved' construction. In particular, for *prime PI-rings* we have Posner's theorem (Posner [1960], Herstein [1968]): if A is such a ring, then the set S of its regular elements is a denominator set, and the ring of (right) fractions A_S (consisting of elements of the form as^{-1} with $s \in S$ and $a \in A$) is a classically simple ring satisfying all identities of A. Using Kaplansky's theorem, we can make this more precise by remarking that A_S is isomorphic to a matrix ring over a division ring, whose dimension over its center is bounded by the integer $[d/2]^2$, where d is the degree of any identity satisfied in A. This theorem, together with many others, can also be derived from the interesting construction of a prime central ring of quotients, obtained by Martindale [1969] (see Sect. 4.3 in Part I).

3.3. The Group of Units and the Adjoint Algebra of an Associative Ring. Let A be a *PI*-algebra and $[A]$ – its *adjoint Lie algebra*, i.e. the same underlying set A with the commutator operation; then, according to Sect. 1.3, the Lie algebra $[A]$ is an algebra with identity. If $A = F(X, \mathfrak{R})$ is the free algebra in a variety $\mathfrak{R}$ of associative algebras, then the Lie subalgebra L generated by the set X relative to the commutator, is free in some variety $\mathcal{L}(\mathfrak{R})$ of Lie algebras. It is easiest to verify this using the following useful internal criterion of the freeness of L in a variety: every mapping of the generating set X into L extends to a homomorphism of the algebra L into itself. In our case every mapping $\varphi : X \to L \subseteq A$ extends to a homomorphism $\bar{\varphi} : A \to A$, which turns out to be a Lie algebra homomorphism. Its restriction $\tilde{\varphi}$ to L is the required homomorphism of L into itself. As a result, we obtain a transformation $\mathfrak{R} \mapsto \mathcal{L}(\mathfrak{R})$ from the set of varieties of associative algebras into the set of varieties of Lie algebras. So far, this correspondence has not been investigated well enough.

Taking the group $U(A)$ of invertible elements (units) of an associative PI-algebra, we often obtain a group which does not satisfy an identity. This is the case for the algebra of $n \times n$ matrices when $n \geq 2$ (see Sect. 1.3, Chapter 1). Nevertheless, in many other situations there is a close link between identities of A, $[A]$ and $U(A)$. For example, if $[A]$ is nilpotent with index n, then $U(A)$ is also nilpotent with index at most n (Zalesskij, Gupta and Levin). M.B. Smirnov showed that if A has no additive 2-torsion and the Lie ring $[A]$ is solvable, then $U(A)$ is a solvable group. For a nil algebra A over an infinite field (and when $U(A)$ is considered with the circle operation $x \circ y = x+y+xy$), the converse holds: if $U(A)$ is solvable then $[A]$ is a solvable Lie algebra (Zalesskij and Smirnov [1982], Gupta and Levin [1983]).

3.4. Operations on Groups. An abstract theory of operations, whose special cases include free and direct products, has been developed in group theory. It appears that the problem of describing the operations on groups from a fixed variety which yield groups with identity relations, has not been posed explicitly. However, a number of so-called verbal products have been constructed (Golovin, Moran), which do produce groups inside the original variety. If $\mathfrak{V}$ is a variety of groups, and $(G_\alpha)_{\alpha \in I}$ – a family of groups, then the *verbal product* $P = \prod_{\alpha \in I}^{\mathfrak{V}} G_\alpha$ is defined as the factor group of the free product $F = \prod_{\alpha \in I}^{*} G_\alpha$ modulo the subgroup $H = V(F) \cap C$, where C is the kernel of the homomorphism from the free product F into the direct product $\prod_{\alpha \in I} G_\alpha$ (the so-called Cartesian subgroup of F). Restriction of the operation to the class $\mathfrak{V}$ (in which case $H = V(F)$, since clearly $F/C \in \mathfrak{V}$) leads to a free product in the variety $\mathfrak{V}$: the group P is then universal in the variety $\mathfrak{V}$, contains the groups G_α as subgroups, and every family of mappings $\varphi_\alpha : G_\alpha \to Q \in \mathfrak{V}$ uniquely extends to a homomorphism $\varphi : P \to Q$ (H. Neumann [1967]). Verbal products (as well as other operations on groups) are studied from the point of view of validity of certain postulates or properties, which hold for free and/or direct products. For example, we will say that a product satisfies Mal'tsev's postulates if the subgroup generated in G by arbitrary subgroups $H_\alpha \subseteq G_\alpha$ is naturally isomorphic to the product (of the same type) of groups H_α. This property is not satisfied by $\mathfrak{V}$-products, where $\mathfrak{V} = \mathfrak{A}^l$ or $\mathfrak{N}_l$ and $l > 1$. The simplest non-trivial products (i.e. ones distinct from free and direct products) satisfying Mal'tsev's postulate is the

$\mathfrak{V}$-product with $\mathfrak{V} = \text{var}\,(\text{Sym}_3)$. The free product in a Burnside variety $\mathfrak{B}_n$, for sufficiently large odd n, is also of Mal'tsev type (S.I. Adyan).

In the group-theoretic situation there arose a question about amalgams with identity. Namely, suppose there is a given family of groups G_α $(\alpha \in I)$ from some variety $\mathfrak{V}$, and a family of their subgroups H_α $(\alpha \in I)$. We will assume that for every pair of indices $\alpha, \beta \in I$ there is an isomorphism $\varphi_{\alpha\beta} : H_\alpha \to H_\beta$. Does there exist a group $G \in \mathfrak{V}$ containing subgroups isomorphic to the G_α's in such a way that $G_\alpha \cap G_\beta = H_\alpha = H_\beta$, where H_α and H_β are identified by means of $\varphi_{\alpha\beta}$? Only partial answers to this question are known. For the variety of all groups, solution is provided by the construction of a free product with a common subgroup (Higman, B.H. Neumann, H. Neumann; see Magnus, Karrass and Solitar [1966]).

Passing to non-trivial varieties, the answer, as a rule, turns out to be negative. For example, if $\mathfrak{V}$ contains a non-abelian finite group and has the amalgamation property, then $\mathfrak{V}$ contains all groups (B.H. Neumann). In the variety of abelian groups however, there is a construction of a direct product with common subgroup. Examples of other varieties satisfying this property are not known at present.

The theory of products has also been developed for Lie algebras, but interesting results in this area have only been obtained in the case of free products of Lie (p-)algebras with a common subalgebra (Shirshov [1962] and Kukin [1974]).

3.5. Contragredient Lie Algebras. Because of several problems of theoretical physics, the so-called *contragredient* (or *Kac-Moody*) *Lie algebras* have attracted considerable attention in recent years. Construction of such algebras uses, as its starting point, a generalized *Cartan matrix* of size l. By this we mean a matrix of integers $A = (A_{i,j})_{1 \le i,j \le l}$ in which $A_{i,i} = 2$ for $i = 1, \ldots, l$, $A_{i,j} \le 0$ when $i \ne j$, and $A_{i,j} = 0$ implies $A_{j,i} = 0$. Special cases of such matrices include Cartan matrices associated with Cartan subalgebras of classical (semi)simple Lie algebras (from the families A_n, B_n, C_n, D_n, together with E_6, E_7, E_8, F_4 and G_2). A Cartan matrix has an associated Dynkin diagram with l vertices labelled with integers $1, 2, \ldots, l$, in which vertex number i is connected with vertex number j by a set of $A_{i,j}A_{j,i}$ edges. A Cartan matrix A is called indecomposable if its Dynkin diagram is connected. We will say that A is symmetrizable if there exists a diagonal matrix D, with positive rational entries, for which DA is a symmetric matrix. The property of being symmetrizable holds automatically whenever the diagram does not contain a cycle.

A *contragredient algebra* $\mathbf{g}(A)$ over the field $\mathbb{C}$, associated with a symmetrizable Cartan matrix, is defined in terms of generators $\{E_i, F_i, H_i \,|\, i = 1, 2, \ldots, l\}$ and defining relations

$$1)\ [H_i, E_j] = A_{i,j} E_j\,, \quad [H_i, F_j] = -A_{i,j} F_j\,,$$
$$[E_i, F_j] = \delta_{i,j} H_j\,, \quad [H_i, H_j] = 0\,;$$

$$2)\ (\operatorname{ad} E_i)^{1-A_{i,j}}(E_j) = 0 = (\operatorname{ad} F_i)^{1-A_{i,j}}(F_j)\ (\text{if } i \neq j)\,.$$

If the bilinear form obtained from symmetrization of the matrix A is positive-definite (it is then said that A is finite), then the algebra $\mathbf{g}(A)$ is finite-dimensional and semisimple. It is clear that in this case $\mathbf{g}(A)$ is an algebra with identity and, moreover, by Ado's theorem $\mathbf{g}(A)$ embeds in a finite-dimensional associative algebra – and hence it is a special Lie algebra. The problem of describing all identities of contragredient algebras of this type is very difficult, and a basis of identities for indecomposable matrices A is known only in the case $l = 1$: we then obtain the algebra $\mathrm{sl}\,(2, \mathbb{C})$. By the theorem of Razmyslov (see Sect. 5.4, Chapter 2) the algebra $\mathbf{g}(A)$, for a finite indecomposable matrix A, is completely determined by its identities.

A Cartan matrix A is called affine (and sometimes Euclidean) if A is indecomposable, singular, and each of its principal minors (i.e. submatrices obtained by deleting row number i and column number i for some i) is a positive matrix. In this situation the Lie algebra $\mathbf{g}(A)$ can be decomposed into

$$\mathbf{g}(A) = G \otimes \mathbb{C}[t, t^{-1}] \oplus \mathbb{C}z\,,$$

where G is some finite-dimensional simple Lie algebra, $\mathbb{C}[t, t^{-1}]$ is the algebra of Laurent series in t, z is a central element, and

$$[g_1 \otimes f_1(t), g_2 \otimes f_2(t)] = [g_1, g_2] \otimes f_1(t) f_2(t) + \operatorname{Res}(f_1 f_2) K(g_1, g_2) z\,,$$

in which $K(\ ,\ )$ is the Killing form on G, and $\operatorname{Res}(f)$ is the coefficient of t^{-1} in the series f. The algebra $\mathbf{g}(A)$ contains a central ideal $\mathbb{C}z$, modulo which $\mathbf{g}(A)$ is isomorphic to $G \otimes \mathbb{C}[t, t^{-1}]$, i.e. the factor algebra satisfies all multilinear identities of the algebra G. It follows that $\mathbf{g}(A)$ is an algebra with a non-trivial identity. More precisely, we can say that $\mathbf{g}(A)$ is a central extension of a special Lie algebra. It is not known, however, whether $\mathbf{g}(A)$ itself is then a special algebra. If this is not the case, then a question of Latyshev about a homomorphic image of a special Lie algebra being special would receive a negative answer. As the simplest example, one should consider $G = \mathrm{sl}\,(2, \mathbb{C})$.[2]

There is a variation of this construction, which makes use of an automorphism of the Dynkin diagram of G, for which our discussion of identities remains valid.

The classes of matrices listed above give rise to the most thoroughly studied Lie algebras. The next best-known class consists of hyperbolic matrices, i.e. finite but not affine indecomposable matrices, for which every indecomposable component of each minor is affine or finite. It is known that the size of a hyperbolic matrix does not exceed 10×10, and that there are only finitely many hyperbolic matrices of size greater than 3×3. Besides, from the point of view of theory of identities there is no reason to isolate other classes of matrices – because the algebra $\mathbf{g}(A)$, where A is not finite, nor affine,

---

[2] Yu.V. Billig recently showed that $\mathbf{g}(A)$ may indeed fail to be special.

necessarily contains the free Lie algebra of rank 2 (Kac [1980]). It is clear that for such A, the algebra $\mathbf{g}(A)$ does not satisfy any non-trivial identities.

The adjoint associative algebra $\operatorname{Ad} L$ of a Lie algebra L with identity (see Sect. 5.4) is not necessarily a PI-algebra. An example is provided by a free solvable algebra of degree at least 3. However, if L is special then $\operatorname{Ad} L$ is a PI-algebra. The question of whether a Lie algebra L is special whenever $\operatorname{Ad} L$ is a PI-algebra is equivalent to the question of Latyshev which was formulated above.

It should be mentioned that the adjoint associative algebra of a finitely generated Jordan algebra with identity is always a PI-algebra.

§ 4. On the Geometry of Defining Relations and Identities in Groups

4.1. Problems Solvable by Geometric Methods. Certain long-standing problems in group theory, connected with the names of O.Yu. Schmidt, J. von Neumann, A. Tarski, S.N. Chernikov, R. Baer and others, have been recently solved by means of a geometric interpretation of deducing consequences from defining relations, and an analogous interpretation of consequences of some identities. Application of the method of diagrams to the analysis of identities and other properties of 'global' character which may be imposed on groups, has not yet found a parallel in the case of varieties. In spite of this, geometric approach to combinatorial study of identities appears very promising and gives an effective way of solving several difficult problems. With this in mind we devote this section to a more detailed discussion of this topic.

We begin with questions which naturally arose in the original period of investigation of infinite groups with various finiteness conditions. An infinite group will be called *quasi-finite*, if all of its proper subgroups are finite. For a prime p, such groups include the p-adic group C_{p^∞} – the union of cyclic groups of order $p, p^2, p^3, \ldots$, embedded in one another. It can also be realized as the group of all complex roots of unity of degrees $p, p^2, p^3, \ldots$. Every locally finite quasi-finite group is isomorphic to the group C_{p^∞} for some p (M.I. Kargapolov). At the same time, the question of O.Yu. Schmidt (1938) about the existence of non-abelian quasi-finite groups remains open.

All known examples of *Noetherian groups* (i.e. groups in which every subgroup is finitely generated) are provided by groups with subnormal series $G = G_0 \triangleright G_1 \triangleright \ldots \triangleright G_n = \{1\}$, in which each factor G_i/G_{i+1} is a finite or cyclic group. Baer's problem asks whether every Noetherian group possesses such a series.

A group in which every descending chain of subgroups $H_1 \supset H_2 \supset \ldots$ becomes stationary is called *Artinian*. They include Chernikov groups, i.e. direct products of finitely many groups of the type C_{p^∞}, and their finite extensions. It is easy to see that an Artinian group is periodic, since it cannot contain infinite cyclic subgroups. V.P. Shunkov showed that a locally finite Artinian group is a Chernikov group. Chernikov's question dealt with existence of other Artinian groups.

The problem of Tarski, posed independently also by A.I. Starostin, consists of the question of existence of the hypothetical 'monster' – an infinite group, whose proper subgroups all have prime orders.

von Neumann's conjecture (interesting to specialists in the theory of dynamical systems, ergodic theory and abstract harmonic analysis) postulating amenability of every group with no free non-cyclic subgroups, dates back to his paper [1929] (one of many equivalent definitions of *amenability* of a group G can be expressed as existence in G of a translation-invariant additive measure μ, defined for every subset of G, such that $\mu(G) = 1$).

In recent years Ol'shanskij constructed groups with new and unexpected properties, which give answers to the questions listed above. Investigation of these groups became possible thanks to the specially developed geometric method of analyzing a large number of defining relations (and in particular, consequences of certain identities). Other results were obtained in this way as well.

4.2. Interpretation of Deriving Consequences from Defining Relations. As was mentioned in Sect. 2.2 of Chapter 1, every group G with generators $a_1, a_2, \dots$ is isomorphic to a factor group of the free group F with free generating set $x_1, x_2, \dots$ modulo the kernel of the homomorphism α, which maps x_i to a_i $(i = 1, 2, \dots)$.

If the group G is given by a homomorphism $F \to G$ (or, as is also said, G is described by a *co-presentation*, i.e. $G \simeq F/N$), then it is not necessary to enumerate all words belonging to the kernel N of that co-presentation. It suffices to identify some set R of words r_i, which generates N as a normal subgroup (in many important cases, R is a finite set). In other words, if $r \in N$ then for some $r_{k_i} \in R$ and $s_i \in F$ we have $r = \prod_{i=1}^{l} s_i r_{k_i}^{\pm 1} s_i^{-1}$. The relations $r_i(a_1, \dots, a_{n_i}) = 1$ in the group G are then called *defining relations* for G (the left-hand side represents a word in the alphabet $A = \{a_1^{\pm 1}, a_2^{\pm 1}, \dots\}$). The abbreviation $G = \langle\, a_1, a_2, \dots ; r = 1,\ r \in R \,\rangle$ is used in this context.

A word $r = \prod_{i=1}^{l} s_i r_{k_i}^{\pm 1} s_i^{-1}$ can be visualized in the following way. We will label one point of the plane with o, and attach to it several segments p_i (see Fig. 3) with endpoints labelled $o_1, \dots, o_l$, and circles $P_1, \dots, P_l$ passing through those points. We will next divide each path p_i into $|s_i|$ edges, where $|s_i|$ is the length of the word s_i, assigning to each edge e a label $\varphi(e)$ from the alphabet A in such a way that by traversing the path p_i we can read off precisely the word s_i. Similarly, we will divide the circumference q_i of each loop P_i into several edges, labelling them in such a way that traversing it in the clockwise direction yields the word $r_{k_i}^{\pm 1}$. It is clear that travelling along the contour of the diagram constructed this way, we obtain a graphic representation of the entire word $w = \prod_i s_i r_{k_i}^{\pm 1} s_i^{-1}$.

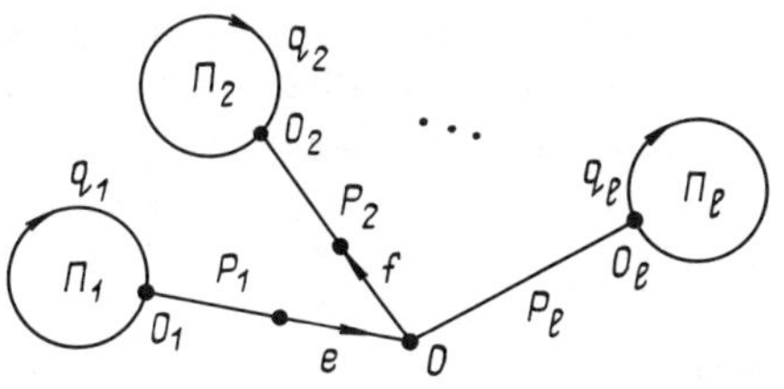

Fig. 3

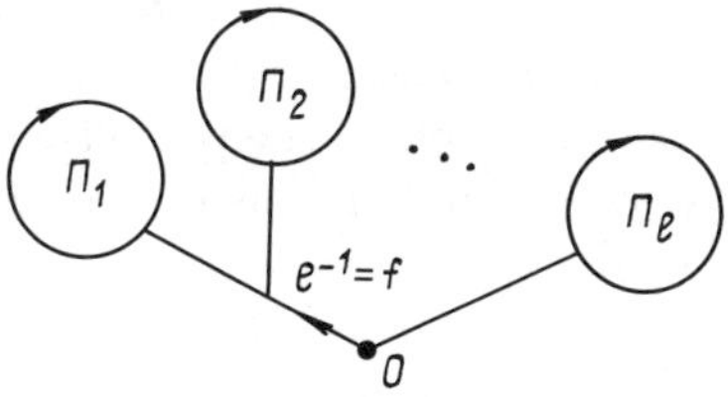

Fig. 4

The word w may not be, in general, reduced, and its graphical representation does not have to coincide with that of the element r in the free group. If labels of two adjacent edges e and f in the diagram's contour are mutual inverses of a letter in A, then by gluing the edges e^{-1} and f together (and labelling the resulting edge with their common name, always assuming that $\varphi(e^{-1}) = \varphi(e)^{-1}$) we obtain a new diagram (Fig. 4) whose contour corresponds to the word resulting from w after two neighboring letters have been cancelled. As a result, we can arrive at a diagram whose contour's labels will yield a word identical to r (see the example in Sect. 4.3 of Chapter 1, and Fig. 2).

Formally defining the notion of an arbitrary *diagram* over a given co-presentation $G = \langle A;\, r = 1, r \in R \rangle$, we first consider a simply connected graph M in the Euclidean plane, consisting of a finite collection of vertices, edges (with two orientations) and regions, for which the usual compatibility conditions hold (i.e. the endpoints of an edge belong to the set of vertices, the boundary of a region consists of edges belonging to the specified set, etc.). Further, to every edge e we assign a letter $\varphi(e)$ from A (its *label*), in such a way that $\varphi(e^{-1}) = \varphi(e)^{-1}$ and that the label of each region (i.e. the word corresponding to its boundary) with a suitable choice of the starting point and direction will yield a word belonging to R.

Above we have essentially showed that to every consequence $r = 1$ of the defining relations there corresponds a diagram over the given co-presentation, whose contour yields the word r. It is easy to see that the converse of this assertion also holds: if Δ is a diagram over a co-presentation α of the group G, then the word read off of its contour equals 1 in G. The reader will easily verify this by applying a process of cutting the diagram Δ into 'petals', reversing the process of gluing described above.

Geometric interpretation of deducing consequences of defining relations is due to van Kampen (van Kampen's lemma; see Lyndon and Schupp [1977]). This approach should not be confused with other well-known connections between co-presentations of groups and 2-dimensional simply connected complexes, in which case the group G can be described as the fundamental group of a 2-dimensional simplicial complex. As we will see later, in applications of van Kampen's lemma it is important to assure that every edge be on the boundary of at most two regions (i.e. the support is not only a 2-dimensional complex, but a surface as well); it is also important for the diagrams to be planar, i.e. realizable in the plane (and sometimes on a torus, or a sphere with a small number of holes, as well as on other surfaces whose Euler characteristic has small absolute value).

4.3. Conditions of Small Cancellation. It is astonishing that this simple observation of van Kampen's remained unnoticed for more than 30 years, until it was combined with various conditions of 'small cancellation' by Weinbaum, Lyndon and their successors. One of the simpler such conditions is the condition $C(\lambda)$, $0 < \lambda < 1$. It is related to *symmetrized co-presentations*, i.e. those for which with every word r_i from R, r_i^{-1} as well as cyclic shifts of r_i belong to R as well (a word vu is called a *cyclic shift* of a word of the form uv). It is clear that every co-presentation can be symmetrized.

For a symmetrized system of reduced words R the condition $C(\lambda)$ holds, by definition, if length $|u|$ of any common initial fragment of two distinct words uv and uw in R is subject to the inequalities $|u| < \lambda|uv|$ and $|u| < \lambda|uw|$. For example, the fundamental group G of an orientable compact surface of genus 2 has one defining relation $a_1a_2a_1^{-1}a_2^{-1}a_3a_4a_3^{-1}a_4^{-1} = 1$. Considering the left-hand side of this relation together with all of its cyclic shifts and their inverses, we can verify that a common beginning of two distinct words from that set may consist of at most a single letter, which implies that this co-presentation of G satisfies the condition $C(\lambda)$ for all $\lambda > 1/8$.

In order for the condition $C(\lambda)$ to be meaningful in the context of diagrams, it is necessary to restrict our attention to reduced diagrams. We will explain what this means. Suppose that some edge e_1 lies on the boundary of two regions P and P' in a diagram Δ, so that $e_1e_2\dots e_n$ and $e_1e_2'\dots e_n'$ represent contours of these regions, traversed in opposite directions. We will assume that $\varphi(e_i') = \varphi(e_i)$ for $i = 2,\dots,n$ – in other words, that the labels on the contours of P and P' form identical words. It is then possible to cut the regions P and P' out of Δ, obtaining a hole with boundary $e_2^{-1}\dots e_n^{-1}e_n'\dots e_2'$, and then contract this hole by gluing e_2 with e_2', etc., and e_n with e_n', preserving the common labels of these edges. It is easy to see that while reducing the number of regions in this way, we will not change the labelling of the outside contour of Δ. After several such cancellations of pairs of regions, every diagram can be made into a *reduced* one, i.e. containing no pairs which are cancellable in the above sense.

It is now easy to notice that the condition $C(\lambda)$ for a co-presentation α implies that in every reduced diagram over α, for any pair of regions the length of any *arc* between them (i.e. a common sub-path of their contours) is less than λ times either of their perimeters. In particular, an interior region (which has no edges in common with the exterior contour of the diagram) cannot be an polygon with n sides for any $n \le \lambda^{-1}$. When $\lambda \le 1/6$, using Euler's formula relating the number of vertices, edges and regions, it isn't difficult to conclude that the perimeter of the diagram grows with the number of regions t at least as fast as ct, where $c = c(\lambda) > 0$. It is clear that in the case of a finite number of defining relations this estimate, together with van Kampen's lemma, allows to give a bound on l and lengths of words s_i depending on the length of a word $r = \prod_{i=1}^{l} s_i r_{k_i}^{\pm 1} s_i^{-1}$, which in turn can be used as a foundation of an algorithm for determining whether words are equal to 1 in G.

Conjugacy of words v and w in G is interpreted by means of *annular diagrams* with two boundaries – inner and outer, with labels v and w (Schupp) which, just as above, leads to an algorithm for determining conjugacy of words in groups with condition $C(\lambda)$ when λ is small.

4.4. Geometric Analysis of Consequences of Certain Identities. We will first present certain combinatorial difficulties which arise in the study of even the apparently simple Burnside's identity $x^n \equiv 1$. It is obvious that a co-presentation of the free Burnside group $B(m, n)$ with m generators can be obtained by imposing the defining relations $v^n = 1$, where v ranges over the set of all words in the alphabet $A = \{a_1^{\pm 1}, \ldots, a_m^{\pm 1}\}$. In this case, however, too many "relations between relations" are introduced; in particular, one defining relation is known to follow from others.

A natural (and successful in the case of large odd n, as has recently transpired) method of isolating a system of independent defining relations is described below. We will first make sure that subsequent relations, defined inductively, would not follow from the preceding ones. Let, for example, $m = 2$, $r_1 = a_1^n$, $r_2 = a_2^n$, $r_3 = (a_1 a_2)^n$, $r_4 = (a_1 a_2^{-1})^n$ etc., and suppose that relations $r_1 = 1, r_2 = 1, \ldots, r_{i-1} = 1$ have already been defined. We will choose a shortest word v_i, whose order in the group $G_{i-1} = \langle a_1, a_2; r_1 = 1, \ldots, r_{i-1} = 1 \rangle$ is infinite. We will set $r_i = v_i^n$, and call v_i a *period of rank i*.

How can one settle Burnside's problem by showing that the group $B(2, n) = G = \langle a_1, a_2; r_i = 1, i = 1, 2, \ldots \rangle$? At this point it is convenient to make use of a combinatorial observation due to Thue: there exists an infinite sequence (a 'word') in the alphabet $\{a_1, a_2\}$ which contains no non-empty sub-words of the form www for any $w = w(a_1, a_2)$. All initial fragments of this word would represent distinct elements of G (i.e. $|G| = \infty$), provided we could show that for every consequence $r = 1$ of the defining relations (with r being a non-empty reduced word), r contains sub-words of the form w^3. In the language of diagrams this would mean that in a reduced diagram Δ there is a region P such that a sufficiently long sub-path p of its contour q ($|p| \geq \frac{3}{n}|q|$) is contained in the outside contour of Δ. In other words, the region P has a sufficiently long exterior arc. In Sect. 4.3 we mentioned that the question of existence of exterior arcs is easily answered in the presence of a condition of type $C(\lambda)$ with small λ. However, for co-presentations of the group $G = B(2, n)$ such a condition does not hold: a subword of length smaller than half the length of r_i may be a part of some period v_j for $j \gg i$. For example, the common beginning of the defining words a_1^n and $(a_1^k a_2)^n$

is a_1^k, with $k \leq n/2$ (when $k > n/2$, the second word may be replaced with $(a_1^{k-n}a_2)^n$). Condition of type $C(1/2)$ alone does not, in principle, yield anything (as was noted by A.I. Gol'berg, every group has a co-presentation satisfying the condition $C(\lambda)$ when $\lambda > 1/5$).

At the same time, an elementary argument shows that a periodic word u with period v_i (i.e. a sub-word of a power of v_i) cannot also have period v_j, $j \neq i$, assuming that $|u| \geq |v_i| + |v_j|$. Choosing a sufficiently large n we therefore see that a condition of type $C(\lambda)$ (with a very small λ) is satisfied for defining words v_i^n and v_j^n, for which v_i and v_j have 'similar' lengths (so that neither $|v_i| \ll |v_j|$ nor $|v_j| \ll |v_i|$ holds). One can ask how can a map be constructed without long exterior arcs of regions, while satisfying such requirements (in particular, ones without a boundary whatsoever – when regions cover the surface of a sphere)? We will attempt to visualise a map on a sphere covered by regions with only two distinct perimeters – 'small' and 'large', placed in such a way that the common boundary of two regions of the same type is small (say, less than 10^{-3} in comparison with their perimeter, and the arc between a small and a large region does not exceed half the perimeter of the small region. This will inevitably result in a map resembling the one in Fig. 5, i.e. there will necessarily be 'long' and 'narrow' semiaxes consisting of small regions, which are 'squeezed' in between pairs of large regions. Technical arguments based on planarity of diagrams show that, indeed, in a diagram Δ over a co-presentation of the group $G = B(m,n)$ some regions will have long exterior arcs, unless Δ does contain long and thin semiaxes of this kind.

The problem therefore becomes: show that Δ doesn't contain long 'semi-axes'. Considering such semi-axis Γ, we will use induction by reducing the parameter i to $i-1$, since Γ is filled with regions whose perimeter is small. The contour of Γ decomposes into a product $p_1q_1p_2q_2$, where words $\varphi(q_1)$ and $\varphi(q_2^{-1})$, which represent labels of q_1 and q_2^{-1}, are periodic words with periods v_k and v_l, $|v_k| \ll |q_1|$, $|v_l| \ll |q_2|$, both $|p_1|$ and $|p_2|$ are less than $|v_k|$ and $|v_l|$, and finally q_1 and q_2^{-1} have no common sub-paths of length comparable with $|q_1|$ or $|q_2|$. We now need to show that such a diagram cannot be constructed.

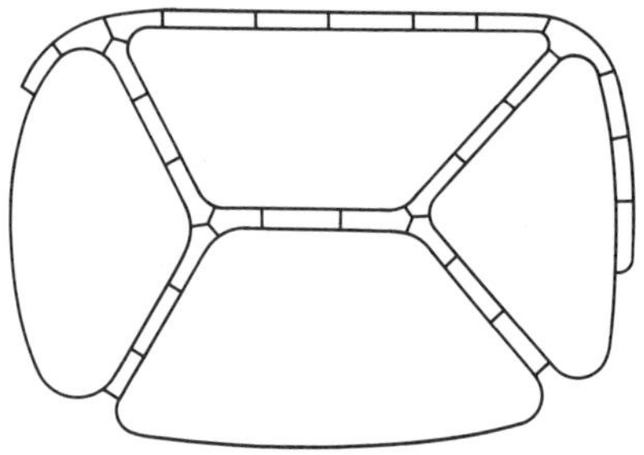

Fig. 5

This task can be reduced to the case $v_k = v_l$ by means of the following device: we attach to Γ its mirror image L with contour $b^1 d^1 b^2 d^2$ (see Fig. 6), so that L is glued to Γ via identifcation of edges of the paths q_1 and d^1; this identification, however, is not a trivial one (which would not yield the desired result, because the word $\varphi(q_1)$ is not assumed to be periodic), but one in which the labels of q_1 and d^1 are shifted to the period v_k.

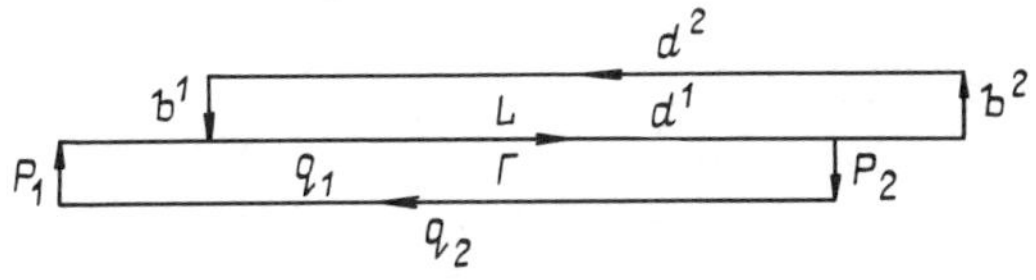

Fig. 6

Conversely, the 'symmetric' case $v_k = v_l$ can be transformed into the 'asymmetric' one with a pair of periods (v_k, v) in which, however, $|v| < |v_k|$ – which allows to carry out a compatible argument by induction on $|v_k| + |v_l|$. Naturally, full formulation of all inequalities and inductive hypotheses requires logical precision, which are not attempting to maintain here. We will only concentrate on the idea of the reduction mentioned above. It is based on the fact that when $v_k = v_l$, one can identify the paths q_1 and q_2^{-1} while preserving labels of their edges, as a result of which the diagram Γ will be transformed into a ring with contours p_1 and p_2 (see Fig. 7). This ring should in fact be 'narrow', due to the inductive hypothesis asserting that in diagrams over co-presentations of the group G_{i-1}, 'almost all' edges of regions are exterior. Since length of the path q_1 is very large in comparison with $|p_1|$ and $|p_2|$, it must be 'wound' around p_1 many times – i.e. the path q_1 is homotopic in E with a path of the form $p_1^s t$, where t is a short path dissecting the diagram E. This means that $\varphi(q_1) = \varphi(p_1)^s \varphi(t)$ in G_{i-1}, and we return to the problem of equality between two periodic words $\varphi(q_1)$ and $\varphi(p_1)^s$ with periods v_k and $v = \varphi(p_1)$, for which $|v| = |p_1| < |v_k|$.

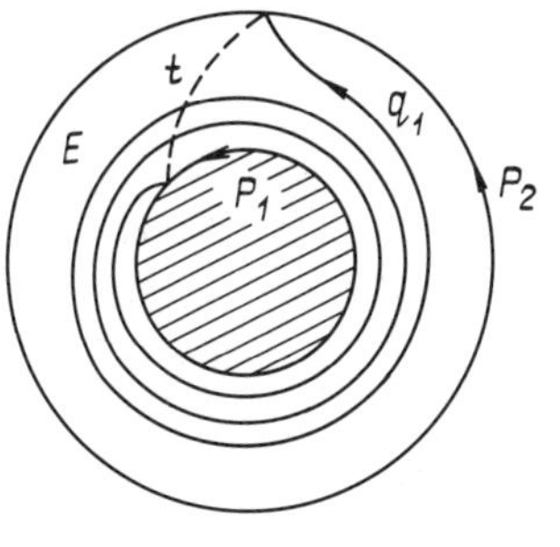

Fig. 7

As the reader has observed, in simplifying the description of a general idea of the proof we have not only omitted technical details, but also the necessity of the condition that n be odd has been lost from our view. In any case, Fig. 8 shows that with n even, there may be arbitrarily long 'semi-axes' between two regions with identical contour labels $(a_1^{n/2}a_2^{n/2})^n$.

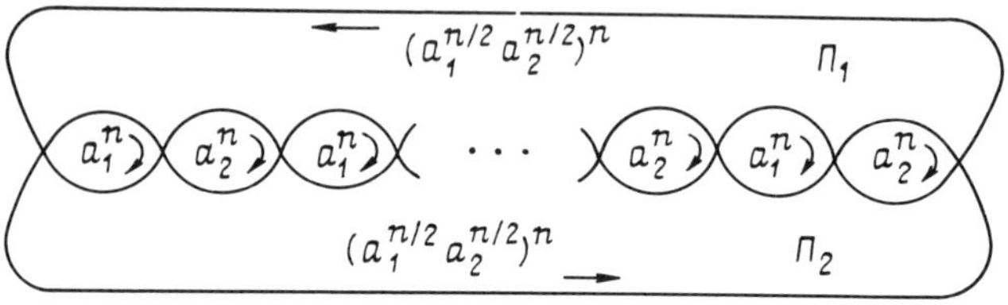

Fig. 8

In order to solve other problems from Sect. 4.1, more complicated relations have to be considered. However, defining words in those cases are almost entirely composed of high powers of their sub-words. And so, in solving the problem of Schmidt and Tarski, we conclude that relations of the form $a = u_i v^r u_i v_i^{r+1} \dots u_i v_i^{r+h}$, where $a = a_1, a_2$ and $r, h \gg 1$, guarantee in effect that every pair of non-commuting elements will generate the whole group. New technical difficulties encountered here are connected with decomposing contours of diagrams into many fragments. Other questions arising in this situation lead to diagrams on a torus, or diagrams with several 'holes', and to problems of equating self-intersecting paths in such diagrams. In effect, the value of the geometric method just described depends on the ability to rely, explicitly and implicitly, on Jordan's lemma and its corollaries for two-dimensional surfaces. Aside from this, it is important that the diagrams adequately reflect not only consequences of relations, but also the process of deriving these consequences in groups.

§ 5. Effect of Identities in Finite and Finite-Dimensional Algebras

5.1. Variety Generated by a Finite Algebra. Let G be a finite algebra. The study of algebras satisfying all identities which can be derived in G is equivalent to the study of the variety $\mathfrak{V}$ generated by G (we write $\mathfrak{V} = \operatorname{var} G$). One of the first questions naturally arising in connection with this deals with the structure of finitely generated algebras from $\mathfrak{V}$. Another question is: let $\operatorname{var} G_1 = \operatorname{var} G_2$, where G_1 and G_2 are finite algebras; what natural conditions imply $G_1 = G_2$? Necessity of such conditions is obvious, since

$\operatorname{var} G \simeq \operatorname{var}(G \times G)$. Finally, what 'parameters' of the algebra G are inherited by algebras in $\operatorname{var} G$?

We will note that Birkhoff's theorem yields a representation of any finite algebra $A \in \operatorname{var} G$ as $A = B/\rho$, where $B \subseteq \underbrace{G \times \ldots \times G}_{t}$ (direct product of a finite number t of isomorphic copies of the algebra G), and ρ is some congruence on B. A more detailed study of the situation is possible after introducing the notion of a subfactor of an algebra (which should not be confused with a factor algebra; see Sect. 2.4 in Chap. 1). We will recall (see Sect. 2.4, Chap. 1) that algebra G is called critical if it does not belong to the variety generated by its proper subfactors. For every algebra G there is an associated set $\mathcal{F}(G)$ of its subfactors. It follows from what was said above that every finite algebra A in $\operatorname{var} G$ can be represented in the form $A = B/\rho$, where $B \subseteq G_1 \times G_2 \times \ldots \times G_t$, with $G_i \in \mathcal{F}(G)$ for $i = 1, 2, \ldots, t$ and $|G_1| \geq |G_2| \geq \ldots \geq |G_t|$. In this case we say that we have given a *representation of the algebra A in the variety* $\operatorname{var} G$. We will introduce a partial order in the set of such representations by comparing the t-tuples $(|G_1|, \ldots, |G_t|)$ lexicographically. Having in mind a representation which is minimal in this sense, we will introduce the concept of a *minimal representation* (which is not always unique). In many classes of algebras minimal representations exhibit a number of desirable properties (for example, all the G_i's are critical, the projection of A on any G_i is surjective etc.), which permit to carry out 'sufficiently categorical' arguments about the structure of the algebra $A \in \operatorname{var} G$. For example, one can show in this way that if G_1 and G_2 are two finite simple algebras (i.e. algebras with no non-trivial congruences), then their identities coincide if and only if these algebras are isomorphic.

5.2. The Class $\mathfrak{C}(e, m, c)$. It appears that the first class of algebras in which varieties generated by a finite algebra were thoroughly investigated, was the class of groups. These studies were originated in connection with the famous theorem of Oates and Powell on existence of a finite basis of identities of a finite group (for more details see Sect. 4.1, Chap. 3). Kovács and Newman [1971] introduced a class of groups $\mathfrak{C}(e, m, c)$ (see also Neumann [1967]). Here e, m and c are integers, and a group G is in $\mathfrak{C}(e, m, c)$ if and only if it satisfies the identity $x^e \equiv 1$, nilpotent subfactors of this group have degree of nilpotency at most c, while principal subfactors have order not exceeding m. A subfactor H/K of a group G is called *principal*, if K is a maximal normal subgroup of G among those which are strictly contained in the normal subgroup H. A not too complicated argument shows that if $G \in \mathfrak{C}(e, m, c)$, then $\operatorname{var} G \subseteq \mathfrak{C}(e, m, c)$. A significantly more difficult proof demonstrates that for any e, m and c the class $\mathfrak{C}(e, m, c)$ is a variety generated by a single finite group. As such a group one can take the direct product of all critical groups from the class $\mathfrak{C}(e, m, c)$.

Finite simple groups are critical (an exercise; see Sect. 5.1). Identities of critical groups are important because every locally finite variety is generated by its critical groups. Unfortunately, critical groups are not uniquely determined by their identities. For example, two non-abelian groups of order 8 – D_4 and Q_8 (the dihedral and quaternion groups) are critical, i.e. the variety generated by their proper subfactors is an abelian variety of exponent 4, and the identities $x^4 \equiv 1$ and $[x, y^2] \equiv 1$ can serve as a basis of identities of either of these groups. Hence $\operatorname{var} D_4 = \operatorname{var} Q_8$, but $D_4 \not\simeq Q_8$. Nevertheless, there is a certain link between these groups, even though it isn't always strong. Let M_1 be the *monolith* of a group G_1 (the smallest normal subgroup distinct from $\{1\}$), $C_1 = C(M_1)$ – its centralizer in G_1, and let M_2, C_2 denote analogous objects in a group G_2. Existence of a monolith in a critical group was mentioned in Sect. 2.4 of Chapter 1. If $\operatorname{var} G_1 = \operatorname{var} G_2$, with G_1 and G_2 – critical groups, then there exist isomorphisms $\alpha : M_1 \to M_2$ and $\beta : G_1/C_1 \to G_2/C_2$ such that for every $x \in G_1$ and $g \in M_1$ we have

$$\alpha(xgx^{-1}) = \beta(xC_1)\alpha(g)\beta(xC_1)^{-1} \tag{3}$$

(the right-hand side is well-defined, since $\beta(C_1) = C_2 = C_{G_2}(M_2)$). Equality (3) defines the notion of similarity, very close to the concept of a homomorphism of modules. This link becomes even stronger in the case of finite Lie algebras.

5.3. Finite Rings. For finite Lie algebras (i.e. finite Lie rings considered over a finite ring of operators) there is a theory, fully analogous to the one described above. In defining a class $\mathfrak{c}(f, m, c)$, corresponding to the class $\mathfrak{C}(e, m, c)$, the first parameter is no longer a number. In accordance with remarks made in the beginning of Sect. 2, it becomes a polynomial (with leading coefficient 1), and an algebra L belongs to $\mathfrak{c}(f, m, c)$ if and only if it satisfies the identity $f(\operatorname{ad} x) \equiv 0$, cardinality of principal subfactors is bounded by m, and the index of nilpotency of nilpotent subfactors does not exceed c. As in the case of groups, $G \in \mathfrak{c}(f, m, c)$ implies $\operatorname{var} G \subseteq \mathfrak{c}(f, m, c)$, and for any choice of parameters f, m, c the class $\mathfrak{c}(f, m, c)$ is a variety generated by a single finite Lie algebra.

Structure of a finite Lie algebra L from a variety $\operatorname{var} G$, where G is another finite Lie algebra, becomes substantially more clear thanks to the existence in L of a *socle series* of bounded length, i.e. a series $L = L_0 \triangleright L_1 \triangleright \ldots \triangleright L_{q-1} \triangleright L_q = \{0\}$ in which, for every i, the ideal L_{i-1}/L_i is equal to the sum of all minimal ideals of the algebra L/L_i.

Length of such series is known to be bounded by the maximum q of cardinalities of critical subfactors of G. Minimal ideals mentioned above are similar to monoliths of critical subfactors. In the context of algebras over a field, cardinality q can be replaced by dimension. For example, for the variety $\operatorname{var} G$ with $G = \langle e, f \mid [e, f] = f,\ 2e = 2f = 0 \rangle$ we can take $q = 2$. If $L \in \operatorname{var} G$, then L contains a series $L = L_0 \triangleright L_1 \triangleright \{0\}$ in which L_0/L_1 is abelian, $L = H_1 \oplus \ldots \oplus H_t$ for one-dimensional ideals H_i of L, and $\dim L/C_L(H_i) = 1$ for every $i = 1, \ldots, t$.

In the case of associative rings, the parallelism described above is also present. The structure theory allows to make several simplifications, which then permit to extend the theory to the case of the so-called almost nilpotent rings, i.e. rings with a nilpotent ideal of finite index. The class $C(e,d,c)$ which is considered in this situation, is defined as follows: an associative ring R belongs to $C(e,d,c)$ if its additive group has exponent e, its primitive subfactors are finite simple rings whose cardinalities divide d, and the index of nilpotency of its nilpotent subfactors does not exceed c.

The technique which originated in group theory, was transferred, through Lie algebras, to other classes of non-associative algebras: Jordan algebras, alternative algebras, and others. Its generalizations are not automatic, and additional complications arise in various places – for example, in defining a module over such more general algebras, etc.

5.4. Structure of Linear Algebras with Identities of a Finite-Dimensional Algebra. Let A be an n-dimensional algebra over a field, $\{e_1,\dots,e_n\}$ – its basis, and $f(x_1,\dots,x_{n+1},y_1,\dots,y_m)$ – a (non-associative) multilinear polynomial, skew-symmetric with respect to the indeterminates $\{x_1,\dots,x_{n+1}\}$. Because of multilinearity, every value of the polynomial f on A is a linear combination of elements of the form

$$f(e_{i_1},\dots,e_{i_{n+1}},a_1,\dots,a_m)\,, \tag{4}$$

where $1 \le i_1,i_2,\dots,i_{n+1} \le n$ and $a_1,\dots,a_m \in A$. Skew-symmetry in the first $n+1$ variables forces (4) to be zero. It follows that

$$f(x_1,\dots,x_{n+1},y_1,\dots,y_m) \equiv 0 \tag{5}$$

is an identity in A.

Definition. For a fixed signature Ω of linear algebras, the set of all identities (5) (with f satisfying the conditions imposed above) is called *the system of Capelli identities of order* $n+1$.

Naturally, a system of Capelli identities can be satisfied in an infinite-dimensional algebra as well. For example, the algebra $N = \langle e_1, e_2, \dots, e_m, \dots \rangle$ with trivial multiplication over a field k satisfies every such system of order $n \ge 2$, but its dimension may be arbitrarily large. At the same time, by extending the domain of scalars (while preserving its commutativity), we can view this algebra as one-dimensional. For example, we can adjoin the endomorphism $\varphi : e_i \mapsto e_{i+1}$ $(i = 1,2,\dots)$ to k. Over the ring of polynomials $K = k[\varphi]$, the module N is cyclic.

The example described above is typical, and at this point we arrive at the important concepts of the adjoint algebra and the centroid. If A is a linear algebra over a field k and Ω is the set of its operations with arity at least 2, then for every n-ary operation $\omega \in \Omega$, and any elements $a_1,\dots,a_{i-1},a_{i+1},\dots,a_n \in$

A, the mapping $x \mapsto \omega(a_1, \dots, a_{i-1}, x, a_{i+1}, \dots, a_n)$ is a linear endomorphism of the vector space A over the field k. The set of all transformations of this form generates an associative subalgebra $\operatorname{Ad} A$ in the algebra $\operatorname{End}_k A$ of all linear endomorphisms of A. We say that $\operatorname{Ad} A$ is the *adjoint associative algebra* of A. The centralizer Z of $\operatorname{Ad} A$ in $\operatorname{End}_k A$ is called the *centroid* of the algebra A. It is clear now that after defining multiplication of elements of Z by elements of A by setting $\varphi a = \varphi(a)$ ($\varphi \in Z$, $a \in A$), A becomes a Z-algebra. In many cases the centroid is non-commutative. In the example discussed above, we have $Z = \operatorname{End} N$. In a number of important situations, however, Z does turn out to be commutative, and when A is a simple algebra, then Z is also a field.

Several ground-breaking results on Ω-algebras with Capelli identities were obtained by Razmyslov [1981]. More precisely, they concern so-called weak Capelli identities. Let V be a subspace of A which is closed under the action of the set Ω' of derived operations, consisting of evaluations of multilinear polynomials of signature Ω (see Chapter 3, Section 1.1). We then say that the pair (A, V) has signature (Ω, Ω'). The pair (A, V) satisfies the system of *weak Capelli identities of degree* $n+1$ if (5) holds whenever elements of V are substituted for $x_1, \dots, x_{n+1}$, and elements of A are substituted for $y_1, \dots, y_m$. A theorem of Razmyslov asserts that the pair (A, V) can be embedded in a pair $(\bar{A}, \bar{V})$ of the same signature, and in such a way that $\bar{V}$ contains elements $e_1, \dots, e_n$ which constitute a basis of $\bar{V}$ in the following sense: $\bar{A}$ is a module over its adjoint associative algebra $\bar{D} = \operatorname{Ad} \bar{A}$. Razmyslov's theorem states that the $\bar{D}$-module $\bar{A}$ can be embedded in a $\bar{D}$-module M such that every element $v \in \bar{V}$ can be written as $v = \sum_{i=1}^{n} \lambda_i e_i$, where λ_i are endomorphisms of the $\bar{D}$-module M for $i = 1, 2, \dots, n$.

Construction of such basis can be illustrated by the example of any infinite-dimensional Lie algebra L with index of nilpotency 2, i.e. $L^3 = \{0\}$. At first, we adjoin to L elements e_1 and e_2 so as to obtain a Lie algebra $\bar{L}$ which is nilpotent of index 2 and satisfies no other relations involving e_1 or e_2. To do this, it is enough to extend the basis by adding to it elements e_1, e_2, $[e_1, e_2]$, $[x_\alpha, e_1]$ and $[x_\alpha, e_2]$, where $\{x_\alpha\}$ is the original basis of the algebra L. This yields the desired algebra ($\bar{A} = \bar{L}$, $A = L$). Now $\bar{D} = \operatorname{Ad} \bar{L}$, and M is the injective hull of the $\bar{D}$-module $\bar{L} \oplus \bar{L}$. We will identify $\bar{L}$ with the first summand in $\bar{L} \oplus \bar{L}$. Then $\{e_1, e_2\}$ is the desired basis. Existence of endomorphisms $\lambda_1, \lambda_2 \in \operatorname{End}_{\bar{D}} M$ for a given $v \in \bar{L}$ follows from the fact that the element $e_1 + \psi e_2 = (e_1, e_2)$, where ψ is the endomorphism interchanging the coordinates, generates a free $\bar{D}$-module. The mapping assigning $v \in L$ to $e_1 + \psi e_2$ extends to a homomorphism of submodules of M and hence, by injectivity of the module M, to an endomorphism $\varphi \in \operatorname{End}_{\bar{D}} M$. We now have $v = \varphi(e_1 + \psi e_2) = \varphi e_1 + \varphi\psi e_2$, so that $\lambda_1 = \varphi$ and $\lambda_2 = \varphi\psi$.

The above example shows that this construction applied blindly, so to speak, sometimes results in algebras more complicated than the original one. Nevertheless, it also has some desirable consequences.

Theorem (Razmyslov [1981]). *Let L be a Lie algebra satysfying a system of weak Capelli identities of order $m+1$; then L contains a nilpotent ideal N with $N^m = \{0\}$, such that $\mathrm{Ad}\,(L/N)$ satisfies all identities of matrices of certain size.*

This approach has interesting applications to the study of identities of simple algebras. The point is that in the case of semiprime algebras (i.e. those without non-zero nilpotent ideals), it is possible to give a more explicit construction of the algebras $\bar{A}$, $\bar{L}$, and of the module M appearing in the statement of Razmyslov's theorem. In particular, we have

Theorem. *If a pair (A, V) satisfies weak Capelli identities of order $m+1$ and A is a prime algebra with classical ring of quotients Q, then $\dim_C V \leq m$; here $Q = CA$, $C = \mathrm{End}\,_D P$, P is the injective hull of the D-module A, and $D = \mathrm{Ad}\,A$.*

It can be also shown that C is a commutative algebra.
This result has the following important application:

Theorem. *Let A_1 and A_2 be two finite-dimensional simple linear Ω-algebras over an algebraically closed field. If $\mathrm{var}\,A_1 = \mathrm{var}\,A_2$ then $A_1 \simeq A_2$.*

The centroid turns out to be useful in the study of simple algebras with identities. For example, the following theorem due to Yu.A. Bakhturin holds: every locally finite-dimensional simple Lie algebra with identity over a field of characteristic zero has finite dimension over its centroid (Bakhturin [1976]).
The importance of systems of Capelli identities is emphasized by the following result:

Theorem. *Let $\mathfrak{V}$ be a variety of linear algebras over a field k of characteristic zero, satisfying the system of all Capelli identities of some fixed order n. Then, modulo this system, $\mathfrak{V}$ is defined by a collection of identities in at most $n-1$ variables.*

Proof. The system of Capelli identities of order n can also be defined by means of Young tableaux (see Sect. 3.4, Chap. 1). Namely, as a basis of this system one should take the set of identities of the form $e_d(v_i) \equiv 0$, where d is any Young tableau with not fewer than n rows. Let now $f(x_1, \ldots, x_m) \equiv 0$ be a multilinear identity of degree m, satisfied in $\mathfrak{V}$. We will view $f(x_1, \ldots, x_m)$ as an element of the Sym_m-module P_m of all multilinear polynomials in m variables. The unit element of the group algebra R of Sym_m over k can be represented as a sum of elements from the left ideals Re_d, generated by elements e_d (see Sect. 3.3, Chap. 1), where d ranges over a set of Young tableaux. We will write $1 \cdot f = \sum_d \alpha_d e_d \cdot f$, where $\alpha_d \in R$. If d has at least n rows, then $e_d f$ is a linear combination of values of those polynomials of the form $e_d(v_i)$, for which $e_d(v_i) \equiv 0$ belongs to the system of Capelli identities of order n. This implies that modulo that system (which is satisfied in $\mathfrak{V}$), the identity $f(x_1, \ldots, x_m) \equiv 0$ is equivalent to a system of identities of the form $\sum_d \alpha_d e_d f \equiv 0$, where the sum involves only tableaux obtained from a single diagram consisting of at most $n-1$ columns. It remains to note that if the variables from an identity $e_d(v_i) \equiv 0$ which appear in the same row of the tableau d are identified, then, by symmetry (see definition of e_d in Sect. 3.3 of Chapter 1), we obtain an identity equivalent to the original one, but involving only indeterminates in the number not exceeding the number of rows in d.

The theorem has been proved.

In studying identities of finite-dimensional algebras, this result allows to restrict attention to a finite set of indeterminates.

Chapter 3
Systems of Identities

The language of identities is expressive enough to reflect important properties of algebraic systems, and, as we have seen, the presence of an identity naturally influences the structure of a concrete algebra. This leads to questions about the structure of identities. Among them we have: the problem of finding all identities of a given algebra or, at the least, of determining whether these identities have a finite basis; comparison of 'strength' of various identities; relations between classes of algebraic systems with given identities (i.e. between varieties); links between identities, and operations on algebras (of the type of extension of groups etc.); determination of natural numeric parameters of identities; identification of identities which are extremal relative to certain properties, etc. In this chapter we concentrate in detail on basic characteristics of identities and varieties defined by them. Our exposition begins with general and 'coarse' properties of systems of identities, and then turns towards finer characteristics, as well as particular identities.

§ 1. The Finite Basis Problem

1.1. On Derivation of Consequences. Questions about links and dependencies between identities, about certain identities being consequences of others, arise in a natural way whenever algebras belonging to natural classes of algebraic systems are being considered: the class of all groups, rings, lattices etc. Losing almost no generality, we will assume that the class $\mathfrak{M}$ in question is a variety of algebras of certain fixed signature Ω. In this case all the information about identities of Ω-algebras from $\mathfrak{M}$ and about subvarieties of $\mathfrak{M}$ is encoded in the $\mathfrak{M}$-free algebra F of countable rank. More precisely, let

$\{x_1, x_2, \ldots\}$ be a free generating set of F. Every element $f \in F$ is constructed from $x_1, x_2, \ldots$ by means of operations from Ω and, as a rule, f is given together with such description. It determines the order in which operations are applied in order to obtain f from a subset $\{x_1, \ldots, x_n\}$. Having such description in mind, we write $f = f(x_1, \ldots, x_n)$. The set of all consequences of a system of identities $\{f_i \equiv g_i\}_{i \in I}$ can be described as the set of all identities $\{f \equiv g\}$, in which f and g is any pair of elements which are equivalent with respect to the smallest congruence $\sim$ on F satisfying 1) $f_i \sim g_i$ for all $i \in I$, and 2) if $v \sim w$ then $\alpha(v) \sim \alpha(w)$ for any homomorphism $\alpha : F \to F$ (i.e. substituting any 'words' for variables in an identity also yields an identity). The minimal congruence with properties (1) and (2) is called verbal (see also Sect. 2.2 in Chap. 1). It is generated by the set $\{f_i \equiv g_i\}_{i \in I}$.

The lattice (see Sect. 6.1) of all subvarieties of a variety $\mathfrak{M}$ is anti-isomorphic to the lattice of all verbal congruences on F. In the most interesting cases of groups or rings, congruences correspond to decompositions into cosets modulo a normal subgroup or an ideal. The verbal subgroup (ideal) corresponding to a set of identities V is denoted by $V(F)$.

Coincidentally, verbal congruences (subgroups, ideals) are of interest not only in free algebras (see definition in Sect. 2.2, Chap. 1). For example, the commutator $[x_1, x_2]$ (or the identity $[x_1, x_2] \equiv 1$) corresponds to the commutant $[A, A]$ of a group A, while in the ring signature, the word $x_1 x_2 \ldots x_n$ corresponds to the forming of the power A^n of a ring A.

The fact that a subvariety $\mathfrak{V}$ of the variety $\mathfrak{M}$ of rings (or groups, etc.) can be distinguished in $\mathfrak{M}$ by means of a finite set of identities, is equivalent to the statement that every chain $\mathfrak{V}_1 \supseteq \mathfrak{V}_2 \supseteq \ldots$ of subvarieties of $\mathfrak{M}$ such that $\bigcap_i \mathfrak{V}_i = \mathfrak{V}$, has to stabilize after finitely many steps; this can also be expressed in terms of chains of verbal ideals (subgroups, etc.) $V_1(F) \subseteq V_2(F) \subseteq \ldots$ for which $\bigcup_i V_i(F) = V(F)$ necessarily stabilizing after a finite number of terms.

A proof of existence of a finite basis of identities in a variety $\mathfrak{V}$ usually becomes easier if it is possible to verify in advance that $\mathfrak{V}$ has finite axiomatic rank, i.e. all of its identities follow from a set of identities involving indeterminates whose number is bounded from above (see Sect. 3.1), since in that situation it suffices to consider an $\mathfrak{M}$-free algebra F_n with a finite set of free generators $\{x_1, \ldots, x_n\}$, and to show that $V(F_n)$ has a finite set of generators (as a verbal congruence, i.e. as an ideal, normal subgroup, etc.).

A typical example is provided by a proof of the existence of a finite basis of identities of any nilpotent associative algebra A. It is clear that in the presence of an identity $x_1 x_2 \ldots x_n \equiv 0$, every other identity of A can be written as $f = \sum_i \alpha_i u_i \equiv 0$, where u_i is a (non-commutative) monomial of degree $< n$. Since each u_i involves fewer than n variables, by substituting zero for all the remaining ones we can replace the identity $f \equiv 0$ with an equivalent system of (normal) identities $f_k \equiv 0$, in which each f_k depends on fewer than n variables. It remains to notice that every ascending chain of verbal ideals in a free nilpotent (of index n) algebra with finite rank stabilizes. This, however, obviously follows from finite-dimensionality of a nilpotent algebra with finitely many generators.

It is easy to see that in the argument just presented, associativity of A was not used. The same assertion can also be easily generalized to the case of nilpotent rings (i.e. algebras over $\mathbb{Z}$), and – with the use of certain commutator relations (see Sect. 4.2, Chap. 1) – to the case of nilpotent groups as well (Lyndon; see H. Neumann [1967]).

Axiomatic rank of subvarieties can be bounded in an even more general case. Quite simple reasoning, dating back to Weyl and using Young diagrams, allows to prove that it is enough to require that a system of Capelli identities of some order be satisfied (see Sect. 5.4, Chap. 2). Unfortunately, simple considerations are very rarely sufficient in proving existence of a finite basis of identities.

1.2. Hereditary Finitely Based (Specht) Varieties. If every subvariety of a variety $\mathfrak{M}$ admits a finite basis of identities, then $\mathfrak{M}$ is called a *hereditary finitely based variety* (in case of linear algebras, they are also called *Specht varieties*). For a hereditary finitely based variety, it is natural to consider in the first place the problem of determining (up to equivalence) all system of identities of its algebras. As was remarked in Sect. 1.1, the nilpotency identity always implies that a variety is a Specht variety. Among varieties of semigroups which are hereditary finitely based we have, for example, the variety of all commutative groups (Perkins [1969]; Volkov and Shevrin [1985]).

The first non-trivial results on heredity of the finite basis property in larger classes of groups and Lie algebras are connected with Cohen's method. For this reason we will now examine in more detail the first such result (Cohen's theorem), according to which the finite basis property holds for every *metabelian group*, i.e. a group G whose commutant G' is abelian (this is equivalent to the identity $[[x_1, x_2], [x_3, x_4]] \equiv 1$, which defines the variety of all metabelian groups; see Cohen [1967]).

To prove Cohen's theorem it suffices to establish the finiteness condition on ascending chains of verbal subgroups of the $\mathfrak{M}$-free group F of countable rank. An even stronger assertion holds – namely, the finiteness condition on ascending chains of normal Φ-subgroups of F (subgroups, invariant under permutations of free generators of F). Cohen makes use of the Magnus embedding (see Sect. 2.1), which allows to reduce the problem to the study of a free module over the free abelian group of countable rank, and of its Φ-submodules – whose definition is similar to the one given above for subgroups. Cohen proves the finiteness condition for Φ-submodules by combining the idea of the proof of Hilbert's basis theorem with the study of partially well-ordered sets and closure operators with the finite basis property (f.b.p.). A closure operator on a set M assigns to each subset $A \subseteq M$ a subset $\bar{A}$ in such a way that: 1) $A \subseteq \bar{A}$, 2) $\bar{\bar{A}} = \bar{A}$, 3) $A \subseteq B \Rightarrow \bar{A} \subseteq \bar{B}$, and 4) if $x \in \bar{A}$, then $x \in \overline{\{a_1, \ldots, a_n\}}$ for some finite subset $\{a_1, \ldots, a_n\} \subseteq A$. The closure operator has the f.b.p. whenever every closed subset is a closure of a finite subset.

For elements of a free module one can define the concepts of the leading term and a generalized exponent. The set of generalized exponents has the f.b.p., and exponents of elements of a Φ-submodule form a closed set. The key point of the proof is the following lemma: assume that a closure operator satisfying f.b.p. is defined on M, and a set P is partially ordered and satisfies the finiteness condition on descending chains; then the closure operator defined in the natural way on $M \times P$ satisfies the f.b.p.

Developing Cohen's method, McKay proved the finite basis property for every variety of groups with the central metabelian identity $[[[x,y],[z,u]],v] \equiv 1$, while Bryant and Newman [1974] showed this for any subvariety of $\mathfrak{N}_c\mathfrak{N}_2 \cap \mathfrak{N}_2\mathfrak{N}_c$ (see Sect. 2.1 for definition of a product of varieties; $\mathfrak{N}_c$ consists of groups G which are nilpotent of degree not more than c, i.e. with the condition $\gamma_{c+1} = \{1\}$ – see Sect. 4.2 of Chapter 1). A.N. Krasil'nikov and A.L. Shmel'kin [1978] established the existence of a finite basis of identities in every group from $\mathfrak{N}_c(\mathfrak{B}_n \cap \mathfrak{A})$ (here $\mathfrak{A}$ stands for the variety of all abelian groups, while $\mathfrak{B}_n$ is defined by the identity $x^n \equiv 1$. This implies, in particular, the presence of a finite basis of identities in every supersolvable group, i.e. a group with finite series $G = G_0 \triangleright G_1 \triangleright \ldots \triangleright G_m = \{1\}$, where G_i are normal subgroups of G whose factors G_i/G_{i+1} are cyclic for $i = 0, 1, \ldots, m-1$.

Cohen's approach applied to identities of all *metabelian Lie algebras* (i.e. Lie algebras with abelian commutant) leads to the consideration of free modules over the polynomial ring $k[x_1, x_2, \ldots]$, and their Φ-transformations – where Φ is the set of isotonic (i.e. order-preserving) mappings φ of the set of positive integers. For an element $u = \sum_i f_i(x_1, \ldots, x_s)y_i$ of such a module we set $\varphi(u) = \sum_i f_i(x_{\varphi(1)}, \ldots, x_{\varphi(s)})y_{\varphi(i)}$. Finite basis property for all subvarieties of the metabelian variety $\mathfrak{M}$ follows from the finiteness condition on ascending chains of Φ-invariant submodules.

One of the outstanding amplifications of this last result is a theorem of Krasil'nikov, according to which, over any field of characteristic zero, every variety of Lie algebras with nilpotent commutant (i.e. belonging to $\mathfrak{N}_c\mathfrak{A}$ for a suitable c) has a finite basis of identities. Significance of this result lies in the fact that, according to the classical theorem of Lie, the commutant of a finite-dimensional solvable Lie algebra over a field of characteristic zero is nilpotent. It follows that identities of any finite-dimensional soluble Lie algebra over a field of characteristic zero admit a finite basis. The proof of this result relies on validity of Capelli identities of certain order in $\mathfrak{N}_c\mathfrak{A}$. Both the question of whether identities of groups with nilpotent commutant are finitely based, and a closely related question of A.L. Shmel'kin about the finite basis property of identities of matrix groups over a field, remain unanswered.

One of the general results on identities of associative algebras over a field k of characteristic zero was obtained by Latyshev [1976] and Genov [1976].

Existence of a finite basis is ensured by the identity

$$[x_1, x_2] \dots [x_{2n-1}, x_{2n}] \equiv 0 .$$

As was shown earlier by Latyshev, this implies that every finitely generated algebra over k satisfying an identity which does not hold in the full matrix algebra $M_2(k)$, generates a Specht variety.

Varieties which do not contain the algebra $M_2(k)$ are called *non-matrix* varieties. The Latyshev-Genov theorem was strengthened by Kemer [1980a]: every non-matrix variety which does not contain a tensor square $G \otimes G$ of an infinite-dimensional Grassman algebra, is a Specht variety[1]. Since $G \otimes G$ does not satisfy identities of the form

$$[x_1, x_2, \dots, x_l][y_1, y_2, \dots, y_m] \dots [z_1, z_2, \dots z_n] \equiv 0 , \qquad (1)$$

every ideal of identities containing the left-hand side of (1), is generated as a verbal ideal (or T-ideal) by finitely many elements. In other words, a variety with an identity (1) is a Specht variety (Latyshev [1976]). A basis of identities for $G \otimes G$ was found by Popov [1981]:

$$[[x, y], [z, u], v] \equiv 0 , \quad [[x, y]^2, z] \equiv 0 .$$

Quite recently, Kemer established Specht property for the standard identity of degree 4 (see Sect. 1.3, Chap. 1). It is also known (Latyshev) that every binomial identity, i.e. an identity $f \equiv 0$ in which f is a linear combination of two monomials, is a Specht identity. Similar results for Lie varieties were obtained by Medvedev [1978].

Modularity of the lattice of congruences (e.g. in case of groups or rings, see Sect. 2.3) allows to deduce that in a variety $\mathfrak{U} \vee \mathfrak{V}$, every subvariety is determined by finitely many identities whenever this property holds for both varieties $\mathfrak{U}$ and $\mathfrak{V}$. However, it is known that – for example – in the case of semigroups, the union of two hereditary finitely based varieties: the variety of all abelian semigroups, and the variety of *orthogonal bands* (operation on an orthogonal band $X \times Y$ is defined by $(x_1, y_1)(x_2, y_2) = (x_1, y_2)$), is not hereditary finitely based, even though it is defined by a single identity $xyzt \equiv xzyt$. The situation in which a variety of semigroups does turn out to be hereditary finitely based is very rare, although such varieties have not yet been completely described. Hereditary finitely based varieties of inverse and orthodox Clifford semigroups have been identified 'modulo' hereditary finitely based group varieties. Similar problem for varieties of groups, however, appears to be extremely difficult (Volkov and Shevrin [1985]).

1.3. Examples of Infinite Systems of Identities. Some General Questions. Systems of identities not reducible to finite ones are relatively easy to find,

---

[1] See the remark in Sect. 1.4 of Chap. 1.

say, in the semigroup signature. It suffices to use certain not too complicated syntactical arguments. First examples of this type were obtained by A.P. Biryukov and Austin. B.H. Neumann's problem on the existence of a finite basis for identities of an arbitrary group, posed in 1935, attracted attention for a long time. Cardinality of the set of varieties of groups was not known either. Solutions of these problems, differing in the methods employed and separated by short intervals, were found by A.Yu. Ol'shanskij [1970], S.I. Adyan (see Adyan [1975]), and Vaughan-Lee [1970].

Ol'shanskij finds a series of finite groups $G_1, G_2, \ldots$ in a locally finite solvable variety $\mathfrak{V}$, such that G_i does not belong to $\mathrm{var}\,(G_k \mid k = 1, \ldots, i-1, i+1, \ldots)$ for any i. This implies, first of all, that $\mathrm{var}\,(G_i \mid i \in I) \neq \mathrm{var}\,(G_i \mid i \in J)$ for two distinct sets I, J of positive integers, and hence demonstrates that the set of (solvable and locally finite) varieties of groups has cardinality continuum. Secondly, there exists an identity $v_i \equiv 1$ of the variety $\mathrm{var}\,(G_j \mid j = 1, \ldots, i-1, i+1, \ldots)$, which fails in G_i, i.e. the system $\{v_i \equiv 1\}_{i=1}^{\infty}$ is independent (none of the identities follows from the others), and as such it cannot be equivalent to any finite system. Finally, local finiteness of the variety $\mathfrak{U}$ defined in $\mathfrak{V}$ by the system $\{v_i \equiv 1\}_{i=1}^{\infty}$ allows to make an argument proving that its axiomatic rank is infinite (see Sect. 3.1).

Adyan's example has another advantage: an independent system $\{v_i\}_{i=1}^{\infty}$ is constructed explicitly. As v_i's, he uses words $[x^{pn}, y^{pn}]^n$, where $n \geq 1003$ is an odd integer and p ranges over all prime integers. The proof is based on a method developed by Novikov and Adyan (see Sect. 4.3, Chap. 1). Vaughan-Lee's solution combines merits of the two examples described above. It represents a group-theoretic interpretation of a system of identities found by him for Lie rings of characteristic 2. It also proves that the set of subvarieties of $\mathfrak{N}_2\mathfrak{N}_2$ has cardinality continuum.

The simplest examples of infinite systems of identities, not equivalent to finite systems, were found by Kleiman [1973] and Bryant [1973]. They include the system discussed in Sect. 1.4 of Chapter 1. There is also an example of varieties $\mathfrak{V}_1 \subset \mathfrak{V}_2$ with continuum of intermediate varieties contained in between, none of them possessing a finite basis of identities (Kleiman [1982]).

Frontiers of the finite basis property appear blurred at this time. For example, the product $\mathfrak{N}_2\mathfrak{N}_1$ is a Specht variety, while $\mathfrak{N}_2\mathfrak{N}_2$ is not. Zorn's lemma can be used to prove the existence of a *'limit' subvariety* in every variety which has no finite basis of identities. This subvariety is not finitely based itself, while at the same time each of its proper subvarieties is defined by a single identity. So far, however, not one explicit example of a limit variety of groups has been found, even though such examples are known for semigroups (Volkov and Shevrin [1985]). An example of a limit variety of Lie rings was obtained by I.B. Volichenko.

To this day it is not known whether there exist non-Specht varieties of associative rings or algebras[2] (this is precisely Specht's problem!). *Alternative*

---

[2] See remark in Sect. 1.4, Chap. 1.

algebras, which are close to being associative, are algebras in which every subalgebra generated by two elements is associative (the 8-dimensional Cayley-Dickson algebra is of this type). The alternative property is equivalent to the identities $(xx)y \equiv x(xy)$ and $y(xx) \equiv (yx)x$ (a theorem of E. Artin; see Zhevlakov et al. [1978]). Examples of solvable varieties over a field of characteristic 2, which have no finite basis of identities were found by Medvedev [1980].

Solution of Specht's problem for Lie algebras depends to a large extent on the characteristic of the base field. For a field of characteristic zero, existence of infinite systems of identities (not reducible to finite ones) has not been settled as yet. In the case of prime characteristic p, first examples were given by Vaughan-Lee (for $p = 2$). The simplest infinite system $u_1 \equiv 0, u_2 \equiv 0, \dots$ not reducible to a finite one was found by Drensky [1974] (for any $p > 0$): $u_n = (x_1x_2 \dots u_{n+2})(x_1x_2)$, where $n = 1, 2, \dots$. This system is not equivalent to any finite one even in the variety $\mathfrak{N}_p\mathfrak{A} \cap \mathfrak{A}^3$, which stands in contrast with Krasil'nikov's theorem for characteristic zero, quoted in Sect. 1.2. In characteristic zero, the identity $(x_1x_2)(x_3x_4)x_5 \equiv 0$, meaning that the second commutant is contained in the center, is a Specht identity; this, however, fails in positive characteristic.

Kleiman [1982, 1983] recently found a method of constructing new solvable varieties, and of studying their free groups; this helped him solve a number of difficult problems of the general theory of group varieties. One of the methods used by him is the reduction of questions about the verbal subgroup $v(F)$ of a free group, where v is a word, to the study of a set of evaluations of a certain other word. This allows to consider simpler dependencies between identities, instead of complicated ones. For example, for every group G with normal subgroup N in which squares of elements are all equal to one, and for a G-invariant subset $M \subset N \setminus \{1\}$, one can construct a group $C(G, N, M)$ as follows: for any word w which is identically 1 on G/N, the statement that $w^4 \equiv 1$ is an identity on G is equivalent to saying that M does not contain evaluations of w on G. An idea, due to V.A. Romankov, of interpreting polynomials with integer coefficients in terms of elements of a free nilpotent group, has also proved to be fruitful. It enabled Kleiman to investigate verbal subgroups of $\mathfrak{U}\mathfrak{V}$-free groups for a nilpotent variety $\mathfrak{U}$ by applying, together with elementary properties of polynomials, a theorem of Yu.V. Matyasevich on diophantine equations. Another characteristic feature of Kleiman's method is the construction of free groups in varieties by means of iterated extensions, in which links between consecutive stages are studied using the Magnus-Shmel'kin technique (see Sect. 2.1).

While solving problems posed by A.I. Mal'tsev, Bokut' and Adyan, Kleiman found an identity $v \equiv 1$ for which there is no algorithm determining whether another identity $w \equiv 1$ is its consequence or not. An answer to Tarski's question was found: there is a (solvable) variety of groups which cannot be defined by any independent system of identities (the same problem was solved for semigroups by A.N. Trakhtman). Examples were found of finitely based varieties $\mathfrak{U}$ and $\mathfrak{V}$ such that their join in the lattice of group varieties does not have a finite basis of identities. It was shown that for every collection of groups $\mathcal{K}$ whose cardinality is less than continuum, there exists continuum of varieties which are solvable of degree 4 and all contain the same set of groups from $\mathcal{K}$. Analogous results imply that there exists continuum of: 1) varieties of periodic groups with identical intersections with the class of locally finite groups; 2) locally finite varieties with identical intersections with the class of solvable groups; and 3) solvable, locally finite varieties with identical intersections with the class of nilpotent groups. Several other subtle examples were also constructed in Kleiman [1973].

§ 2. Operations on Varieties

2.1. Products of Group Varieties. Multiplication of varieties was devised by H. Neumann for the purpose of studying identities of group extensions. A product $\mathfrak{U}\mathfrak{V}$ consists of all groups G which contain a normal subgroup $N \in \mathfrak{U}$, for which the factor group G/N belongs to the variety $\mathfrak{V}$. Product of varieties is always a variety (see Sect. 3.1, Chap. 2). The verbal subgroup of any group G, corresponding to that product, is obtained according to the rule $U(V(G))$, where U and V are, respectively, the operators of identifying $\mathfrak{U}$- and $\mathfrak{V}$-verbal subgroups. Multiplication of group varieties is associative (but not commutative), i.e. all varieties under this operation form a semigroup with zero (the variety $\mathcal{O}$ of all groups), and one (the trivial variety $\mathcal{E}$).

Perhaps the most important example is the variety of solvable groups with degree of solvability $\leq l$. It turns out to be the l-th power of the variety $\mathfrak{A}$ of all abelian groups. When $l = 2$, we obtain the variety of all metabelian groups. Multiplying $\mathfrak{N}_{c_1}\mathfrak{N}_{c_2}\ldots\mathfrak{N}_{c_t}$ (see Sect. 3.2) yields, by definition, the variety of all *polynilpotent groups* corresponding to the sequence $(c_1, c_2, \ldots, c_t)$.

The fundamental tool in the study of products of varieties and free groups in them, is provided by wreath products. To define the *unrestricted* (or *complete*) *wreath product* $\bar{W} = A \bar{\wr} B$ of groups A and B, we first construct the group A^B of all functions $f : B \to A$ with component-wise multiplication; then, for $f \in A^B$ and $b \in B$, we define the function $f^b(x) = f(b^{-1}x)$. Finally, the group $\bar{W}$ consists of pairs (f, b), where $f \in A^B$ and $b \in B$, whose product is defined by the rule

$$(f_1, b_1)(f_2, b_2) = (f_1 \cdot f_2^{b_1}, b_1 b_2)$$

(in other words, $\bar{W}$ is the semidirect product of the groups A^B and B, in which B acts regularly on A^B by means of conjugation). The notion of a *restricted wreath product* $W = A \wr B$ is obtained in exactly the same way, when we use only the functions in A^B with finite support $f^{-1}(A \setminus \{1\})$. It is easy to verify that the groups W and $\bar{W}$ satisfy the same identities. They both belong to the product $\mathfrak{U}\mathfrak{V}$, provided that $A \in \mathfrak{U}$ and $B \in \mathfrak{V}$.

The rôle of wreath products in the investigation of group extensions and products of varieties is explained by the Frobenius-Kaluzhnin-Krasner theorem on the existence of isomorphic embedding of any extension G of a group A by a group B (i.e. A is normal in G and $G/A \simeq B$) into the wreath product $A\bar{\wr}B$. Using wreath products, B.H. Neumann, P.M. Neumann and A.L. Shmel'kin proved the following theorem (H. Neumann [1967]).

Theorem. *Every variety of groups $\mathfrak{V} \neq \mathcal{O}, \mathcal{E}$ decomposes into a product of indecomposable varieties in a unique way.*

Existence of such a decomposition follows from a theorem of Levi on chains of automorphically admissible subgroups of a free group.

In order to find groups generating a product of varieties, Baumslag, B.H. Neumann and P.M. Neumann introduced the concept of a discriminating set of groups in a variety $\mathfrak{M}$ (H. Neumann [1967]).

Definition. A set S of groups from $\mathfrak{M}$ is called *discriminating* for $\mathfrak{M}$ if, for any finite set of words $v_1(x_1, \ldots, x_n), \ldots, v_m(x_1, \ldots, x_n)$ which are not identities in $\mathfrak{M}$, there exists a group $G \in S$ and elements $g_1, \ldots, g_n \in G$ such that $v_i(g_1, \ldots, g_n) \neq 1$ for $i = 1, \ldots, m$.

For example, the infinite cyclic group $\mathbb{Z}$ discriminates the variety $\mathfrak{A}$ (i.e. $\{\mathbb{Z}\}$ is a disriminating set for $\mathfrak{A}$); an $\mathfrak{M}$-free group of countable rank is also always discriminating, while no single finite group can be discriminating for any variety other than $\mathcal{E}$. A discriminating set for $\mathfrak{M}$ is, obviously, a generating set of $\mathfrak{M}$. If, on the other hand, $\mathfrak{M} = \operatorname{var}(G \mid G \in S)$, then the set of finite direct products of groups from S discriminates the variety $\mathfrak{M}$.

Theorem. *If a variety $\mathfrak{U}$ is generated by a set of groups S, and a variety $\mathfrak{V}$ is discriminated by a set T, then the product $\mathfrak{U}\mathfrak{V}$ is discriminated by the set of all wreath products $A\bar{\wr}B$, where $A \in S$ and $B \in T$.*

It follows that the product $\mathfrak{U}\mathfrak{V}$ is generated by finite groups (finite p-groups, for a given prime p), if the same property holds for both $\mathfrak{U}$ and $\mathfrak{V}$. In particular, $\mathfrak{A}^l$ is generated by its p-groups for every prime p.

A free group F in the variety $\mathfrak{U}\mathfrak{V}$ with a free generating set $\{x_i\}_{i \in I}$ has a constructive description as a subgroup of the wreath product $W = A \wr B$, where A and B are free groups in the varieties $\mathfrak{U}$ and $\mathfrak{V}$ with generators, respectively, $\{a_i\}_{i \in I}$ and $\{b_i\}_{i \in I}$, such that $x_i = a_i b_i$ in W (in which case A is identified with the set of those functions $f : B \to A$ for which $f(b) = 1$ whenever $b \neq 1$).

Remeslennikov and Sokolov [1971] determined criteria for an element of W to belong to the subgroup F in the above *Magnus embedding.* Generalization of this embedding to the product of varieties $\mathfrak{U}\mathfrak{V}$, where $\mathfrak{U} \neq \mathfrak{A}$, was found by Shmel'kin. In his generalization, direct wreath products is replaced by the $\mathfrak{U}$-verbal wreath product introduced by him (see H. Neumann [1967]).

The power $\mathfrak{A}^l$ is one of the most important varieties in group theory. Since free solvable groups $F_n(\mathfrak{A}^l)$ are finitely approximated, A.I. Mal'tsev's observation implies that there exists an algorithm for recognizing equality of words in $F_n(\mathfrak{A}^l)$. Using the Magnus embedding, Sokolov and Remeslennikov proved finite approximability relative to conjugation in the free solvable group: two elements of this group are conjugates if and only if their images under all epimorphisms of $F_n(\mathfrak{A}^l)$ onto finite groups are conjugates. This facilitated an algorithmic solution of the question about conjugacy of two elements of the free solvable group. Kolmakov, in order to transfer this theorem to free poly-nilpotent groups, needed to construct embeddings of such groups into so-called twisted wreath products.

Remeslennikov found an example of a group defined in the variety $\mathfrak{A}^5$ by a finite number of generators and relations, for which there is no algorithm for recognizing equality of words. Later, Kharlampovich [1981] constructed a solvable group of solvability degree 3, in which the problem of recognizing equality of words cannot be decided algorithmically. This group is defined by a finite number of generators and relations even in the class of all groups.

Groups from the varieties $\mathfrak{N}_c\mathfrak{A}$ and $\mathfrak{A}\mathfrak{N}_c$ are distinguished among others by their properties. In the first case, this is reflected in the Kolchin-Mal'tsev theorem, which implies that every solvable matrix group contains a subgroup of finite index whose commutant is nilpotent. In the second case, theorems of Hall assert finite approximability of every finitely generated group G containing an abelian normal subgroup A such that the factor group G/A is nilpotent, and the ability to define every finitely generated group from $\mathfrak{A}\mathfrak{N}_c$ by a finite number of relations in that variety (P. Hall [1954]).

A strong result was obtained by Romanovskij [1977].

Theorem. *For any group defined in $\mathfrak{A}^l$ by generators $x_1, \dots, x_n$ and $m < n$ relations between them, there is a subset of $n - m$ elements from $\{x_1, \dots, x_n\}$ which freely generates a solvable group of degree l.*

In view of the equality $\bigvee_{l=1}^{\infty} \mathfrak{A}^l = \mathcal{O}$ in the lattice of varieties, the theorem yields a solution of Lyndon's problem: a group with generators $x_1, \dots, x_n$ and $m < n$ relations between them contains an (absolutely) free subggroup with basis $\{x_{i_1}, \dots, x_{i_{n-m}}\}$ for some $1 \leq i_1 < i_2 < \dots < i_{n-m} \leq m$. In settling this and an many other questions about varieties $\mathfrak{A}^l$, the machinery of group rings and modules over them plays a strong supporting rôle.

2.2. Multiplication of Varieties of Linear Algebras. Product of varieties of associative rings or Lie algebras is defined in a completely analogous way to that of group varieties (Sect. 2.1), and is applied in the study of identities of extensions. In the case of assoviative algebras, however, this multiplication is no longer associative, and does not play as prominent a part as it does for groups. The situation is different for Lie algebras.

If L is a free Lie algebra of countable rank, then the ideal of identities of the variety $\mathfrak{U}\mathfrak{V}$ is the ideal of L generated by the ideal $U(V(L))$ of the algebra $V(L)$. When k is an infinite field, then for every derivation $\delta : G \to G$ of any Lie algebra G the verbal ideal $V(G)$ is invariant under δ, and hence

– obviously – $U(V(L))$ is an ideal of L. For example, if k is a field of characteristic zero, then considering in $V(L)$ multilinear polynomials which generate this set of identities (see Sect. 3.1 in Chap. 1) we see that for every $v(x_1, \ldots, x_n) \in V(L)$ and $g_1, g_2, \ldots, g_n \in G$, the equality

$$\delta(v(g_1, \ldots, g_n)) = \sum_{i=1}^{n} v(g_1, \ldots, \delta(g_i), \ldots, g_n)$$

holds, which demonstrates the invariance asserted above.

For a triple of varieties $\mathfrak{U}, \mathfrak{V}, \mathfrak{W}$ of Lie algebras over an infinite field k, calculation of identities for $(\mathfrak{U}\mathfrak{V})\mathfrak{W}$ and $\mathfrak{U}(\mathfrak{V}\mathfrak{W})$ yields the same ideal of identities $U(V(W(L)))$, i.e. varieties form a semigroup, denoted $v(k)$ (an example due to Bakhturin shows that over the field of two elements, multiplication of varieties is non-associative). In case of a field of characteristic zero, Parfyonov proved that $v(k)$ is a free semigroup with zero and one, i.e. that every variety other than the variety $\mathcal{O}$ of all Lie algebras (zero) and the variety consisting of the zero algebra alone (one), uniquely decomposes into a product of indecomposable varieties. Parfyonov's work pioneered the general theory of Lie identities. Bakhturin generalized that result to the case of an arbitrary infinite field. One of the proofs of this theorem uses the *monotonicity theorem*, which is of independent interest: for any two ideals R and S of a free (not one-dimensional) Lie algebra L over k, and any variety $\mathfrak{V}$ distinct from one, the inclusion $V(R) \subseteq V(S)$ implies $R \subseteq S$ (Bakhturin [1985]). Formulation of this theorem is fully analogous to a theorem of Bronshtein for group varieties.

G.P. Kukin proved a number of general embedding theorems for Lie algebras in terms of products of varieties. For example, every countably generated Lie algebra from any variety $\mathfrak{M}$ over a field of characteristic zero can be isomorphically embedded into an algebra with two generators from the variety $\mathfrak{M}\mathfrak{A}^2$. In particular, a countably generated solvable Lie algebra embeds in a solvable Lie algebra generated by two elements (a similar phenomenon was earlier noticed for groups by Kargapolov, Myerzlyakov and Remeslennikov). Also, a recursively defined Lie algebra (or group) from $\mathfrak{M}$ embeds in a finitely related algebra (resp. group) from $\mathfrak{M}\mathfrak{A}^2$ (with certain restrictions on the base field in the algebra case). Kukin obtained important results of algorithmic character as well. For example, $\mathfrak{N}_3\mathfrak{A}$ already contains a finitely related Lie algebra in which the problem of equality of words is not algorithmically solvable (see Kukin [1980]).

A.I. Mal'tsev generalized the definition of a product $\mathfrak{U}\mathfrak{V}$ of subvarieties of a variety $\mathfrak{M}$ to varieties of systems with arbitrary signature (see A.I. Mal'tsev [1967]). $\mathfrak{U}\mathfrak{V}$ is the subclass consisting of algebras $A \in \mathfrak{M}$ in which there is a congruence θ such that every equivalence class of θ which is an $\mathfrak{M}$-subalgebra belongs to $\mathfrak{U}$, while the factor algebra A/θ belongs to $\mathfrak{V}$. Mal'tsev noted, however, that already in the class of all semigroups a product of varieties is not necessarily a variety. He also gave conditions on $\mathfrak{M}$ which are sufficient for $\mathfrak{U}\mathfrak{V}$ to be a variety.

2.3. Intersection, Union and Other Operations. Among the most natural operations on varieties of the same signature there are, of course, the intersection $\mathfrak{U} \wedge \mathfrak{V}$ and *join*, or *union* $\mathfrak{U} \vee \mathfrak{V}$, by which we understand the smallest variety containing $\mathfrak{U}$ and $\mathfrak{V}$. The set of subvarieties of a variety $\mathfrak{M}$ constitutes a lattice $l(\mathfrak{M})$ with respect to $\wedge$ and $\vee$, which is anti-isomorphic to the lattice of verbal congruences of the $\mathfrak{M}$-free algebra F of countable rank.

For classical systems – groups, rings and algebras – the lattice $l(\mathfrak{M})$ satisfies the *modularity condition*: for $\mathfrak{W} \subseteq \mathfrak{V}$ and any $\mathfrak{U}$, the equality

$$\mathfrak{W} \vee (\mathfrak{V} \wedge \mathfrak{U}) = \mathfrak{V} \wedge (\mathfrak{W} \vee \mathfrak{U}) \tag{2}$$

holds (and so does its dual). The modular identity is easily explained when we notice that to a union of varieties there corresponds an intersection of verbal subgroups (ideals) of F, and to an intersection – a union (sum) of verbal subgroups (ideals). In other cases – say, when $\mathfrak{M}$ is the class of all semigroups – the lattice $l(\mathfrak{M})$ is not modular. In order for $l(\mathfrak{M})$ to be modular, it suffices that congruences in $\mathfrak{M}$-algebras be interchangeable with one another. An elegant theorem of A.I. Mal'tsev says that this last property is equivalent to the existence of an element $f(x_1, x_2, x_3)$ in F for which $f(x_1, x_1, x_2) \equiv x_2$ and $f(x_1, x_2, x_2) \equiv x_1$ are identities in $\mathfrak{M}$. For example, if $\mathfrak{M}$ is the variety of all groups, then $f = x_1 x_2^{-1} x_3$.

Distributivity of the lattice $l(\mathfrak{M})$, i.e. the identity

$$\mathfrak{W} \vee (\mathfrak{V} \wedge \mathfrak{U}) = (\mathfrak{W} \vee \mathfrak{V}) \wedge (\mathfrak{W} \vee \mathfrak{U})$$

together with its dual, is only satisfied in 'small' varieties $\mathfrak{M}$. In this case subvarieties of $\mathfrak{M}$ can be easily described in full, since – for example – decomposition of every subvariety into a union of indecomposable ones is unique. If $\mathfrak{M}$ is the variety of linear algebras over a field of characteristic zero, then for $l(\mathfrak{M})$ to be distributive it is necessary and sufficient that, for every n, the natural representation of the group Sym_n in the n-linear component of the algebra $F_n(\mathfrak{M})$ (see Sect. 3.4, Chap. 1) contain no distinct equivalent irreducible representations. Varieties of associative algebras with distributive lattice of subvarieties were described by Anan'in and Kemer – these are the varieties with a non-zero identity of the form $\alpha y[x, y] + \beta [x, y] y \equiv 0$ for some scalars α, β.

While a basis of identities of the intersection $\mathfrak{U} \wedge \mathfrak{V}$ of two varieties is obtained as a union of the identities which define $\mathfrak{U}$ and $\mathfrak{V}$, the problem of finding a basis for the union $\mathfrak{U} \vee \mathfrak{V}$ is far from trivial. Taking as $\mathfrak{U}$ the variety of all associative algebras, and as $\mathfrak{V}$ – the variety of all Jordan algebras, the union will yield a variety defined by the identity

$$(xy, t, z) + (zx, t, y) + (zy, t, z) \equiv 0,$$

where $(u, v, w) = (uv)w - u(vw)$ is the associator of three elements. Identities of that union, as well as identities of the smallest variety containing all commutative and all associative algebras, or all Jordan and all alternative ones, together with identities of the union of a number of other common varieties were found by Dorofeev [1976].

For subvarieties of concrete varieties, there are also specific natural operations. For example, multiplication of verbal ideals of the free associative algebra defines a corresponding operation on subvarieties, under which they become a free semigroup (Bergman and Lewin [1975]).

In the case of varieties of groups or Lie rings, commutation $[U, V]$ of verbal subgroups (ideals), and the operation on varieties corresponding to it, deserve attention. The variety $\mathfrak{N}_c$ of all nilpotent groups (Lie algebras) with index of nilpotency $\leq c$ is obtained from the unit variety by commutation applied c times: $[\dots[\mathcal{E},\mathcal{E}],\mathcal{E}],\dots,\mathcal{E}]$, while the variety of all centrally metabelian groups (Lie algebras) is equal to $[[[\mathcal{E},\mathcal{E}],[\mathcal{E},\mathcal{E}]],\mathcal{E}] = [\mathfrak{A}^2,\mathcal{E}]$.

An even richer assortment of operations is found in varieties of multiground algebras. For example, by analogy with multiplication of group varieties, Plotkin (see Plotkin and Vovsi [1983]) introduced multiplication of varieties of linear representations of groups (see also Sect. 1.3, Chap. 1). This resulted in a free semigroup, on which the semigroup of group varieties acts freely according to the following rule: a representation (L, Γ) (where L is a linear representation, and Γ – the group acting on it) belongs to the variety $\mathfrak{M}\mathfrak{V}$ (where $\mathfrak{M}$ is a variety of representations and $\mathfrak{V}$ – a variety of groups), if the pair $(L, V(\Gamma))$ belongs to $\mathfrak{M}$.

§ 3. Ranks of Systems of Identities

3.1. Basis Rank and Axiomatic Rank. There are two chains of varieties associated with each variety $\mathfrak{V}$. One of them is descending:

$$\mathfrak{V}^{(1)} \supseteq \mathfrak{V}^{(2)} \supseteq \dots; \quad \bigwedge_{i=1}^{\infty} \mathfrak{V}^{(i)} = \mathfrak{V}\,. \tag{3}$$

Here $\mathfrak{V}^{(i)}$ is determined by all identities of the variety $\mathfrak{V}$, which can be written in terms of not more than i variables. In other words, $\mathfrak{V}^{(i)}$ contains those algebras whose subalgebras generated by i generators belongs to $\mathfrak{V}$. If $\mathfrak{V}^{(k)} = \mathfrak{V}^{(k+1)} = \dots = \mathfrak{V}$ (i.e. $\mathfrak{V}$ can be defined by identities involving k or fewer variables), then the minimal such $r_a(\mathfrak{V}) = k$ is called the axiomatic rank of the variety $\mathfrak{V}$. If no such k exists, $r_a(\mathfrak{V}) = \infty$.

Another, increasing, series is

$$\mathfrak{V}_1 \subseteq \mathfrak{V}_2 \subseteq \dots; \quad \bigvee_{i=1}^{\infty} \mathfrak{V}_i = \mathfrak{V}\,, \tag{4}$$

and is defined as follows: $\mathfrak{V}_i$ is determined by identities which are valid in all algebras from $\mathfrak{V}$ generated by i elements, i.e. $\mathfrak{V}_i$ is generated by the $\mathfrak{V}$-free algebra of rank i. The minimal integer i for which $\mathfrak{V}_i = \mathfrak{V}_{i+1} = \ldots = \mathfrak{V}$ is the basis rank $r_b(\mathfrak{V})$ of $\mathfrak{V}$. Once again, we write $r_b(\mathfrak{V}) = \infty$ if such integer does not exist.

Associativity is defined by an identity in three variables, and the axiomatic rank of the variety of all associative algebras equals three (and not less, because the Cayley-Dickson algebra is non-associative, but alternative – i.e. binary-associative). Similar statement also holds for the variety of all Lie algebras, which is strictly contained in the variety of binary-Lie algebras.

At the same time, basis ranks of the varieties of all groups, all associative or Lie algebras over a given field, are equal two, since in these varieties free algebras with two generators contain isomorphic copies of countably generated free algebras of those same varieties. For example, the commutant of the free group $F(x_1, x_2)$ is a free group of countable rank with basis $[x_1^i, x_2^j]$, where i, j are non-zero integers.

Considering the examples cited above, Shestakov's theorem asserting that basis rank of the variety of all alternative algebras is infinite (see Zhevlakov et al. [1978]), is quite unexpected. This means that every finitely generated alternative algebra satisfies a non-trivial identity (i.e. one which does not follow from the alternative identities). For proper subvarieties of the varieties of all groups, all associative or Lie algebras, similar situations are no longer so rare.

Finiteness of the axiomatic rank of any locally finite variety $\mathfrak{V}$ of finite signature is equivalent to its being finitely based. Indeed, as a basis of identities in r variables one can choose all defining relations of the $\mathfrak{V}$-free algebra F_r of rank r, and the operation tables (multiplication, addition table etc.) for F_r are finite. It is also clear that a locally finite variety has finite basis rank if and only if it is generated by a single finite algebra.

Injection rank of a variety $\mathfrak{V}$ is defined by the ability to embed any countably generated $\mathfrak{V}$-algebra into one generated by r-generators. For example, injection rank of the variety of all groups equals two (Higman, B.H. Neumann, H. Neumann), as is the rank of the variety of all Lie algebras (Shirshov). At the same time, for non-unit solvable varieties of groups (or Lie algebras), the injection rank is infinite. For varieties of classical algebras (groups, rings etc.), it would be interesting to know if embeddability of $F_\infty(\mathfrak{M})$ in $F_r(\mathfrak{M})$ is sufficient for the injection rank not to exceed r, and whether embeddability of $F_{r+1}(\mathfrak{M})$ in $F_r(\mathfrak{M})$ implies embeddability of $F_\infty(\mathfrak{M})$ in $F_r(\mathfrak{M})$.

3.2. Ranks of Group Identities. The existence of group varieties with infinite axiomatic rank was mentioned in Sect. 1.3. The simplest one of them is given by the system of identities from Sect. 1.4 of Chapter 1, and represents the product of Burnside varieties $\mathfrak{B}_4\mathfrak{B}_2$. This example also shows that a product of finitely based group varieties does not necessarily have a finite basis of identities. The difficulty involved in this example lies in explaining why none of the identities follows from the preceding ones.

It is far easier to provide examples of varieties with infinite basis rank. It suffices to consider the product $\mathfrak{A}_p\mathfrak{A}_p$, where $\mathfrak{A}_p$ is the variety of abelian groups satisfying the identity $x^p \equiv 1$. The point of the matter is that a wreath product of two infinite groups from $\mathfrak{A}_p$ belongs to $\mathfrak{A}_p^2$, and is center-free – hence not nilpotent. On the other hand, every finite p-group satisfies some nilpotency identity. It follows that $\mathfrak{A}_p^2$ is not generated by a single finite group, and so – in view of Sect. 3.1 – its basis rank cannot be finite (we will recall here that, according to a theorem of O.Yu. Schmidt, a product of locally finite group varieties is a locally finite variety).

At the same time, basis rank of the variety $\mathfrak{A}^l$ of all solvable groups of index $\leq l$ equals two. For a broad class of products of varieties, finiteness of the basis rank was proved by Baumslag and the Neumanns, who used the notion of discrimination (see Sect. 2.1). Totally different considerations were needed in finding the formula $r_b(\mathfrak{N}_c) = c - 1$ (Kovács, Newman, Pentony, Levin) in case of the variety of all nilpotent groups with index of nilpotency $c > 1$ (Kovács, Newman and Pentony [1968]).

The variety $\mathfrak{M}$ of metabelian groups (see Sect. 1.2) is defined by the identity $[[x_1, x_2], [x_3, x_4]] \equiv 1$, and hence it coincides with $\mathfrak{M}^{(4)}$. Indeed, $r_a(\mathfrak{M}) = 4$. The variety $\mathfrak{M}^{(3)}$ is given by the identity $[[x_1, x_2], [x_1, x_3]] \equiv 1$ (MacDonald), and the second commutant of a $\mathfrak{M}^{(3)}$-group is contained in its center. $\mathfrak{M}^{(2)} \neq \mathfrak{M}^{(3)}$, and it is not known whether $\mathfrak{M}^{(2)}$ is a solvable variety. It is defined by the identity $[[x_1, x_2][x_1^{-1}, x_2]] \equiv 1$ (Higman [1959]). It would be interestring to describe $r_a(\mathfrak{A}^l)$ as a function of l.

Equalities and strict inclusions in the chain (3) can, in general, be distributed in a quite random manner: relying on Kleiman's method (see Sect. 1.3), S.V. Aivazyan showed that for every subset $S \subseteq \{8, 9, \ldots, n, \ldots\}$, there exists a group variety $\mathfrak{V}$ such that varieties $\mathfrak{V}^{(n)}$ and $\mathfrak{V}^{(n+1)}$ (with $n \geq 8$) coincide if and only if $n \in S$.

For a locally finite group variety $\mathfrak{V}$, the variety $\mathfrak{V}^{(1)}$ coincides with some Burnside variety $\mathfrak{V}_n$, because it is defined by an identity in one variable $x^n \equiv 1$. Consequently, the question whether $\mathfrak{V}^{(1)}$ is locally finite reduces to the bounded Burnside problem. Making this question more complicated, B.H. Neumann asks whether for every locally finite variety of groups $\mathfrak{V}$ there exists an integer d, for which $\mathfrak{V}^{(d)}$ is locally finite. We will note that so far there are no examples of this situation in which, say, $\mathfrak{V}^{(2)}$ would fail to be locally finite.

Birkhoff's theorem allows to reformulate Kostrikin's theorem on boundedness of orders of groups with m generators and an identity $x^p \equiv 1$ (p – prime; see Sect. 2.2, Chap. 2) in the following way: all locally finite groups from $\mathfrak{B}_p$ form a variety – so-called Kostrikin's variety $\mathfrak{R}_p$. When $p \geq 5$, Razmyslov's theorem (see Sect. 2.2, Chap. 2) implies that this variety has infinite basis rank, because it is locally nilpotent but not nilpotent. The question of finiteness of axiomatic rank, i.e. of existence of a finite basis of the variety $\mathfrak{R}_p$, is open (in this case, that problem is also equivalent to the problem of 'essential infinite basedness' – see Sect. 4.3).

When $p = 2$, it is clear that $\mathfrak{N}_2 = \mathfrak{B}_2 = \mathfrak{A}_2$ is the variety of abelian groups satisfying the identity $x^2 \equiv 1$. It is not hard to verify that for $p = 3$ we have $\mathfrak{B}_3 = \mathfrak{N}_3$. Indeed, in Sect. 4.3 of Chapter 1 we derived the identity $[x, y^{-1}xy] \equiv 1$ from $x^3 \equiv 1$, which means that every element commutes with its conjugates – i.e. it lies in an abelian normal subgroup. If $G \in \mathfrak{B}_3$ and $g_1, \dots, g_n$ generate that group, then arguing by induction we see that G/N is finite for an abelian normal subgroup N which contains g_n. Schreier's formula (see Sect. 4.2, Chap. 1) implies that the number of generators of N does not exceed $|G/N|(n-1)+1$. It follows that $|N| \leq 3^{|G/N|(n-1)+1}$, and the group G is finite. In fact, $G \in \mathfrak{N}_3$, and an exact Levi-van der Waerden estimate is valid (see Sect. 2.2, Chap. 2). Local finiteness and solvability of the variety $\mathfrak{B}_6$ has been established by M. Hall, while Razmyslov showed that $\mathfrak{B}_n$ is solvable if and only if $n = 1, 2, 3, 6$. In particular, $r_b(\mathfrak{B}_4) = \infty$ which, by the way, follows from the inclusion $\mathfrak{B}_4 \supset \mathfrak{A}_2^2$ (see Sect. 2.2, Chap. 2).

3.3. Ranks of Varieties of Lie Algebras. Under the correspondence between group varieties and varieties of Lie algebras over $\mathbb{Q}$ described in Sect. 4.2 of Chapter 1, the variety $\mathfrak{N}_c$ of all nilpotent groups of degree $\leq c$ is associated with the variety of Lie algebras of the same type, $\mathfrak{A}^l$(groups) $\mapsto$ $\mathfrak{A}^l$(Lie algebras), and for any $c_1, \dots, c_t$ we have $\mathfrak{N}_{c_1}\mathfrak{N}_{c_2}\dots\mathfrak{N}_{c_t}$(groups) $\mapsto$ $\mathfrak{N}_{c_1}\mathfrak{N}_{c_2}\dots\mathfrak{N}_{c_t}$(Lie algebras). Parallels of this kind gave rise to several interesting problems. These theories, manifesting similarities in some of their aspects, significantly diverge in regard to basis ranks.

In case of Lie algebras over an infinite field k, $r_b(\mathfrak{A}^2) = 2$, since every proper subvariety of $\mathfrak{A}^2$ is locally nilpotent (see Sect. 6.2), and there exists a two-dimensional Lie algebra $G = \left\{ \begin{pmatrix} a & b \\ 0 & 0 \end{pmatrix} \mid a, b \in k \right\}$, which is not nilpotent. However, already $r_b(\mathfrak{A}^3) = \infty$, because the free solvable Lie algebra L of rank n in this variety satisfies an identity of the form

$$\sum_{\sigma \in \mathrm{Sym}_{n+2}} \varepsilon_\sigma (y_1 z x_{\sigma(1)}) \dots (y_{n+2} z x_{\sigma(n+2)}) \equiv 0 \,.$$

This identity does not hold in $\mathfrak{A}^3$. The most general result states that the product $\mathfrak{N}_{c_1}\dots\mathfrak{N}_{c_t}$ has infinite basis rank, except possibly when $t = 1$, or $t = 2$ and $c_2 = 1$. If the base field is finite, then the only exception is $t = 1$. When the field is infinite, then $r_b(\mathfrak{N}_c\mathfrak{A}) \leq c+1$. More precisely, over an infinite field k the variety $\mathfrak{N}_c\mathfrak{A}$ is generated by the Lie algebra of $(c+1) \times (c+1)$ upper triangular matrices. The basis rank of any nilpotent ring of degree c does not exceed c. As to the variety $\mathfrak{N}_c$ itself, its basis rank over an arbitrary commutative ring R has been computed by Bakhturin, who showed that $r_b(\mathfrak{N}_c) = c - 1$ in all cases, except the one in which $c = 2p$ and R contains an element of prime additive order p. In that situation basis rank is equal to c (Bakhturin [1985]). It is instructive to compare this result with the analagous one for groups.

It easily follows from the theorem of Baumslag and the Neumanns (see Sect. 2.1) that the variety of Lie algebras $(\mathfrak{A}^m)_i$ from the chain (4) corresponds to the variety of groups $\mathfrak{A}^m$ (as in Sect. 4.2, Chap. 1). In view of what was said about basis rank, we see that for $m \geq 3$, $\mathfrak{A}^m$ has infinitely many preimages. Making use of the fact that the correspondence is bijective on the set of nilpotent varieties, we see that for any $l \geq 3$, basis rank of the group variety $\mathfrak{N}_c \cap \mathfrak{A}^m$ grows without bound as c increases. At the same time it is known that $r_b(\mathfrak{N}_c \cap \mathfrak{A}^2) = 2$ whenever $c \geq 2$ (H. Neumann [1967]).

Finally, an important difference between this and the group theoretic situation lies in the fact that a free solvable Lie algebra of countable rank and with index of solvability 3 cannot be approximated by algebras with bounded number of generators.

Generally speaking, knowledge of the basis rank is quite sketchy. For example, there are indications that basis rank of a product $\mathfrak{U}\mathfrak{V}$ is finite only in the case when $\mathfrak{U}$ is nilpotent and $\mathfrak{V}$ – abelian.

V.V. Stovba [1986] proved that over an infinite field, every subvariety $\mathfrak{V}$ of $\mathfrak{N}_c\mathfrak{N}_d$ with finite axiomatic rank is finitely based. In proving that rank of a $\mathfrak{V}$-free algebra is finite, he considered the action of the factor algebra $G = L/R$ modulo the nilpotent radical on factors R^i/R^{i+1}. In this setting, R^i/R^{i+1} is a module over an associative Noetherian ring U, where U is the universal enveloping algebra of G. By introducing cleverly chosen filtrations in the ring and the module, and passing to the associated graded ring and module, the problem is reduced to a commutative situation which in the end yields the required result – finiteness condition on chains of verbal ideals in L (see Sect. 1.1). Special cases of this result include the theorem of Krasil'nikov on finite basis property of a solvable finite-dimensional Lie algebra, mentioned in Sect. 1.2. The point here is that a finite dimensional algebra satisfies all Capelli identities of some degree (see Sect. 5.4, Chap. 2), and varieties with this property always have finite axiomatic rank.

§ 4. Identities of Finite Algebras

In a concrete finite algebra, it usually isn't difficult to find several characteristic identities. For example, one can easily verify that relations $x^6 \equiv 1$ and $[x^2, y^2] \equiv 1$ hold in the permutation group Sym_3, while the group of quaternion units $Q_8 = \{\pm 1, \pm i, \pm j, \pm k\}$ satisfies identities $x^4 \equiv 1$ and $[x^2, y] \equiv 1$. But even for groups of such low order, it isn't easy to show that these identities imply all of their other identities. The question of B.H. Neumann about existence of a finite basis of identities for every finite group has remained open for over a quarter of the century.

4.1. Identities of Finite Groups. B.H. Neumann's problem was settled affirmatively in 1963 by Oates and Powell [1964]; Kovács and Newman soon shortened the proof which, nevertheless, remains saturated with ideas. As was remarked in Sect. 3.1, in order to settle the finite basis problem for a finite group G, it is enough to show that all identities of $\operatorname{var} G$ follow from identities which can be written using a bounded number of variables. However, the decisive advance was made thanks to the notions of a critical group and a Cross variety. Definitions given below can be carried over in an obvious way to the context of associative rings, Lie rings and other algebras; they have been successfully exploited there as well.

A finite group G is called *critical* if it does not belong to the variety $\operatorname{var}_{-} G$ generated by all of its proper subfactors – i.e. by all factor groups of subgroups of G, with the exception of $G/\{1\}$. For example, the (non-abelian) groups Sym_3 and Q_8 mentioned above are critical, since $\operatorname{var}_{-} \mathrm{Sym}_3 = \mathfrak{A}_6$ and $\operatorname{var}_{-} Q_8 = \mathfrak{A}_4$ are abelian varieties. It immediately follows from the definition that every locally finite variety of groups is generated by its critical groups.

As an example, we will again turn to the identities

$$x^6 \equiv 1, \quad [x^2, y^2] \equiv 1 \tag{5}$$

of the group Sym_3. These identities define a locally finite variety. Indeed, in any group G with n generators, the subgroup H generated by squares of all elements of G has index not greater than 2^n, since G/H satisfies the identity $x^2 \equiv 1$; if relations (5) hold in G, then H is an abelian group with period 3, and finiteness of H (and hence of G) now follows from Schreier's formula, as in Sect. 3.2.

The variety $\operatorname{var} \mathrm{Sym}_3$ clearly contains the cyclic groups $\mathbb{Z}_2$ and $\mathbb{Z}_3$ of orders 2 and 3. In order to show that all identities of the group Sym_3 are consequences of (5), it is therefore sufficient to establish that the identities (5) are not satisfied in any critical group other than $\mathbb{Z}_2$, $\mathbb{Z}_3$ and Sym_3.

Suppose then that G is a critical group in which relations (5) hold. Let, as before, H be the subgroup generated by squares of elements of G. Since H is an abelian group of period 3, it can be viewed as a vector space over the field k of integers modulo 3, which becomes a G/H-module with respect to the conjugation action of G on H, because this action restricted to H is trivial.

A critical group is always monolithic (see Sect. 2.4, Chap. 1). In case $H = \{1\}$ or $H = G$, we therefore immediately obtain isomorphisms $G \simeq \mathbb{Z}_2$ or, respectively, $G \simeq \mathbb{Z}_3$. Further, we note that $C = H$, where C is the centralizer of H in G, for otherwhise G would contain an abelian normal subgroup K, properly containing the subgroup H, and non-triviality of its primary decomposition would contradict the fact that G is a monolithic group. It follows that H is a faithful $k[G/H]$-module. G being monolithic, together with Maschke's theorem, implies that this module is irreducible.

Because G/H is an abelian group with period 2, a faithful irreducible representation can exist only when G/H is cyclic, i.e. $|G/H| = 2$, and H is a one-dimensional space over k – i.e. $|H| = 3$. As a result, the non-abelian group G had order 6, i.e. $G \simeq \mathrm{Sym}_3$, as required.

Definition. *A variety* $\mathfrak{V}$ *is a Cross variety* if: 1) $\mathfrak{V}$ is locally finite, 2) identities of $\mathfrak{V}$ have a finite basis, and 3) the number of non-isomorphic critical groups in $\mathfrak{V}$ is finite.

Thanks to condition 3), the variety $\mathfrak{V}$ satisfies the minimal condition on subvarieties – and hence each of its subvarieties is also a Cross variety. When G is a finite group, to prove existence of a finite basis for the variety $\operatorname{var} G$ it is therefore enough to find a finite set of identities $v_1 \equiv 1, \ldots, v_l \equiv 1$ such that: 1) all of them hold in G, 2) the variety defined by this set of identities is locally finite, and 3) orders of critical groups satisfying the identities $v_1 \equiv 1, \ldots, v_l \equiv 1$ have a common bound. Initially, this program was realized by Cross for a finite group with nilpotent commutant. Definition of a Cross variety is important, because in the course of this reasoning it transpires that it is equivalent to the the notion of a variety generated by a finite group.

The example presented above barely hints at the actual difficulties which were overcome by Oates and Powell in obtaining the following remarkable result:

Theorem. *All identities of any finite group G are consequences of their finite subset.*

In order to simplify the proof of this theorem, Kovács and Newman introduced an important class $\mathfrak{C}(e, m, c)$. It consists of groups satisfying the identity $x^e \equiv 1$, whose nilpotent factors have degree of nilpotency at most c, and principal factors have orders not exceeding m (see Sect. 5.2, Chap. 2). It turns out that there exists a finite set of identities of a finite group G, which defines a variety contained in some class $\mathfrak{C}(e, m, c)$. One of such identity relations is the *'principal centralizer identity'*: let $v_2 = [[x_1, x_2] y_{1,2} x_1^{-1} x_2 y_{1,2}^{-1}]$, where $y_{1,2}$ is also an independent variable; furthermore, by recursion, let

$$v_n = [\ldots [[v_{n-1}, y_n x_n y_n^{-1}], y_{1,n} x_1^{-1} x_n y_{1,n}^{-1}], \ldots, y_{n-1,n} x_{n-1}^{-1} x_n y_{n-1,n}^{-1}] \,.$$

It is easy to notice that $v_n \equiv 1$ holds in a group of order $\leq n$. On the other hand, presence of the identities $v_n \equiv 1$ and $x^e \equiv 1$ in a group implies that orders of its principal factors are bounded by e^n. This phenomenon was used not only by Kovács and Newman – it makes the task of finding a basis of identities of a finite group easier in general.

The class $\mathfrak{C}(e, m, c)$ contains only finitely many critical groups, and is a locally finite variety. Orders of critical groups from $\mathfrak{C}(e, m, c)$ can be bounded by means of a subtle criterion for "non-criticality" of a group which was discovered by Oates and Powell, together with results of Gaschütz. Being unable to describe here all stages of the elegant and instructive proof, for further details we refer the reader to H. Neumann [1967].

4.2. Identities of Finite Rings. We will reiterate that the concepts of a critical algebra and a Cross variety are applicable not only to groups. After group theory, they proved useful in the context of associative rings. Taking into account properties specific to rings, Kruse [1973] and L'vov [1973] independently proved that identities of a finite associative ring are finitely based and, moreover, the fact that var A is a Cross variety whenever A is finite. It also turned out that for a variety $\mathfrak{M}$ of rings to be a Cross variety, it is necessary and sufficient that it satisfy an identity $nx \equiv 0$ for some positive n, and that all nilpotent rings from $\mathfrak{M}$ have bounded index of nilpotency. Identities of a finite alternative ring are also finitely based (L'vov).

Identities of a finite Lie ring have a finite basis as well (Bakhturin, Ol'shanskij). Here, the biggest difficulties arise in attempting to estimate a bound on the number of variables in identities, which in turn limit the parameters f, m and c (see Sect. 5.3, Chap. 2). There is clearly no natural analogue of a principal centralizer (Sect. 4.1) in this situation. This significantly complicates proofs, and makes it necessary to perform a number of subtle reductions, which transform the problem into that of limiting the cardinalities of principal factors by means of certain universal formulæ (not identities!) involving a bounded number of variables. Such formulæ are valid in the Lie algebra whose finite basis property is being considered and, as in the group case, the proof is concluded by applying ideas of Cross (see Sect. 4.1 and Bakhturin [1985]).

4.3. On Critical Algebras. Since every locally finite variety is generated by its critical algebras, the study of the latter represents a natural task of the theory of varieties. According to Sect. 5.1 of Chap. 2, critical algebras include finite simple groups and rings. As a whole, however, classes of critical objects have not been surveyed very well as yet. The sets of critical nilpotent groups and rings are particularly amorphous. We will only state two results of general character. Bryant has proved the following

Theorem. *For critical groups* G_1 *and* G_2, *the equality* $\operatorname{var} G_1 = \operatorname{var} G_2$ *implies* $\operatorname{var}_- G_1 = \operatorname{var}_- G_2$, *where* $\operatorname{var}_- G$ *is the variety generated by proper factors of the group* G *(see Sect. 4.1).*

Yu.N. Mal'tsev obtained an analogue of this theorem for associative rings. We will remark that examples of such a situation are quite common; two non-isomorphic non-abelian groups, say, of order eight – the dihedral group and the quaternion group – are critical, and satisfy the same identities. Finally, the ring of matrices $M_n(A)$ over a finite associative ring A is critical if and only if A itself is a critical ring.

4.4. Finite Algebras with Infinite Bases of Identities. It may turn out that identities of every finite ring are finitely based. Another conjecture has also been advanced: that a positive solution of the finite basis problem will follow from modularity of the lattice of congruences of a finite algebra (of any finite signature). This, however, was disproved by S.V. Polin [1976]: there is a finite linear algebra A over a given field k, which does not have a finite basis of identities; A satisfies the identity $x(yz) \equiv 0$. Another example of an algebra with modular lattice of congruences, but without finite basis of identities, was found by Bryant; his example is a group with a distinguished point.

At the same time, for any fixed signature almost all finite algebras of this signature are finitely based, i.e. the ratio of the number of algebras of cardinality n with finite basis of identities to the number of all algebras with n elements tends to 1 when $n \to \infty$. This fact was discovered by Murskij [1975], who also found that for groupoids (i.e. in signature with only one binary operation), this ratio is of the order $1 - n^{-6}$.

In connection with the finite basis problem, a great deal of work was done in the class of finite semigroups. The smallest semigroup with no finite basis of identities can be represented by six matrices under ordinary matrix multiplication (Perkins [1969]):

$$\begin{pmatrix} 0 & 0 \\ 0 & 0 \end{pmatrix}, \begin{pmatrix} 1 & 0 \\ 0 & 0 \end{pmatrix}, \begin{pmatrix} 0 & 1 \\ 0 & 0 \end{pmatrix}, \begin{pmatrix} 0 & 0 \\ 1 & 0 \end{pmatrix}, \begin{pmatrix} 0 & 0 \\ 0 & 1 \end{pmatrix}, \begin{pmatrix} 1 & 0 \\ 0 & 1 \end{pmatrix}. \tag{6}$$

The large diversity of semigroups of order less than six has thwarted attempts to find a unified approach to the finite basis problem in that case. Many authors have contributed to this task, in the end determining the existence of finite bases of identities for all semigroups of these orders.

Variety generated by the semigroup (6) is *essentially infinitely based*, in the sense that no locally finite variety containing this semigroup has a finite basis (this was shown by Sapir, who also found a general condition under which a finite semigroup is essentially infinitely based; see Volkov and Shevrin [1985]). This means that for every n there exists a not locally finite variety of semigroups, all of whose n-generated semigroups are locally finite and satisfy all identities of the semigroup (6). Similar examples are not known for groups, and the question of their existence is equivalent to the problem of B.H. Neumann from Sect. 3.2.

The semigroup (6) satisfies identities which are extremal in many respects. For example, a variety generated by a finite inverse semigroup is finitely based if and only if it does not contain the semigroup (6).

For a finite semigroup S, the variety $\operatorname{var} S$ is, as a rule, far removed from being a Cross variety. The class of finitely based finite semigroups is not closed under passing to sub-semigroups, homomorphic images etc. Finally, here exist finite semigroups which have no independent bases of identities (Sapir; see Volkov and Shevrin [1985]).

§ 5. Numerical Characteristics of Varieties and Identities of Concrete Algebras

5.1. Series Related to Varieties of Linear Algebras. Aside from the basis and axiomatic ranks which were discussed in Sect. 3, important numerical characteristics of varieties of linear algebras are derived from the series $c(\mathfrak{M}, t)$, $H_n(\mathfrak{M}, t)$ and $H(\mathfrak{M}, t_1, \ldots, t_n)$.

Every free algebra $F = F(x_1, \ldots, x_n)$ in a variety $\mathfrak{M}$ of linear algebras over an infinite field is equipped with a natural gradation (see Sect. 3.2, Chap. 1) and, as with any finitely generated graded algebra, there is a Hilbert-Poincaré series $H_n(\mathfrak{M}, t) = \sum_{i=1}^{\infty} (\dim F_i) t^i$ associated with F. It isn't difficult to calculate that when, for example, F is the free associative algebra, then the coefficients of this series are $h_i = n^i$, while imposing in addition the commutative law makes them become $h_i = \binom{n+i-1}{i}$; in the first case, growth of h_i is exponential, in the second – polynomial. Since a free Lie algebra embeds in a free associative algebra (Sect. 4.2, Chap. 1), the *Hilbert-Poincaré series of any variety* of Lie algebras, as with a variety of associative algebras, converges in some neighborhood of zero.

Considering a finer gradation of F which takes into account degrees of a monomial with respect to each variable, we have $F = \bigoplus_\alpha F_\alpha$, where $\alpha = (m_1, \ldots, m_n)$ and, setting $d_\alpha = \dim L_\alpha$, we obtain the series

$$H(\mathfrak{M}, t_1, \ldots, t_n) = \sum_{(m_1, \ldots, m_n)} d_{m_1, \ldots, m_n} t^{m_1} \ldots t^{m_n} .$$

Finally, recalling that over a field of characteristic zero every identity is equivalent to a system of multilinear identities, important numerical data are provided by the dimension c_n of the n-linear part of the $\mathfrak{M}$-free algebra, together with the series $c(\mathfrak{M}, t) = \sum_{n=1}^{\infty} c_n t^n$; keeping in mind that free algebras in associative or Lie varieties are factor algebras of (absolutely) free associative algebras modulo verbal ideals, the series $c(\mathfrak{M}, t)$ is also referred to as the *codimension series of the variety* $\mathfrak{M}$. In the variety of all associative algebras, one basis of the n-linear part obviously consists of the monomials $x_{\sigma(1)} \cdots x_{\sigma(n)}$ for all permutations σ, i.e. $c_n = n!$. For the variety of all Lie algebras one can derive the formula $c_n = (n-1)!$. This means that for every variety of associative or Lie algebras, growth of codimensions does not exceed $n!$.

Properties of the series defined above are closely related with properties of the variety $\mathfrak{M}$. The former usually include: rationality of the series, rate of growth of the coefficients, radius of convergence, etc. We say that growth of the variety is polynomial if there is a constant M and a positive integer s for which $c_n < Mn^s$. Examples of such varieties clearly include all nilpotent varieties (for which the number of non-zero c_n's is finite).

Theorem (Kemer). *Every associative variety with polynomial growth of codimensions is finitely based.*

Benediktovich and Zalesskij [1980] described verbal ideals of a free Lie algebra with polynomial growth of codimensions.

As was mentioned earlier (Sect. 3.4, Chap. 1), in a metabelian algebra we have

$$yzx_1, \ldots, x_n = yzx_{\sigma(1)} \cdots x_{\sigma(n)}$$

for every permutation σ. This implies that $\mathfrak{A}^2$ has polynomial growth. Growth of $\mathfrak{N}_2\mathfrak{A}$, however, is not polynomial, and results of Benediktovich and Zalesskij, together with those of Mishchenko [1982], show that a variety $\mathfrak{V}$ of Lie algebras over a field of characteristic zero has polynomial growth if an only if for some c we have $\mathfrak{V} \subseteq \mathfrak{N}_c\mathfrak{A}$, but $\mathfrak{V} \not\supset \mathfrak{N}_2\mathfrak{A}$. The codimension series of every variety with polynomial growth has the desirable property of being rational, i.e. expressible as a quotient of two polynomials (Drensky).

Further, we will say that growth of $c_n(\mathfrak{V})$ is subexponential if there are constants M and d such that $c_n(\mathfrak{V}) \leq Md^n$ for $n = 1, 2, \ldots$. Among varieties with this property we have all special varieties, since for all varieties of associative algebras such bound on their growth is always valid (Regev; see Sect. 3.2, Chap. 2). In particular, varieties generated by a finite-dimensional Lie algebra have subexponential growth. Moreover, varieties generated by infinite-dimensional Lie algebras of vector fields on a finite-dimensional manifold also have this property (Mishchenko), together with some other varieties. At the same time, Volichenko observed that already the variety $\mathfrak{A}\mathfrak{N}_2$ has exponential growth, and Mishchenko [1984] showed that varieties whose growth is exponential constitute a very large class which submits to detailed investigation more readily than the class of all varieties. Apart from this, Mishchenko proved that there are no varieties with intermediate growth between polynomial and exponential ones. Moreover, if $c_n(\mathfrak{V}) < Md^n$ for some $d < 2$, then $\mathfrak{V}$ has polynomial growth.

Several questions regarding the series $c(\mathfrak{V}, t)$ remain open; in particular, there is a conjecture that there are finitely many varieties with almost polynomial growth, i.e. varieties whose growth is not polynomial, but whose proper subvarieties have polynomial growth; these include $\operatorname{var} \mathrm{sl}_2(k)$, $\mathfrak{N}_2\mathfrak{A}$, $\mathfrak{G}$ (see Sect. 6.3). Granting that this conjecture is true, the problem of verifying polynomial growth of a variety would become algorithmic in nature.

The series $H_n(\mathfrak{V}, t)$ have only been calculated for concrete varieties. One approach to this computation, using representation theory of the group $\mathrm{GL}(n, k)$, was given by Drensky [1984], who obtained exhaustive results for $\mathfrak{V} = \operatorname{var} \mathrm{sl}_2(k)$ when $\operatorname{char} k = 0$. As a special case for $n = 2$, the formula

$$H(\mathfrak{V}, t_1, t_2) = t_1 t_2 (1 + t_1 + t_2)(1 - t_1^2)^{-1}(1 - t_2^2)^{-1}$$

is due to Bakhturin. We end with one more formula:

$$H_m(\mathfrak{A}^2, t) = 1 + t + (mt - 1)(1 - t)^{-m}, \quad m = 2, 3, \ldots .$$

5.2. Other Numerical Characteristics of Varieties. One of the natural characteristics of a locally finite variety $\mathfrak{M}$ is its *order function* f, where $f(n)$ is the cardinality of the $\mathfrak{M}$-free algebra of rank n. For example, in the varieties $\operatorname{var} \mathrm{Sym}_3$ and $\operatorname{var} Q_8$ (see beginning of Sect. 4), the order functions are respectively equal $2^n \cdot 3^{2^n(n-1)+1}$ and $2^{n(n+3)/2}$. The difference in the rates of growth is explained by an observation due to P.M. Neumann: a locally finite variety of groups is nilpotent if and only if $\log f(n)$ is a polynomial in n. Otherwise, $\log f(n)$ grows faster than any polynomial. For a variety generated by a single finite group G (or any finite algebra), Birkhoff's theorem (Sect. 2.3, Chap. 1) implies that growth of the function $f(n)$ is bounded from above by the quantity c^{c^n}, where $c = |G|$.

The order function can be explicitly computed for many varieties. It is easy to infer from Schreier's formula that for a product $\mathfrak{V}_1\mathfrak{V}_2$ of two locally finite group varieties, the function f can be computed according to the rule $f(n) = f_2(n)f_1((n-1)f_2(n)+1)$. In general, however, it is difficult to describe the behavior of $f(n)$. On the one hand, there exist distinct varieties with identical order functions; on the other, there is an example of a continuum of locally finite varieties, linearly ordered by inclusion like the interval $[0,1]$ (and hence their order functions are also distinct). The question of existence of varieties with order functions whose growth is arbitrarily rapid remains open.

A number of quantitative parameters are associated with Cross varieties (see Sect. 4.1 and 4.2). We will remark that the proof of the theorem on finite basis of a finite Lie ring involves induction on socular height of a variety (see Sect. 5.2, Chap. 2). Finite algebras from a variety generated by a finite group or a finite ring always have bounded socular height. For example, socular height of the variety generated by the alternating group A_5 equals two.

Peculiarities of varieties of different signatures lie in the diversity of numerical characteristics. For associative algebras it is important to know the degrees of standard identities satisfied in them, or the minimal n such that $M_n(k)$ does not belong to the given variety. For locally nilpotent (locally solvable) varieties, growth functions of the nilpotency index (solvability index) of n-generated free algebras are of special interest. In case of semigroups, it is interesting to study the difference between lengths of words in an identity $u \equiv v$ and the number of variables that they include, or such a parameter as the minimal number of identities which define a given variety (as a consequence of this, we will turn the reader's attention to a simple exercise: every finitely based variety of groups or rings can be described by means of a single identity).

5.3. On Identities of Certain Concrete Algebras. As was mentioned in Chapter 2, characteristic identities of those algebras which arise in the structure theory play an important part in the general theory of varieties as well. The problem of finding bases of identities of such algebras has therefore always attracted attention, and particular difficulties in doing so are encountered in proving that all identities of a given algebra can be derived from those that were already found.

The greatest progress has been made in the study of identities of simple associative rings. For the algebra $M_2(k)$ when k is a field of characteristic zero, existence of a finite basis of identities was proved by Razmyslov (such basis was described in Sect. 1, Chap. 1). Solution of the general finite basis problem for identities of the algebra $M_n(k)$ over a field of characteristic zero is provided by Yakovlev's generalization, mentioned in Sect. 1.4 of Chap. 1. In particular, when $m \geq n^2 + n$, every multi-linear identity of degree m in $M_n(k)$ follows from multi-linear identities of this algebra with degree lower than m. The announced proof of this fact uses Razmyslov's theorem on trace identities (in its convolution variant), as well as the highly developed representation theory of symmetric groups – in particular, the branching theorem. The proof also gives an algorithm for finding a basis of identities for every n, even though it fails to provide general formulas.

It can be easily verified that the algebra $T_n(k)$ of upper triangular $n \times n$ matrices satisfies the identity

$$(x_1x_2 - x_2x_1)\dots(x_{2n-1}x_{2n} - x_{2n}x_{2n-1}) \equiv 0\,. \tag{7}$$

Indeed, the value of each commutator $xy - yx$ is a strictly upper triangular matrix, while a product of n strictly upper triangular matrices of size n is zero.

Yu.N. Mal'tsev [1971] showed that (7) represents a basis of identities for $T_n(k)$ in the case of base field of characteristic zero. For an arbitrary ring k, the ideal of identities I_n of $T_n(k)$ coincides with I_1^n, where I_1 is the ideal of identities of the ring k (U.E. Kalyulaid, A. Popov).

Over an infinite field k, the identity

$$(x_1x_2)(x_3x_4)\dots(x_{2n-1}x_{2n}) \equiv 0$$

forms a basis of identities of the algebra $T_n(k)$ viewed as a Lie algebra, i.e. $\operatorname{var} T_n(k) = \mathfrak{N}_{n-1}\mathfrak{A}$. Even earlier, an analogous result was obtained by Romanovskij for identities of the group of non-singular upper triangular matrices. In case of a field of characteristic 0, a single identity $[[x_1,x_2],\dots,[x_{2n-1},x_{2n}]] \equiv 1$ suffices, while in characteristic p one has to include additional identities of the form $[[x_1,y_1],\dots,[x_i,y_i]]^{p^{\alpha(i)}} \equiv 1$ for $i = 1,\dots,c$, where $\alpha(i) = \min\{\alpha \mid \alpha \in \mathbb{Z},\ \alpha > 0,\ ip^\alpha \geq c\}$. For the semigroup of all upper triangular matrices over a finite field, the finite basis problem remains open (if the field is infinite, then for $n > 1$ there are no semigroup identities satisfied in that algebra).

Grassman algebra G – the exterior algebra of an infinite-dimensional vector space – plays a special rôle in a number of questions about identities. It is readily checked that G satisfies the identity $(xy - yx)z \equiv z(xy - yx)$. This can be easily explained: in the standard form of any commutator $xy - yx$, only monomials of even length are present – and they belong to the center of G. It turns out that this identity implies all other identities valid in the exterior algebra (obviously, for the exterior algebra of a finite-dimensional space additional nilpotency identities have to be included).

Quite a lot is known about identities of concrete infinite-dimensional Lie algebras. Let W_n be the Lie algebra of derivations of the ring of polynomials in n indeterminates $y_1,\dots,y_n$ over a field k. We will suppose that

$s_m(x_1, \ldots, x_m, x_{m+1}) \equiv 0$ is a standard identity (in the Lie algebra sense – see Sect. 1.3, Chap. 2), which is not satisfied in W_n. Every element of W_n has the form

$$\sum_{i=1}^{n} f_i(y_1, \ldots, y_n) \frac{\partial}{\partial y_i}, \tag{8}$$

where $f_1, \ldots, f_n$ are polynomials in $y_1, \ldots, y_n$. By multi-linearity of the standard polynomial s_m, we can assume that it becomes zero when $x_1, \ldots, x_{m+1}$ are replaced with derivations of the form $u_p = g_p \frac{\partial}{\partial y_{i_p}}$, where g_p is a monomial without a coefficient for $p = 1, \ldots, m+1$. The degree of the derivation (8) will means the maximum of degrees of the polynomials $f_1, \ldots, f_n$. We now write down a homogeneous derivation of minimal degree $s \geq 0$, which occurs as a value of the polynomial s_m on $u_1, \ldots, u_{m+1}$, as above:

$$s_m(u_1, \ldots, u_{m+1}) = f_1 \frac{\partial}{\partial y_1} + \ldots + f_n \frac{\partial}{\partial y_n} .$$

First, we will assume that $s > 0$. Then there are i, j such that $\frac{\partial f_i}{\partial y_j} \neq 0$. Consider the commutator of $\frac{\partial}{\partial y_j}$ with both sides of the above equality. On the right we will have a non-zero homogeneous derivation $\sum_{t=1}^{n} \frac{\partial f_t}{\partial y_t} \frac{\partial}{\partial y_t}$ of degree $s-1$. Recalling that commutation in a Lie algebra is a derivation, we see that the left side becomes an expression of the form

$$\sum_{l=1}^{m+1} s_m\left(u_1, \ldots, \left[u_l, \frac{\partial}{\partial y_j}\right], \ldots, u_{m+1}\right),$$

which is a sum of values of the standard polynomial s_m on homogeneous elements of W_n. Comparing the two sides we conclude that s cannot me minimal.

It remains to consider the case $s = 0$. Since a single commutation of homogeneous derivations decreases the degree by one, we arrive at the equality $\sum_{i=1}^{m+1} \deg g_i = m$. Let p_j be the number of elements u_p of degree j. Then, obviously,

$$p_0 + p_1 + \sum_{i>1} p_i = m+1, \text{ and } p_1 + \sum_{i>1} i p_i = m .$$

It follows that $\sum_{i>1}(i-1)p_i = p_0 - 1$. By skew-symmetry of the standard polynomial with respect to $x_1, \ldots, x_m$, the values $u_1, \ldots, u_m$ which are substituted for them should be pairwise distinct. We will remark that dimension of the space of derivations of degree 0 is n, and of those of degree not exceeding 1 equals $n + n^2$. Taking u_{m+1} into account, we get $p_0 \leq n+1$ and $p_0 + p_1 \leq n^2 + n + 1$. In this case

$$\sum_{i>1} p_i \leq \sum_{i>1} p_i = p_0 - 1 \leq n . \tag{9}$$

Furthermore,

$$m = p_0 + p_1 + \sum_{i>1} p_i \leq n^2 + n + 1 + n = (n+1)^2 \ .$$

This calculation shows that when $m > (n+1)^2$, all values of the polynomial $s_m(x_1, \dots, x_{m+1})$ on W_n are zero. A more precise analysis of the inequality (9) shows that W_n in fact satisfies the identity

$$s_{(n+1)^2}(x_1, \dots, x_{(n+1)^2+1}) \equiv 0 \ .$$

For $n = 1$ we arrive at the identity of the algebra W_1

$$s_4(x_1, x_2, x_3, x_4, x_5) \equiv 0 \ . \tag{10}$$

The presence of non-trivial identities in Lie algebras of vector fields was first established by E.I. Sumenkov in 1976 (see *Third all-Soviet symposium on theory of rings, algebra and modules*, Summary of communications, Tartu, 1976, p. 92).

The algebra W_n has the same identities as the algebra $\mathrm{Vect}(M)$ of vector fields on a smooth n-dimensional manifold M. It would be interesting to find a basis of those identities. Even for $n = 1$ it is not known whether the single identity (10) is such a basis. Important results on identities of the algebra $\mathrm{Vect}(M)$ are due to Razmyslov, as well as Kirillov [1983] and his students. We will also remark at this point that the Lie algebra G of an n-dimensional Lie group $\mathfrak{G}$ is a Lie subalgebra of left-invariant fields from $\mathrm{Vect}(M)$, i.e. it is an algebra with identity. Since G is finite-dimensional, its identities are significantly more numerous than those of $\mathrm{Vect}(M)$.

Because finite simple groups can be distinguished by means of identities (see Sect. 5.1, Chap. 2), it is interesting to find their bases of identities explicitly. Solution of this difficult problem is made somewhat easier by applying the principal centralizer identity (Sect. 4.1). So far, no visible general patterns have been found here, and bases of identities have been explicitly described only for simple groups from the family $\mathrm{PSL}\,(2, 2^n)$, as well as for $\mathrm{PSL}\,(2,7)$, $\mathrm{PSL}\,(2,9)$ and $\mathrm{PSL}\,(2,11)$ (Cossey, Oates and MacDonald, Street, Southcott). These proofs rely on the study of the structure of critical groups from $\mathrm{var}\, G$, where G is one of the groups listed above, as well as on delicate investigation of relationships between two-dimensional characters and values of words on two letters in the group $\mathrm{PSL}\,(2,q)$.

The task of finding bases of identities in concrete semigroups is even harder. For example, the monoid of all transformations of the set $\{1, \dots, n\}$ into itself, does not have a finite basis of identities for $n \geq 3$ (Volkov). In a number of natural examples, bases of identities are known – with the exception of cyclic semigroups.

§ 6. Small and Extremal Varieties

6.1. Atoms in Lattices of Varieties. In every signature, the unit variety is defined by the identity $x \equiv y$. Standard application of Zorn's lemma allows to conduct an argument showing that subvarieties of a given non-unit variety include *atoms*, i.e. varieties which themselves do not contain any non-unit proper subvarieties. For example, every non-trivial group contains a non-trivial cyclic subgroup. Consequently, every group variety $\mathfrak{V} \neq \mathcal{E}$ contains a group of prime order p. Such groups in fact generate all the *atoms* $\mathfrak{A}_p = \mathfrak{A} \cap \mathfrak{B}_p$ *in the lattice of varieties* of groups. Similar simple considerations usually permit to find atoms in all more or less interesting lattices of varieties. Nevertheless, it is useful to know what algebras one is bound to encounter in this or that variety.

Clearly, one natural atomic variety of Lie algebras over a field is the variety $\mathfrak{A}$ of vector spaces, i.e. algebras with trivial multiplication. It isn't difficult to show that in the lattice of varieties of associative rings, atoms are determined by identities $px \equiv 0, xy \equiv 0$, or $px \equiv 0, x^p \equiv x$ (*Fermat's identity*), where p is a prime – that is, every variety contains the group $\mathbb{Z}_p$ with either zero multiplication, or multiplication modulo p. In particular, identities $\{2x \equiv 0, x^2 \equiv x\}$ define the *variety of Boolean rings.*

A partially ordered set in which for every two elements a and b there exists the least upper bound $a \vee b$ and the greatest lower bound $a \wedge b$, is called a *lattice.* The class of all lattices can be axiomatized by the associative, commutative, absorption and idempotent laws, thus becoming a variety. Variety $\mathcal{D}$ of all distributive lattices, introduced by means of identities in Sect. 2.3, is the unique atom in the class of *lattice varieties.* This can be explained by the fact that a non-unit lattice always contains a 2-element sub-lattice L; the latter generates all of $\mathcal{D}$, since every distributive lattice is isomorphic to a sub-lattice of the Cartesian power of L, i.e. is embeddable in the lattice of all subsets of some set.

All atoms in the lattice of semigroup varieties have also been determined (Kalicki and Scott [1961]). Aside from group varieties $\mathfrak{A}_p$ mentioned above, they include: *left (right) null semigroups* with the identity $xy \equiv x$ or $xy \equiv y$, *semilattices* (with identities $x^2 \equiv x$ and $xy \equiv yx$), and semigroups nilpotent with index two ($xy \equiv zt$) (see Volkov and Shevrin [1985]).

6.2. Classification of Certain Identities. The problem of describing all consequences of an identity $f \equiv g$ (or, equivalently, all subvarieties of the variety defined by it) can be completely solved only for identical relations which are sufficiently strong[3]. In other words, the lattice of subvarieties is possible to describe only for comparatively small varieties. This is the reason

[3] With the exception of a variety with particularly 'poor' signature – the variety of *unars*, i.e. algebras with a single unary operation. The lattice of unar varieties is distributive, countable and easily described.

why determination of lattices of small varieties always forms a part of the investigation of identities of algebras with a given signature.

It is an easy exercise to enumerate all possible identities of abelian groups given by $\{[x, y] \equiv 1, x^n \equiv 1\}$ for some $n \geq 0$. Similarly, it is not difficult to find all subvarieties in the variety $\mathfrak{N}_2$ of nilpotent groups with index 2. Identities which define them have the form $[x, y, z] \equiv 1$, $[x, y]^k \equiv 1$ and $x^l \equiv 1$, where $k|l$, and $k, l \geq 0$.

Commutator identities are usually manipulated by means of certain relations of Lie type, which hold in all groups (see equations (16) in Chap. 1). Bases of identities for all subvarieties of $\mathfrak{N}_3$ have been found by Jónsson and Remeslennikov: for $k_1, k_2, k_3, k_4 \geq 0$ satisfying $k_4|k_3|k_2|k_1$ (together with certain other restrictions), they have the form

$$x_1^{k_1} \equiv 1, \ , [x, y]^{k_2} \equiv 1, \ [x, y, z]^{k_3} \equiv 1, \ , [x, y, y]^{k_4} \equiv 1$$
$$\text{and } [x, y, z, t] \equiv 1 .$$

The lattice of subvarieties of $\mathfrak{N}_4$ is significantly more complicated. Its description required greater effort (Fitzpatrick and Kovács [1983]). This can be partially explained by lack of distributivity in that lattice. Significant further progress in the direction of determining the lattice of subvarieties $l(\mathfrak{N}_c)$ for increasing c is hardly possible.

A similar picture can be painted for the varieties of Lie or associative rings with small index of nilpotency. In the case of linear algebras over a field of characteristic zero, however, representation theory of the symmetric group (see Sect. 3.4, Chap. 1) provides a practical algorithm for finding all varieties of bounded index of nilpotency k, and the values of k which can be achieved depend on our computational capabilities.

Finite basis property of metabelian groups makes the task of determining the lattice of subvarieties of $\mathfrak{A}^2$ realistic. The greatest progress in this area was achieved by Bryce [1970], who gave a description of metabelian varieties which are not decomposable into a union of non-nilpotent ones. Classification of all nilpotent metabelian group varieties has not been completed as yet. All proper subvarieties of the variety of metabelian Lie algebras over a field of characteristic zero can be easily listed: they are $\mathfrak{A}^2 \cap \mathfrak{N}_c$, where $c = 0, 1, 2, \ldots$ (see Sect. 3.4, Chap. 1). The lattice of varieties of metabelian Lie algebras over a field of characteristic $p > 0$, even though distributive, has a more complex structure. It involves additional identities which typically have the form $x_1 x_2 \ldots x_2 x_n \ldots x_n x_1 \ldots x_1 \equiv 0$, where the number of copies of each variable is a power of p (see Bakhturin [1985]). The analogous problem for a finite field has not been solved.

Even fewer subvarieties have been found in lattices of varieties of semigroups and other algebras with 'bad' congruences. Problems of this type are either easily solved (for example, all varieties of commutative monoids have been described), or prove to be extremely difficult.

6.3. Extremal Varieties. What can be said about a group identity, if it holds in some finite non-abelian group or, conversely, about an identity which is not satisfied in the two-dimensional Lie algebra of matrices of the form $\begin{pmatrix} * & * \\ 0 & 0 \end{pmatrix}$ over $\mathbb{C}$? It is often possible to give a valid answer to questions of this type even in the cases when the lattice of all varieties in question is not easy to study. It is merely necessary to find minimal or maximal varieties which are 'responsible for' the given property.

For example, the structure of those non-abelian finite groups whose proper subgroups are abelian was determined long ago. This easily yields a list of minimal non-abelian locally finite group varieties: $\mathfrak{B}_p \cap \mathfrak{N}_2$ (with p odd), $\mathfrak{B}_4 \cap \mathfrak{N}_2$, and $\mathfrak{A}_p\mathfrak{A}_q$, where p, q are distinct primes. This list also turns out to be exhaustive among all (locally) solvable varieties, as well as many others (generally speaking, however, there exist other almost abelian varieties, as evidenced by example of Sect. 4.3, Chap. 1).

A more difficult problem was posed by Kovács and Newman: give a description of *almost Cross varieties*, i.e. varieties which are not generated by finite groups, but whose every proper subvariety has the form $\operatorname{var} G$ with $|G| < \infty$. Such description would give a direct indication of the minimal collection of groups which should be contained in any variety $\mathfrak{V}$ which is not generated by a single group. Kovács and Newman [1971] gave the following list of known almost Cross varieties: $\mathfrak{A}$, $\mathfrak{A}_p\mathfrak{A}_p$, $\mathfrak{A}_p\mathfrak{A}_q\mathfrak{A}_r$ (with p, q, r – distinct primes), $\mathfrak{A}_p(\mathfrak{N}_2 \cap \mathfrak{B}_q)$ $(p \neq 2)$, and $\mathfrak{A}_p(\mathfrak{N}_2 \cap \mathfrak{B}_q)$ $(p, q \neq 2$ – distinct primes). Ol'shanskij [1971] confirmed the Kovács-Newman conjecture, proving that this list of almost Cross varieties is complete. Hence every solvable variety either contains one of them, or is generated by a finite group.

This gives rise to a simple algorithm for checking whether an identity

$$f(x_1, \ldots, x_n) \equiv 1 \tag{11}$$

determines, in the variety $\mathfrak{A}^l$ of all solvable groups of a given index l, a subvariety generated by a single finite group, or not. First, one can determine whether identity (11) implies an identity $x^n \equiv 1$ for some n. This step is obvious, because n is the greatest common divisor of exponents of indeteminates in the word $f = x_1^{k_1} \ldots x_n^{k_n}$, obtained from f by a permutation of letters. In the list given above, one has to consider all varieties of period n (i.e. those with identity $x^n \equiv 1$), and compute the value of the word f in their free groups with n free generators (this is easily achieved by means of embedding in a wreath product). If (and only if) values of f are distinct from 1 in all of those groups, then f defines a Cross variety. For simple identities, no calculations are necessary at all. For example, $\mathfrak{B}_p \cap \mathfrak{A}^l$ is a Cross variety for any l and a prime p, since the above list doesn't include any varieties of prime period.

The situation is different when the assumption of solvability is dropped. Namely, in a variety of prime period $\mathfrak{R}_p$ (see Sect. 3.2), Razmyslov [1972] exhibited the first example of a non-solvable almost Cross variety $(p \geq 5)$. In order to do so, he first produces an almost Cross variety of Lie algebras in characteristic p, satisfying the Engel identity of degree $p-2$. Groups obtained from these algebras by the Campbell-Hausdorff formula (more precisely, by means of its truncated form; see Sect. 4.2, Chap. 1), generate an almost Cross variety.

In the case of solvable Lie rings, once again, a complete description of almost Cross varieties has been found (Bakhturin and Ol'shanskij; see Bakhturin [1985]). For a finite field of characteristic p the only such variety is Artamonov's variety $\mathfrak{M}$ (which was considered by them for other reasons): 1) $(xy)(zt) \equiv 0$, 2) $xyz \dots z \equiv 0$ and 3) $xyx \dots xy \dots y \equiv 0$, where the number of copies of z in (2) is p, and (3) involves p copies of each of the letters x and y. Artamonov observed that all proper subvarieties of $\mathfrak{M}$ are nilpotent, while $\mathfrak{M}$ itself is not. This implies that $\mathfrak{M}$ is an almost Cross variety. It is more difficult to show that there are no other solvable almost Cross varieties. As in the case of groups, the problem can be translated into the language of representations, but this proves to be far from sufficient for obtaining a similar solution.

Free algebra of the variety $\mathfrak{M}$ has a very clear description as a subalgebra of a semidirect product of two abelian algebras. Similarly as for groups, this phenomenon can be used to obtain a simple algorithm for identifying identities of finite algebras inside $\mathfrak{A}^l$.

A complete answer was obtained in the case of associative rings by L'vov: every almost Cross variety of rings is defined either by the identity $xy \equiv 0$, or by the system $px \equiv x^p \equiv xy - yx \equiv 0$. L'vov also described all almost nilpotent varieties of rings.

A large amount of work in this area was done by Artamonov, who described all chain varieties of linear algebras, i.e. varieties for which the lattice of subvarieties forms a chain, thus being particularly uncomplicated (the variety $\mathfrak{M}$ discussed above is a chain variety; Artamonov [1978]).

Kargapolov and Churkin [1971], as well as Groves [1971,1972], fully explained the special rôle of the variety $\mathfrak{A}^2$ of metabelian groups:

Theorem. *Every solvable variety $\mathfrak{V}$ (i.e. subvariety of some $\mathfrak{A}^l$) either contains $\mathfrak{A}^2$ as a subvariety, or – conversely – is contained in a product $\mathfrak{B}_m\mathfrak{N}_c\mathfrak{B}_m$ for some m, c.*

Groves extended this result to any subvariety $\mathfrak{V}$ of a product of solvable and locally finite varieties, and obtained other variants of the theorem.

Bakhturin [1985] focused attention on extremal property of the variety $\mathfrak{A}^2$ of metabelian Lie algebras over an infinite field: a solvable variety $\mathfrak{V}$ is locally nilpotent if and only if $\mathfrak{A}^2 \not\subset \mathfrak{V}$. To show this, he proved a preliminary result stating that every variety not containing all metabelian algebras has to satisfy an Engel identity. Quite recently however, Zel'manov solved a long-standing problem of Kostrikin, showing that over a field of characteristic zero an Engel identity implies nilpotency (see Sect. 2.3, Chap. 2). Consequently, every variety of Lie algebras over a field of characteristic zero is nilpotent if and only if it does not contain all metabelian algebras.

Various extremal properties of the group variety $\mathfrak{A}_p\mathfrak{A}$, where p is a prime, have also been noted. If, for example, a finitely generated solvable group G contains a subgroup which is not finitely generated, then $\operatorname{var} G \supseteq \mathfrak{A}_p\mathfrak{A}$ for some p. Moreover, $\mathfrak{A}_p\mathfrak{A}$ has no *semigroup identities* (i.e. identities $f \equiv g$, where f and g are semigroup words), since the wreath product $\mathbb{Z}_p \wr \mathbb{Z}$ contains an absolutely free semigroup (Belyaev, Sesekin). At the same time, varieties $\mathfrak{N}_c\mathfrak{B}_n$ do satisfy non-trivial semigroup identities (A.I. Mal'tsev). Among solvable group varieties, $\mathfrak{A}_p\mathfrak{A}$ is distinguished as a variety 'almost satisfying semigroup identities', since whenever $\mathfrak{V} \subseteq \mathfrak{A}^l$, then $\mathfrak{V} \supseteq \mathfrak{A}_p\mathfrak{A}$ or $\mathfrak{V} \subseteq \mathfrak{N}_c\mathfrak{B}_n$ for some p, c, n.

Volichenko [1980] proved that there exists a unique variety $\mathfrak{G}$ of Lie algebras over a field of characteristic zero, with the property that any variety $\mathfrak{V}$ satisfies some standard identity (see Sect. 1.3, Chap. 2) if and only if $\mathfrak{V} \not\supseteq \mathfrak{G}$. Such variety $\mathfrak{G}$ is defined in $\mathfrak{A}\mathfrak{N}_2$ by the identity $(x_1x_2)((x_1x_3)(x_2x_4)) \equiv 0$.

Variety generated by the Grassman algebra (see Sect. 1.1, Chap. 2) is extremal in certain respects. For example:

Theorem (Kemer). *A variety $\mathfrak{M}$ of associative algebras satisfies all Capelli identities of some degree (see Sect. 5.4, Chap. 2) if and only if $\mathfrak{M}$ does not contain the Grassman algebra.*

Kemer [1980a] also showed that a variety of associative algebras over a field of characteristic zero is non-matrix (see Sect. 1.2) if and only it satisfies an identity of the form

$$\prod_{i=1}^{n}[[x_{i,1}, x_{i,2}], [x_{i,3}, x_{i,4}], x_{i,5}] \equiv 0\,,$$

where [,] is the ring commutator.

A list of 'extremal' examples can be continued well beyond this. We will mention the semigroup (6) from Sect. 4.3, limit varieties from Sect. 1.2, and others. One can also note old theorems which assert that a variety of lattices satisfies the modular identity if and only if it doesn't include a lattice in Fig. 9, and that a variety of modular lattices is distributive if and only if it does not contain the lattice in Fig. 10.

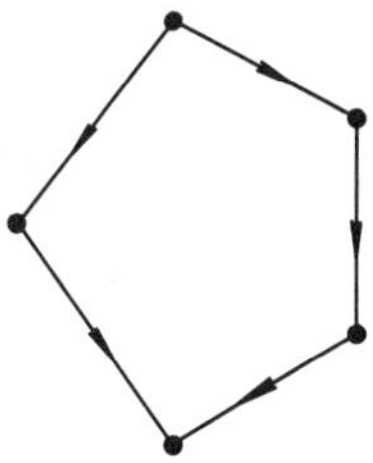

Fig. 9

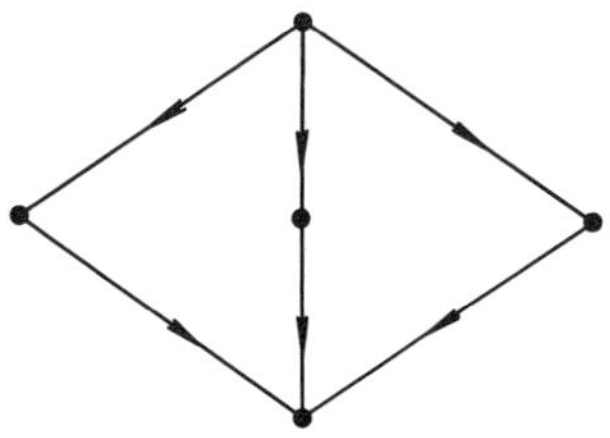

Fig. 10

Chart of the infinite ocean of algebraic varieties, with scattered islands of important examples, compels one to think that correctly posed problems about properties of varieties and identities of algebras should lead to a search for varieties and identities which are extremal relative to some natural property, rather than to a description of lattices of all varieties in question.

Comments on the Literature

We will first give short annotation of recommended books on the subject. The monograph Adyan [1975] contains a complete and updated variant of the negative solution of the Burnside problem for groups with identity $x^n \equiv 1$, n odd and $n \geq 665$. An example of a group variety without a finite basis of identities, as well as solutions of certain other problems from group theory can also be found there. Monograph by Bakhturin [1985] provides a general outline of the theory of varieties of Lie algebras, and their links with varieties of groups and associative rings. This exposition should be accessible even to beginning algebraists. Chapter 10 of the monograph by Jacobson [1956] is devoted to the theory of PI-algebras. In his second book (Jacobson [1975]), the reader will find yet another approach to this subject, as well as a detailed discussion of the structure of finite-dimensional division rings. Questions of the theory of identities of non-associative rings are clarified in the monograph by Zhevlakov et al. [1978]. Solution of the bounded Burnside problem is presented in Kostrikin [1986]. That monograph also contains newer results on Engel Lie algebras. Questions pertaining to links between abstract groups and Lie algebras are explained in Magnus, Karrass and Solitar [1966]. A.I. Mal'tsev [1970] presents a general approach to the theory of varieties and quasivarieties of algebraic systems (see also Kurosh [1973]). The book by H. Neumann [1967] is the first tome entirely devoted to the theory of varieties. Problems which were posed in it ('Hanna Neumann's problems') had a great influence on the development of this theory (because of the rate at which these problems were being solved, see also Kovács and Newman [1974]). Problems associated with structure theory of associative PI-algebras are touched upon in Herstein [1968]. Monographs by Procesi [1973] and Rowen [1980] are entirely devoted to the subject of PI-algebras. The book by Plotkin and Vovsi [1983] contains elements of the theory of varieties of representations.

In the list of references we find surveys on theory of groups, semigroups, rings, lattices and universal algebras, forming an introduction on which the reader can base details of development and more specific questions of the theory of identities. VINITI surveys by Bakhturin, Slin'ko and Shestakov [1981], Beidar et al. [1984], Noskov, Remeslennikov and Roman'kov [1979], and Plotkin [1971] have a purely encyclopædic nature (they also contain references to earlier surveys in the area). The book by Volkov and Shevrin

[1985], as well as the report *Ordered sets and lattices* [1985], Bratislava, Komenskij University have a similar character. Among other surveys, we will note Artamonov [1978] and Amitsur [1974].

The collection *Solvable and infinite simple groups. Collection of articles.* Series: Matematika. Novoe v zarub. nauke. Vol. 21. Mir: Moscow contains, in particular, an article of P. Hall on the construction of finitely defined infinite simple groups.

Apart from books and surveys, references also include some original papers which are of historical interest as first sources on various chapters of the theory of identities (Birkhoff [1935], Burnside [1902], B.H. Neumann [1937], Novikov [1959], Specht [1950]). We also list articles containing important results on problems such as the existence of finite bases of identities (Genov [1976], Kemer [1984], Latyshev [1976], Ol'shanskij [1970], Oates and Powell [1964], Vaughan-Lee [1970], as well as Cohen [1967] and a closely related important work by Higman [1952]), questions on periodic groups and algebraic algebras (Braun [1982], M. Hall [1958], Kleiman [1983], Ol'shanskij [1982, 1983], Razmyslov [1978], Shirshov [1957], Zel'manov [1983 and 1984b]), and other problems pertaining to the theory of varieties (Baumslag et al. [1964], Formanek [1972], Jónsson [1979], Kukin [1980], McKenzie [1973], Razmyslov [1973b], Shmel'kin [1965]).

References*

Artamonov, V.A. [1973] Chain varieties of linear algebras. Tr. Mosk. Mat. O.-va *29*, 51–78. Zbl. 289.17005. English translation: Trans. Mosc. Math. Soc. *29*, 47–76 (1976)

Artin, M. [1969] On Azumaya algebras and finite dimensional representations of rings. J. Algebra *11* (4), 532–563. Zbl. 222.16007

Bakhturin, Yu.A. [1976] Simple Lie algebras satisfying a non-trivial identity. Serdica *2* (3), 241–246. Zbl. 357.17013

Benediktovich, I.I., Zalesskij, A.E. [1980] T-ideals of free Lie algebras with polynomial growth of the sequence of codimensions. Izv. Akad. Nauk SSSR, Ser. Fiz.-Mat. Nauk *1980* (3), 5–10 (Russian). Zbl. 434.17009

Bergman, G.M., Lewin, J. [1975] The semigroup of ideals of a fir is (usually) free. J. Lond. Math. Soc., II. Ser *11* (1), 21–31. Zbl. 275.16003

Block, R.E., Wilson, R.L. [1984] The restricted simple Lie algebras are of classical or Cartan type. Proc. Natl. Acad. Sci. USA *81* (16), 5271–5274. Zbl. 542.17003

Bryant, R.M. [1973] Some infinitely based varieties of groups. J. Aust. Math. Soc. *16* (1), 29–32. Zbl. 272.20019

Bryant, R.M., Newman, M.F. [1974] Some finitely based varieties of groups. Proc. Lond. Math. Soc., II. Ser. *28* (2), 237–252. Zbl. 278.20020

Bryce, R.A. [1970] Metabelian groups and varieties. Philos. Trans. R. Soc. Lond., Ser. A *266* (1176), 281–355. Zbl. 203,321

* For the convenience of the reader, references to reviews in Zentralblatt für Mathematik (Zbl.), compiled using the MATH database, and Jahrbuch über die Fortschritte der Mathematik (Jrb.) have, as far as possible, been included in this bibliography.

Dorofeev, G.V. [1976] Join of varieties of algebras. Algebra Logika *15* (3), 267–291. English translation: Algebra Logic *15*, 165–181 (1977). Zbl. 362.08004

Drensky, V.S. (= Drenskij, V.S.) [1974] On identities in Lie algebras. Algebra Logika *13* (3), 265–290. Zbl. 298.17011. English translation: Algebra Logic *13*, 150–165 (1975)

Drensky, V.S. [1984] Codimensions of T-ideals and Hilbert series of relatively free algebras. J. Algebra *91* (1), 1–17. Zbl. 552.16006

Dubnov, Ya.S., Ivanov, V.K. [1943] On reduction of the degree of affine polynomials. Dokl. Akad. Nauk SSSR *41*, 95–98 (Russian). Zbl. 60,33

Farkaš, D.R., Snider, R.L. [1974] Group algebras whose simple modules are injective. Trans. Am. Math. Soc. *194*, 241–248. Zbl. 287.16004

Fitzpatrick, P., Kovács, L.G. [1983] Varieties of nilpotent groups of class four. I. J. Aust. Math. Soc. *A35* (1), 59–73. Zbl. 524.20013

Goldie, A.W. [1974] Lectures on quotient rings and rings with polynomial identities. Math. Inst. Giessen, Univ. Giessen: Giessen. Zbl. 312.16001

Golod, E.S. [1964] On nil algebras and finitely approximated p-groups. Izv. Akad. Nauk SSSR, Ser. Fiz.-Mat. Nauk *28* (2), 273–276. Zbl. 215,392. English translation: Transl., II. Ser., Am. Math. Soc. *48*, 103–106 (1965)

Groves, J.R.J. [1971] On varieties of soluble groups. I. Bull. Aust. Math. Soc. *5* (1), 95–109. Zbl. 216,84

Groves, J.R.J. [1972] On varieties of soluble groups. II. Bull. Aust. Math. Soc. *7* (3), 437–441. Zbl. 241.20023

Gupta, N.D., Levin, F. [1983] On the Lie ideals of a ring. J. Algebra *81* (1), 225–231. Zbl. 514.16024

Hall, M. Jr. [1959] The theory of groups. MacMillan: New York. Zbl. 84,22

Hall, P. [1954] Finiteness conditions for soluble groups. Proc. Lond. Math. Soc., II. Ser. *4* (16), 419–436. Zbl. 56,256

Hartley, B. [1977] Injective modules over group rings. Q. J. Pure Appl. Math. *28* (1), 1–29. Zbl. 345.16010

Heinecken, H. [1963] Liesche Ringe mit Engelbedingung. Math. Ann. *149* (3), 232–236. Zbl. 108,32

Higgins, P.J. [1954] Lie rings satisfying the Engel condition. Proc. Camb. Philos. Soc. *50* (1), 8–15. Zbl. 55,26

Higman, G. [1959] Some remarks on varieties of groups. Q. J. Pure Appl. Math. *10* (39), 165–178. Zbl. 89,13

Isaacs, I.M., Passman, D.S. [1964] Groups with representations of bounded degree. Can. J. Math. *16* (2), 299–309. Zbl. 124,267

Kac, V.G. [1980] On simplicity of certain infinite-dimensional Lie algebras. Bull. Am. Math. Soc. *2*, 311–314. Zbl. 427.17012

Kalicki, J., Scott, D. [1955] Equational completeness of abstract algebras. Indagationes Math. *17* (5), 650–659. Zbl. 73,245

Kargapolov, M.I., Churkin, V.A. [1971] On varieties of solvable groups. Algebra Logika *10* (6), 651–657. English translation: Algebra Logic *10*, 395–398 (1973). Zbl. 258.20033

Kargapolov, M.I., Merzlyakov, Yu.I., Remeslennikov, V.N. [1960] On completions of groups. Dokl. Akad. Nauk SSSR *134* (3), 518–520. Zbl. 97,14. English translation: Sov. Math., Dokl. *1*, 1099–1101 (1961)

Kemer, A.R. [1980a] On non-matrix varieties. Algebra Logika *19* (3), 255–283. English translation: Algebra Logic *19*, 157–178 (1981). Zbl. 467.16025

Kemer, A.R. [1980b] Capelli identities and nilpotency of the radical of a finitely generated PI-algebra. Dokl. Akad. Nauk SSSR *255* (4), 793–796. English translation: Sov. Math., Dokl. *22*, 750–753. Zbl. 489.16011

Kharlampovich, O.G. [1981] A finitely solvable group with unsolvable word problem. Izv. Akad. Nauk SSSR, Ser. Fiz.-Mat. Nauk *45* (4), 852–873. Zbl. 485.20023. English translation: Math. USSR, Izv. *19*, 151–169 (1981)

Kirillov, A.A. [1983] On identities of the Lie algebra of Hamiltonian vector fields on a plane. Prepr., Inst. Prikl. Mat. Im. M. V. Keldysha Akad. Nauk SSSR, Mosk. *121* (Russian)

Kleiman, Yu.G. (= Klejman, Yu.G.) [1973] On the basis of the product of varieties of groups. Izv. Akad. Nauk SSSR, Ser. Fiz.-Mat. Nauk *37* (1), 95–97. English translation: Math. USSR, Izv. *7*, 91–94 (1974). Zbl. 279.20022

Kleiman, Yu.G. [1982] On identities in groups. Tr. Mosk. Mat. O.-va *44*, 62–108. Zbl. 495.20013. English translation: Trans. Mosc. Math. Soc. 1983 (2), 63–110 (1983)

Kostrikin, A.I. [1958] On the Burnside problem. Dokl. Akad. Nauk SSSR *119* (6), 1081–1084 (Russian). Zbl. 84,255

Kóvacs, L.G., Newman, M.F. [1971] On non-cross varieties of groups. J. Aust. Math. Soc. *12* (2), 129–144. Zbl. 221.20037

Kóvacs, L.G., Newman, M.F., Pentony, P.F. [1968] Generating groups of nilpotent varieties. Bull. Am. Math. Soc. *74* (5), 968–971. Zbl. 186,37

Krasil'nikov, A.N., Shmel'kin, A.L. [1981] On the Specht property and basic rank of certain variational products. Algebra Logika *20* (5), 546–554. English translation: Algebra Logic 20, 357–363 (1982). Zbl. 529.20016

Kruse, R. [1973] Identities satisfied by a finite ring. J. Algebra *26* (2), 298–318. Zbl. 276.16014

Kukin, G.P. [1974] On free products of restricted Lie algebras. Mat. Sb., Nov. Ser. *95* (1), 53–83. English translation: Math. USSR, Sb. *24*, 49–78 (1976). Zbl. 327.17005

Kushkulei, A.Kh. (= Kushkulej, A.Kh.) [1978] On identities of finite-dimensional representations. In: Work collection on algebra, Moscow 1978. 134–157: (Russian). Zbl. 462,20033

Latyshev, V.N., Shmel'kin, A.L. [1969] On a problem of Kaplansky. Algebra Logika *8* (4), 447–448. Zbl. 206,47. English translation: Algebra Logic *8*, 257 (1971)

Leron, U., Vapne, A. [1970] Polynomial identities of related rings. Isr. J. Math. *8* (2), 127–137. Zbl. 205,345

Levin, F. [1970] Generating groups for nilpotent varieties. J. Aust. Math. Soc. *11* (1), 28–32. Zbl. 191,19

L'vov, I.V. [1973] On varieties of associative rings. I. Algebra Logika *12* (3), 269–297. English translation: Algebra Logic *12*, 150–167 (1974). Zbl. 288.16008

Lyndon, R.C., Schupp, P.E. [1977] Combinatorial group theory. Springer-Verlag: New York, Heidelberg, Berlin. Zbl. 368.20023

Mal'cev, A.I. (= Mal'tsev, A.I.) [1953] Nilpotent semigroups. Uch. Zap. Ivanov. Ped. Inst. *4*, 107–111 (Russian). Zbl. 87,255

Mal'cev, A.I. [1967] On multiplication of classes of algebraic systems. Sib. Mat. Zh. *8* (2), 346–365. English translation: Sib. Math. J. *8*, 254–267 (1968). Zbl. 228.08007

Mal'cev, Yu.N. (= Mal'tsev, Yu.N.) [1971] Basis of identities of the algebra of upper triangular matrices. Algebra Logika *10* (4), 393–400. Zbl. 247.16005. English translation: Algebra Logic *10*, 242–247 (1973)

Markov, V.T. [1973] On the dimension of non-commutative affine algebras. Izv. Akad. Nauk SSSR, Ser. Fiz.-Mat. Nauk *37* (2), 284–288. Zbl. 255.16007. English translation: Math. USSR, Izv. *7*, 281–285 (1974)

Martindale, W.S. III [1969] Prime rings satisfying a generalized polynomial identity. J. Algebra *12* (4), 576–584. Zbl. 175,31

Medvedev, Yu.A. [1978] Finite basis property of varieties with a two-term identity. Algebra Logika *17* (6), 705–726. Zbl. 425.16016. English translation: Algebra Logic *17*, 458–472 (1979)

Medvedev, Yu.A. [1980] Example of a variety of solvable alternative algebras over a field of characteristic 2, which has no finite basis of identities. Algebra Logika *19* (3), 300–313. English translation: Algebra Logic *19*, 191–201 (1981). Zbl. 468.17005

Mishchenko, S.P. [1982] Varieties of Lie algebras with slow growth of the sequence of codimensions. Vestn. Mosk. Univ., Ser. I *19* (5), 63–66. English translation: Mosc. Univ. Math. Bull. *37* (5), 78–82. Zbl. 517.17006

Mishchenko, S.P. [1984] On the Engel problem. Mat. Sb., Nov. Ser. *124* (1), 56–67. Zbl. 546.17004. English translation: Math. USSR, Sb. *52*, 53–62 (1985)

Murskij, V.L. [1975] Finite basis property and other properties of 'almost all' finite algebras. Probl. Kybern. *30*, 43–56 (Russian) Zbl. 415.08005

Novikov, P.S., Adyan, S.I. [1968] On infinite periodic groups I, II, III. Izv. Akad. Nauk SSSR, Ser. Fiz.-Mat. Nauk *32* (1), 212–244, (2), 251–524 and (3), 709–731. Zbl. 194,33. English translation: Math. USSR, Izv. *2*, 209–236, 241–480, 665–685 (1969)

Ol'shanskij, A.Yu. [1971] Solvable almost cross varieties of groups. Mat. Sb., Nov. Ser. *85* (5), 115–131. Zbl. 234.20012. English translation: Math. USSR, Sb. *14*, 115–129 (1972)

Passman, D.S. [1972] Group rings satisfying a polynomial identity. J. Algebra *20* (1), 103–117. Zbl. 226.16015

Perkins, P. [1969] Bases for equational theories of semigroups. J. Algebra *11* (2), 298–314. Zbl. 186,34

Platonov, V.P. [1967] Linear groups with identity relations. Dokl. Akad. Nauk BSSR *11* (7), 581–582 (Russian). Zbl. 252.20036

Polin, S.V. [1976] On identities of finite algebras. Sib. Mat. Zh. *17* (6), 1356–1366. Zbl. 353.17003. English translation: Sib. Math. J. *17*, 992–999 (1977)

Popow, A. [1981] Identities of tensor product of two copies of Grassmann algebra. Compt. Rend. Bulg. Acad. Sc. *34*, 1205–1208

Posner, E.C. [1960] Prime rings satisfying a polynomial identity. Proc. Am. Math. Soc. *11* (2), 180–183. Zbl. 215,381

Procesi, C. [1972] On a theorem of M. Artin. J. Algebra *22* (2), 309–315. Zbl. 238.16015

Procesi, C., Small, L.W. [1968] Endomorphism rings of modules over PI-algebras. Math. Z. *106* (3), 178–180. Zbl. 175,317

Razmyslov, Yu.P. [1972] On an example of non-solvable almost cross varieties of groups. Algebra Logika *11* (2), 186–205. Zbl. 248.20033. English translation: Algebra Logic *11*, 108–120 (1973)

Razmyslov, Yu.P. [1973a] The existence of a finite basis of identities of the matrix algebra of order 2 over a field of characteristic zero. Algebra Logika *12* (1), 83–113. English translation: Algebra Logic *12*, 47–63 (1974). Zbl. 282.17003

Razmyslov, Yu.P. [1974] On the Jacobson radical of PI-algebras. Algebra Logika *13* (3), 337–360. English translation: Algebra Logic *13*, 192–204 (1975). Zbl. 354.16008

Razmyslov, Yu.P. [1981] Algebras satisfying identical relations of Capelli type. Izv. Akad. Nauk SSSR, Ser. Fiz.-Mat. Nauk *45* (1), 143–166. Zbl. 465.16009. English translation: Math. USSR, Izv. *18*, 125–144 (1982)

Razmyslov, Yu.P. [1983] Central polynomials in irreducible representations of a semisimple Lie algebra. Mat. Sb., Nov. Ser. *122* (1), 97–125. Zbl. 526.17002. English translation: Math. USSR, Sb. *50*, 99–124 (1985)

Regev, A. [1971] Existence of polynomial identities in $A \otimes_F B$. Bull. Am. Math. Soc. *77* (6), 1067–1069. Zbl. 225. 16012

Remeslennikov, V.N., Sokolov, V.G [1971] Finite approximability of groups with respect to conjugacy. Sib. Mat. Zh. *12* (5), 1085–1099. English translation: Sib. Math. J. *12*, 783–792 (1972). Zbl. 282.20022

Romanovskij, N.S. [1977] Free subgroups in finitely presented groups. Algebra Logika *16* (1), 88–97. Zbl. 384.20030. English translation: Algebra Logic *16*, 62–68 (1978)

Schelter, W.F. [1978] Non-commutative affine PI-rings are catenary. J. Algebra *51* (1), 12–18. Zbl. 375.16015

Shestakov, I.P. [1976] Centers of alternative algebras. Algebra Logika *15* (3), 343–362. English translation: Algebra Logic *15*, 214–226 (1977). Zbl. 402.17019

Shirshov, A.I. [1962] On a conjecture in the theory of Lie algebras. Sib. Mat. Zh. *3* (2), 297–301 (Russian). Zbl. 104,260

Stovba, V.V. [1982] On the finite basis property of certain varieties of Lie and associative algebras. Vestn. Mosk. Univ., Ser. I *1982* (2), 54–58. English translation: Mosc. Univ. Math. Bull. *37* (2), 69–74. Zbl. 494.17011

Stovba, V.V. [1986] On the property of being a finite basis of certain solvable varieties of Lie algebras of finite axiomatic rank. Mat. Sb., Nov. Ser. *129* (1), 104–120. Zbl. 594.17008. English translation: Math. USSR, Sb. *57*, 111–129 (1987)

Volichenko, I.B [1980] On a variety of Lie algebras, connected with standard identities I, II. Izv. Akad. Nauk BSSR, Ser. Fiz.-Mat. Nauk *Nauk 1980* (1), 23–30, Zbl. 421.17009 and (2), 22–27 (Russian). Zbl. 432.17005

von Neumann, J. [1929] Zur allgemeinen Theorie des Masses. Fundam. Math. *13*, 73–116. Jrb. 55,151

Zalesskij, A.E., Smirnov, M.B. [1982] Associative rings satisfying the condition of Lie solvability. Izv. Akad. Nauk BSSR, Ser. Fiz.-Mat. Nauk *Nauk 1982* (2), 15–20 (Russian). Zbl. 499.16022

Zel'manov, E.I. [1984a] On the theory of Jordan algebras. Proc. Int. Congr. Math., Warszawa, 1983 Vol. *I*, 455–463. Zbl. 575.17009

Recommended Literature

Adyan, S.I. [1975] The Burnside problem and identities in groups. Nauka: Moscow. Zbl. 306.20045. English translation: Springer-Verlag: Berlin, Heidelberg, New York (1979). Zbl. 417.20001

Amitsur, S.A. [1974] Polynomial identities. Isr. J. Math. *19* (1-2), 183–199. Zbl. 297.16009

Artamonov, V.A. [1978] Lattices of varieties of linear algebras. Usp. Mat. Nauk *33* (2), 135–167. Zbl. 384.17002. English translation: Russ. Math. Surv. *33* (2), 155–193

Bakhturin, Yu.A. [1985] Identities in Lie algebras. Nauka: Moscow. Zbl. 571,17001. English translation: VNU Science Press: Utrecht (1987)

Bakhturin, Yu.A., Slin'ko, A.M., Shestakov, I.P. [1981] Non-associative rings. Itogi Nauki Tekh., Ser. Algebra, Topologiya, Geom. *18*, 3–72. English translation: J. Sov. Math. *18*, 169–211 (1982). Zbl. 486.17001

Baumslag, G., Neumann, B.H., Neumann, H., Neumann, P.M. [1964] On varieties generated by a finitely generated group. Math. Z. *86* (2), 93–122. Zbl. 125,14

Beidar, K.I., Latyshev, V.N., Markov, V.T., Mikhalev, A.V., Skornyakov, L.A., Tuganbaev, A.A. [1984] Associative rings. Itogi Nauki Tekh., Ser. Algebra, Topologiya, Geom. *22*, 3–115. Zbl. 564.16002. English translation: J. Sov. Math. *38*, 1855–1929 (1987)

Birkhoff, G. [1935] On the structure of abstract algebras. Proc. Camb. Philos. Soc. *31*, 433–454. Zbl. 13,1

Braun, A. [1982] The radical in a finitely generated PI-algebra. Bull. Am. Math. Soc. *7* (2), 385–386. Zbl. 493.16011

Burnside, W. [1902] On an unsettled question in the theory of discontinuous groups. Q. J. Pure Appl. Math. *33*, 230–238. Jrb. 33,149

Cohen, D.E. [1967] On the laws of a metabelian variety. J. Algebra *5* (3), 267–273. Zbl. 157,348

Formanek, E. [1972] Central polynomials for matrix rings. J. Algebra *23* (1), 129–132. 242.15004

Genov, G.K. [1976] The Specht property of certain varieties of associative algebras over a field of characteristic zero. C. R. Acad. Bulg. Sci. *29* (7), 939–941 (Russian). Zbl. 358.16013

Hall, M. Jr. [1959] Solution of the Burnside problem for exponent six. Ill. J. Math. *2*, 764–786. Zbl. 83,248

Hall, P. [1933] A contribution to the theory of groups of prime-power order. Proc. Lond. Math. Soc., II. Ser. *36*, 29–95. Zbl. 7,291

Herstein, I.N. [1968] Noncommutative rings. Carus Math. Monographs 15, J. Wiley: New York. Zbl. 177,58

Higman, G. [1952] Ordering by divisibility in abstract algebras. Proc. Lond. Math. Soc., II. Ser. *2*, 326–336. Zbl. 47,34

Jacobson, N. [1964] Structure of rings (revised edition). Am. Math. Soc. Colloq. Publ. 37, AMS: Providence. Zbl. 73,20

Jacobson, N. [1975] *PI*-algebras: An introduction. Lect. Notes Math., Vol.441, Springer-Verlag: New York, Heidelberg, Berlin. Zbl. 326.16013

Jónsson, B. [1979] Congruence varieties. In: Grätzer, G., Universal algebra. Springer-Verlag: New York, Berlin, Heidelberg. Zbl. 412.08001

Kemer, A.R. [1984] Varieties and $\mathbb{Z}_2$-graded algebras. Izv. Akad. Nauk SSSR, Ser. Fiz.-Mat. Nauk *48* (5), 1042–1059. English translation: Math. USSR, Izv. *25*, 359–374 (1985). Zbl. 586.16010

Kleiman, Yu.G. (= Klejman, Yu.G.) [1983] On certain problems in the theory of group varieties. Izv. Akad. Nauk SSSR, Ser. Fiz.-Mat. Nauk *47* (1), 37–74. Zbl. 516.20014. English translation: Math. USSR, Izv. *22*, 33–65 (1984)

Kostrikin, A.I. [1986] Around Burnside. Nauka: Moscow. Zbl. 624.17001. English translation: Springer-Verlag: Berlin, Heidelberg, New York (1990)

Kóvacs, L.G., Newman, M.F. [1974] Hanna Neumann's problems on varieties of groups. Lect. Notes Math., Vol.372, 417–431, Springer-Verlag: New York, Heidelberg, Berlin. Zbl. 306.20028

Kukin, G.P. [1980] On embeddings of recursively presented Lie algebras and groups. Dokl. Akad. Nauk SSSR *251* (1), 37–39. English translation: Sov. Math., Dokl. *21*, 378–381. Zbl. 447.20026

Kurosh, A.G. [1973] Lectures on general algebra. 2nd ed. Nauka: Moscow. Zbl. 271.08001. English translation: Pergamon Press: London, New York (1965)

Latyshev, V.N. [1976] Partially ordered sets and non-matrix identities of associative algebras. Algebra Logika *15* (1), 53–70. English translation: Algebra Logic *15*, 34–45 (1977). Zbl. 358.16012

Magnus, W., Karrass, A., Solitar, D. [1966] Combinatorial group theory. J. Wiley: New York, London. Zbl. 138,256

Mal'cev, A.I. (= Mal'tsev, A.I.) [1970] Algebraic systems. Nauka: Moscow. Zbl. 223. 08001. English translation: Springer-Verlag: Berlin, Heidelberg, New York (1973). Zbl. 266.08001

McKenzie, R. [1973] Some unsolved problems between lattice theory and equational logic. In: Proc. Univ. Houston Lattice Theory Conf., Houston 1973, 564–573. Zbl. 329.06002

Neumann, B.H. [1937] Identical relations in groups. I. Math. Ann. *114*, 506–525. Zbl. 16,351

Neumann, H. [1967] Varieties of groups. Springer-Verlag: New York, Berlin, Heidelberg. Zbl. 251.20001

Noskov, G.A., Remeslennikov, V.N., Roman'kov V.A. [1979] Infinite groups. Itogi Nauki Tekh., Ser. Algebra, Topologiya, Geom. *17*, 65–157. Zbl. 429.20001. English translation: J. Sov. Math. *18*, 669–735 (1982)

Novikov, P.S. [1959] On periodic groups. Dokl. Akad. Nauk SSSR *127* (4), 749–752. Zbl. 119,22. English translation: Transl., II. Ser., Am. Math. Soc. *45*, 19–22 (1965)

Oates, S., Powell, M.B. [1964] Identical relations in finite groups. J. Algebra *1* (1), 11–39. Zbl. 121,272

Ol'shanskij, A.Yu. [1970] On the finite basis problem for groups. Izv. Akad. Nauk SSSR, Ser. Fiz.-Mat. Nauk *34* (2), 376–384 (Russian). Zbl. 215,105

Ol'shanskij, A.Yu. [1982] Groups of bounded period with subgroups of prime order. Algebra Logika *21* (5), 553–618. English translation: Algebra Logic *21*, 369–418 (1983). Zbl. 524.20024

Ol'shanskij, A.Yu. [1983] On a geometric method in the combinatorial group theory. Proc. Int. Congr. Math., Warszawa 1983, Vol. *I*, 415–424 (1984). Zbl. 565.20015

Plotkin, B.I. [1971] General group theory. Itogi Nauki Tekh., Ser. Algebra, Topologiya, Geom. *1970*, 5–73. Zbl. 281.20001. English translation: J. Sov. Math. *1*, 527–570 (1973)

Plotkin, B.I., Vovsi, S.M. [1983] Varieties of group representations. General theory, related topics and applications. Zinatne: Riga (Russian). Zbl. 527.20004

Procesi, C. [1973] Rings with polynomial identities. M. Dekker: New York. Zbl. 262.16018

Razmyslov, Yu.P. [1973b] On a problem of Kaplansky. Izv. Akad. Nauk SSSR, Ser. Fiz.-Mat. Nauk *37* (3), 483–501. English translation: Math. USSR, Izv. *7*, 479–496 (1974). Zbl. 314.16016

Razmyslov, Yu.P. [1978] On a problem of Hall-Higman. Izv. Akad. Nauk SSSR, Ser. Fiz.-Mat. Nauk *42* (4), 833–847. Zbl. 394.20030. English translation: Math. USSR, Izv. *13*, 133–146 (1979)

Rowen, L.H. [1980] Polynomial identities in ring theory. Academic Press: New York, London. Zbl. 461.16001

Schelter, W.F. [1978] Non-commutative affine PI-rings are catenary. J. Algebra *51* (1), 12–18. Zbl. 375.16015

Shestakov, I.P. [1976] Centers of alternative algebras. Algebra Logika *15* (3), 343–362. English translation: Algebra Logic *15*, 214–226 (1977). Zbl. 402.17019

Shirshov, A.I. [1957] On rings with identity relations. Mat. Sb., Nov. Ser. *43* (2), 277–283 (Russian). Zbl. 78,24

Shmel'kin, A.L. [1965] Wreath products and varieties of groups. Izv. Akad. Nauk SSSR, Ser. Fiz.-Mat. Nauk *29* (1), 149–170 (Russian). Zbl. 135,47

Specht, W. [1950] Gesetze in Ringen. I.. Math. Z. *52*, 557–589. Zbl. 32,389

Vaughan-Lee, M.R. [1970] Uncountably many varieties of groups. Bull. Lond. Math. Soc. *2* (6), 280–286. Zbl. 216,84

Volkov, M.V., Shevrin, L.N. [1985] Identical relations in semigroups. Izv. Vyssh. Uchebn. Zaved., Mat 1985 *11*, 3–47. English translation: Sov. Math. *29* (11), 1–64. Zbl. 629.20029

Zel'manov, E.I. [1983] Lie algebras with algebraic adjoint representation. Mat. Sb., Nov. Ser. *121* (4), 545–561. English translation: Math. USSR, Sb. *49*, 537–592 (1984). Zbl. 544.17003

Zel'manov, E.I. [1984b] Lie algebras with finite gradation. Mat. Sb., Nov. Ser. *121* (3), 353–392. Zbl. 546.17005. English translation: Math. USSR, Sb. *52*, 347–385 (1985)

Zhevlakov, K.A., Slin'ko, A.M., Shestakov, I.P., Shirshov, A.I. [1978] Rings that are nearly associative. Nauka: Moscow. Zbl. 445,17001. English translation: Academic Press: New York (1982)

Author Index

Subject Index

Encyclopaedia of Mathematical Sciences

Editor-in-chief: R. V. Gamkrelidze

Algebra

Volume 11: **A. I. Kostrikin, I. R. Shafarevich** (Eds.)

Algebra I

Basic Notions of Algebra

1989. V, 258 pp. 45 figs. ISBN 3-540-17006-5

This book is wholeheartedly recommended to every student or user of mathematics. Although the author modestly describes his book as 'merely an attempt to talk about' algebra, he succeeds in writing an extremely original and highly informative essay on algebra and its place in modern mathematics and science. From the fields, commutative rings and groups studied in every university math course, through Lie groups and algebras to cohomology and category theory, the author shows how the origins of each algebraic concept can be related to attempts to model phenomena in physics or in other branches of mathematics. Comparable in style with Hermann Weyl's evergreen essay **The Classical Groups,** Shafarevich's book is sure to become required reading for mathematicians, from beginners to experts.

Contents: Preface. – What is Algebra? – Fields. – Commutative Rings. – Homomorphisms and Ideals. – Modules. – Algebraic Aspects of Dimension. – The Algebraic View of Infinitesimal Notions. – Noncommutative Rings. – Modules over Noncommutative Rings. – Semisimple Modules and Rings. – Division Algebras of Finite Rank. – The Notation of a Group. – Examples of Groups: Finite Groups. – Examples of Groups: Infinite Discrete Groups. – Examples of Groups: Lie Groups and Algebraic Groups. – General Results of Group Theory. – Group Representations. – Some Applications of Groups. – Lie Algebras and Nonassociative Algebra. – Categories. – Homological Algebra. – *K*-theory. – Comments on the Literature. – References. – Index of Names. – Subject Index.

Springer-Verlag
Berlin
Heidelberg
New York
London
Paris
Tokyo
Hong Kong
Barcelona